Smart Building Materials and Designs for Sustainable Built Environment

Smart Building Materials and Designs for Sustainable Built Environment

Guest Editor

Kishor Zingre

MDPI

Basel • Beijing • Wuhan • Barcelona • Belgrade • Novi Sad • Cluj • Manchester

Guest Editor
Kishor Zingre
Architecture and Built
Environment
University of Northumbria
Newcastle
UK

Editorial Office
MDPI AG
Grosspeteranlage 5
4052 Basel, Switzerland

This is a reprint of the Special Issue, published open access by the journal *Buildings* (ISSN 2075-5309), freely accessible at: https://www.mdpi.com/journal/buildings/special_issues/CL1Q16ZPF3.

For citation purposes, cite each article independently as indicated on the article page online and as indicated below:

Lastname, A.A.; Lastname, B.B. Article Title. *Journal Name* **Year**, *Volume Number*, Page Range.

ISBN 978-3-7258-7442-2 (Hbk)
ISBN 978-3-7258-7443-9 (PDF)
https://doi.org/10.3390/books978-3-7258-7443-9

Contents

About the Editor

Kishor Zingre

Kishor Zingre joined Northumbria University in Newcastle, UK, as an Assistant Professor in the department of Architecture and Built Environments. Primarily working within the fields of sustainability and building thermal physics, his research focuses on the decarbonisation of built environments using cool materials and bio-inspired designs and employs experimental methods (measurements, thermography, and laser scanning) and computational simulations. The international impact of his work includes the following: (i) development of novel heat transfer models and thermal comfort analysis for pedestrians in street canyons and (ii) contributions to the housing policies in Singapore (for cooling the city by 2oC). National impact of his work includes (i) the submission of four (3*) outputs to REF2021; (ii) working with the National Trust to achieve their sustainability goals for Coleshill and Cambo villages; (iii) carrying out energy assessments of Newcastle City Council schools; and (iv) contributions towards the development of North of Tyne Combined Authority's Green Growth Skills programme, which will have a real-world impact for the Northeast in terms of (a) skills development; (b) jobs creation, and (c) achieving the central government's sustainability goals. NU impact includes assisting carbon managers with decarbonising city campus buildings by diagnosing various building defects (air leakage through envelops, thermal bridges, etc.) and providing recommendations for energy-saving measures. The collaborative and interdisciplinary nature of his work has allowed him to foster an international network with academics across four continents and make significant contributions with demonstrable impact (60 research outputs). He has delivered invited talks at 25+ national and international conferences.

buildings

MDPI

Article

One-Pot Synthesis of Carbon-Based Composite Foams with Tailorable Structure

Florina S. Rus [1,*], Cristina Mosoarca [1], Nicolae Birsan [1], Mihai Petru Marghitas [1,2], Raul Bucur [1], Dan Rosu [1], Emanoil Linul [2] and Radu Banica [1,*]

[1] National Institute of Research and Development for Electrochemistry and Condensed Matter (INCEMC), Str. Dr. A. Păunescu Podeanu, Nr. 144, 300569 Timişoara, Romania

[2] Department of Mechanics and Strength of Materials, Politehnica University of Timisoara, 1 Mihai Viteazu Blvd., 300222 Timişoara, Romania; emanoil.linul@upt.ro

* Correspondence: rusflorinastefania@gmail.com (F.S.R.); radu.banica@yahoo.com (R.B.)

Abstract

Dehumidification plays a vital role across industrial, commercial, and residential settings, where controlling moisture is essential for maintaining air quality, protecting materials, and ensuring comfort. Calcium chloride ($CaCl_2$) is a widely used, low-cost desiccant, but it suffers from a critical drawback: under humid conditions, particles tend to agglomerate, which reduces their ability to absorb water. In addition, when the salt dissolves in hydration water, its contact surface with moist air decreases, and corrosive liquid leakage can occur. Embedding $CaCl_2$ into hydrophilic porous matrices offers a solution by dispersing particles more effectively, preventing agglomeration, increasing the contact area, and retaining liquid within the pore network to suppress leakage. In this study, we introduce a novel approach for fabricating carbon-based foams impregnated with $CaCl_2$, produced through the thermal decomposition of glucose under self-induced pressure. These foams exhibit a composite architecture that integrates $CaCl_2$ and calcium carbonate, enabling controlled porosity through selective dissolution. Importantly, the in situ transformation of $CaCl_2$ into calcite refines the internal structure, improving both stability and acids absorption performance. FTIR confirmed the strong hydrophilicity of the foam walls, which enhances water vapor uptake while preventing leakage of saturated salt solutions. The carbon matrix further suppresses salt particle agglomeration during moisture absorption, resulting in high efficiency. These multifunctional foams not only capture water vapor and volatile acids but also show potential as phase change materials. Mechanical testing revealed tunable behavior among the fabricated foams, ranging from high-stiffness structures with superior energy absorption (e.g., C2) to more compliant foams with extended strain capacity (e.g., A2), illustrating their versatility for practical applications.

Keywords: carbon foams; dehumidification; calcium chloride; calcite; moisture adsorption; phase change materials; thermal management; porous materials; mechanical tuning; volatile acid capture

1. Introduction

Aerogels are among the most intriguing types of porous materials. They combine extreme lightness with impressive structural and functional traits. These materials are produced when the liquid part of a gel is replaced by gas, keeping the fragile solid structure intact. This process leads to solids that often have porosities above 90 to 99.8%. The first mention of silica aerogels comes from S.S Kistler in 1931, who showed that removing liquid under supercritical conditions could stop pore collapse and maintain a highly porous structure [1]. Since that time, researchers have made aerogels from many inorganic and organic sources, including silica [2] alumina, titania [3], polymers [4], carbon [5], as well as hybrid materials [6,7]. Their low thermal conductivity, high porosity, and adjustable chemistry have sparked interest in areas like insulation, energy storage [8], catalysis [9], and environmental cleanup [10].

Aerogels' unique properties come from their hierarchical pore structure and very large internal surface area. Depending on the precursor and method used, pore sizes can be classified as micro- (less than 2 nm), meso- (2 to 50 nm), or macro- (over 50 nm). The mix of these pore types affects both the material's adsorption capacity and transport behavior. Foams represent a distinct class of porous materials, characterized by significantly larger pore sizes and thicker cell walls [11]. In certain cases, the cell walls may incorporate aerogel, provided a continuous 3D meso- or microporous structure is present. Unlike aerogels, foams can exhibit either closed or open porosity, with each type serving different application domains. Impregnation of highly porous materials with hygroscopic substances acts on increasing their capacity to retain water vapor, making them particularly useful in construction applications. Carbon foams are generally biodegradable and exhibit low toxicity; however, the precursors employed in their synthesis are typically synthetic chemicals [12]. Furthermore, the production process often involves costly equipment and high energy consumption. Impregnation is usually performed in a secondary stage after the carbon matrix has been formed, although a one-step synthesis would be preferable.

Carbon-based porous masses are generally obtainable from more cost-effective precursors compared to other types of aerogels. They are typically made through sol–gel polymerization of resorcinol-formaldehyde or other organic materials, followed by drying and pyrolysis in inert conditions [8,13]. This process turns the organic gel into a conductive carbon network while keeping the open porous structure. Additional activation steps, like treating with KOH or CO_2, can further increase the surface area; however, the KOH activation method is not yet fully understood. Some authors suggest that metallic potassium forms within the carbon structure at temperatures below 500 °C, through the reduction of potassium ions by carbon [14]. However, this mechanism appears unlikely, as no potassium vapor has been detected in the system. Natural carbon sources such as anthracite contain non-negligible amounts of ash, and the transition metals present are likely reduced by carbon prior to potassium. This sequence may influence the activation process by promoting carbon oxidation and leading to the formation of K_2CO_3 as a secondary reaction product. The water present in the system originates primarily from the thermal decomposition of KOH [15]. These traits make carbon aerogels suitable for electrochemical applications, including electrodes for supercapacitors, lithium-ion batteries, and electrocatalytic devices [16,17]. For instance, carbon aerogels used as supercapacitor electrodes can provide high specific capacitances due to their large surface area for charge storage and hierarchical pores that help with ion transport [17,18].

Besides energy storage, carbon aerogels are also useful as catalyst supports because of their high conductivity and adjustable surface chemistry [18]. They create a stable structure for dispersing nanoparticles such as platinum, iron oxide, or cobalt catalysts, improving performance in fuel cells, Fischer–Tropsch synthesis, or environmental catalysis [19]. Their excellent ability to adsorb makes them promising materials for capturing greenhouse gases like CO_2, volatile organic compounds, and heavy metals from water solutions [20,21]. Their thermal stability and lightweight nature also enable uses in thermal insulation, sound damping, and electromagnetic shielding [22].

 https://doi.org/10.3390/buildings16010056

Recently, the focus has shifted to more sustainable methods for producing aerogels, particularly those made from biomass. Renewable sources, such as cellulose, starch, lignin, and glucose, can be converted into highly porous gels that yield biopolymer aerogels upon drying [23]. When subjected to pyrolysis, these biopolymer aerogels turn into carbon aerogels or foams, providing an eco-friendly alternative to polymers made from petroleum [24]. Glucose, in particular, is a flexible building block because it can undergo hydrothermal carbonization or polymerization to create gels that can be freeze-dried into aerogels with open structures [25]. After carbonization, glucose-derived aerogels maintain much of their porosity and develop conductive carbon structures [26]. These bio-based aerogels not only reduce environmental impact but also offer unique pore architectures and surface properties due to their natural structure [27].

Several methods have been shown to convert glucose into carbon aerogels or foams. In hydrothermal synthesis, concentrated glucose solutions are heated under pressure, leading to polymerization and dehydration reactions that produce hydrothermal carbons. Following drying and pyrolysis, this results in carbon-rich networks with hierarchical porosity [28]. Other methods mix glucose with crosslinkers to form hydrogels, which can produce ultralight aerogels after freeze-drying. Carbonizing these structures at high temperatures leads to lightweight carbon foams with specific surface areas suitable for adsorption and energy storage [29]. Using glucose and other biomass-derived sources gives a promising route to sustainable aerogels that match the humidity properties of those made from synthetic sources [30].

A key application area for both synthetic and bio-derived foams is removing moisture and chemical pollutants. Their high surface area and open-pore structure allow for efficient uptake of water vapor, volatile organics, and ions. Moisture adsorption is important for humidity control, water harvesting, and dehumidification technologies. Dehumidification plays a critical role in maintaining air quality, protecting sensitive materials, and ensuring comfort across industrial, commercial, and residential environments. Conventional moisture-absorbing materials often suffer from limited capacity, poor mechanical integrity, or leakage during saturation, prompting the need for multifunctional alternatives with enhanced performance and structural control.

Carbon-based foams have emerged as promising candidates due to their lightweight architecture, high surface area, and chemical tenability. However, most existing synthesis routes involve multi-step procedures or lack control over pore morphology and functional integration. In this context, one-pot synthesis strategies offer a streamlined pathway to engineer composite foams with tailored properties, enabling simultaneous control over porosity, chemical functionality, and mechanical behavior. For example, foams treated with hygroscopic salts or polar groups can absorb water up to several times their own weight and can be regenerated through mild heating or solar exposure [31]. These systems are promising for collecting atmospheric water in dry areas. Additionally, foams have been studied as desiccants in packaging, electronics protection, and controlled environments where moisture management is essential [32]. This study introduces a novel one-pot approach for fabricating carbon-based composite foams via the thermal decomposition of glucose under self-induced pressure. The incorporation of calcium chloride ($CaCl_2$) and calcium carbonate ($CaCO_3$) into the carbon matrix allows for selective dissolution and in situ transformation into calcite, yielding a hierarchically porous structure with enhanced moisture and volatile acid capture capabilities. The resulting foams exhibit strong hydrophilicity, phase change material (PCM) behavior, and tunable mechanical properties, making them suitable for a wide range of environmental and energy-related applications.

 https://doi.org/10.3390/buildings16010056

Foams also work effectively to absorb chemical pollutants. Activated carbon foams can capture significant amounts of CO_2 due to their micropore structure and surface chemistry [33]. Additionally, they have been shown to remove dyes, heavy metals, and organic solvents from water, making them valuable for treating wastewater [34]. Their adsorption works through both physical filling of pores and chemical interactions with functional groups on the aerogel surface. Compared to traditional activated carbons, foams provide a tunable and interconnected pore network that can improve diffusion rates and adsorption kinetics [35].

These findings show that foams are a versatile material platform, linking fundamental science with practical technologies. Carbon foams, whether made from synthetic polymers or renewable glucose, demonstrate the potential of combining lightweight design with electrical conductivity and environmental friendliness. Ongoing development is likely to boost progress in energy storage, catalysis, pollutant capture, and water harvesting. However, challenges remain in scaling production, ensuring mechanical durability, and optimizing cost-performance balance. Still, the unique combination of properties makes foams essential candidates for creating next-generation functional materials [36].

Porous carbon-based materials have attracted significant attention due to their versatility in energy storage, thermal insulation, and environmental control. Within this broad class, composite foams derived from carbohydrate precursors offer a sustainable route to functional materials with tunable porosity and mechanical resilience. Recent advances (2024–2025) have highlighted the growing role of carbon-based aerogels and foams in sustainable construction. Novel aerogel composites have been developed for building insulation with high thermal performance [37], while circular initiatives such as CARBIP [38] emphasize eco-friendly, reusable insulation systems. Reviews of aerogel insulation for exterior walls [39] and studies on low-carbon architecture [40] further underline their relevance in passive climate regulation and energy-efficient building design. These developments provide a timely context for our investigation of glucose–$CaCl_2$-derived foams as potential building materials.

In this study, we present a simple and affordable way to create carbon-based foams, produced by the thermal decomposition of glucose under its own pressure. What makes these foams special is not only the elegance of their synthesis, but also the possibilities they open up. By introducing a sacrificial calcium salt, we can guide the formation of pores in a controlled manner and later dissolve the salt to fine-tune the structure. This gives us a material that is lightweight, customizable, and surprisingly versatile. Even more intriguing is the in situ transformation of the calcium salt into calcite, achieved through a gentle vapor treatment with ammonia and carbon dioxide. This step adds a mineral backbone to the foam, strengthening it and enhancing its resistance to heat and chemical stress capable of standing up to the demanding conditions of modern buildings. The porous carbon framework acts like a natural shield against heat transfer, making these foams excellent candidates for thermal insulation. Their ability to trap air in the pores reduces energy loss, while their stability across a wide temperature range ensures reliable performance in both hot summers and cold winters. At the same time, the foams interact with water vapor and gases in a way that helps regulate moisture indoors, preventing condensation and improving air quality. This multifunctionality means they are more than just insulators—they are active participants in creating healthier, more sustainable living spaces.

 https://doi.org/10.3390/buildings16010056

2. Materials and Methods

2.1. Synthesis of Carbon-Based Foams

Anhydrous glucose (G) and calcium chloride (C) were mixed at mass ratios 4:1 (C1), 3:2 (C2), and 3.33:1.66 (C3). The mixtures were transferred into Berzelius beakers placed at the bottom of 100 mL Teflon-lined stainless-steel autoclave vessels. The autoclaves were sealed and heated at 220 °C for 24 h. Calcium chlorine anhydrous was procured from Sigma Aldrich and NH_4CO_3 were obtained from Reactivul Bucuresti. Glucose ($C_6H_{12}O_6$) was procured from SC Chimreactiv SRL (Bucharest, Romania).

After cooling, the resulting carbonaceous foams were extracted and cut into cylindrical specimens with a diameter of 18 ± 0.2 mm and a height of 5 ± 0.3 mm using a CNC system.

Calcite Loading Procedure

Foam specimens were subjected to a secondary vapor-phase treatment in an atmosphere containing CO_2, NH_3, and H_2O. These vapors were generated in situ by the thermal decomposition of 58 mg of ammonium bicarbonate (NH_4HCO_3) placed inside the autoclave, ensuring no direct contact with the samples. The treatment was conducted at 180 °C for 12 h. Subsequently, the samples were heated in air at 200 °C for 60 h to sublimate residual byproducts formed during the in situ conversion of $CaCl_2$ to $CaCO_3$ (calcite). For clarity, the overall experimental workflow is summarized in Figure 1, which outlines the synthesis, treatment, and characterization steps in a sequential manner.

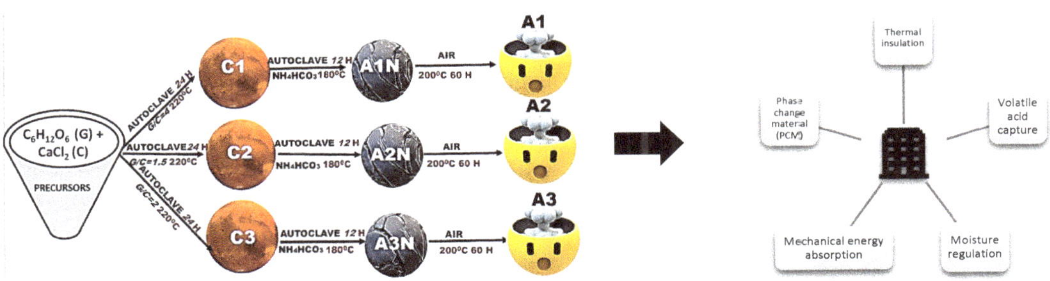

Figure 1. Workflow representation of sample synthesis.

2.2. Physico-Chemical Characterization

X-ray diffraction (XRD) was performed using an X'Pert PRO MPD diffractometer (PANalytical, Almelo, The Netherlands) with Ni-filtered Cu Kα radiation (λ = 1.54 Å).

Fourier-transform infrared (FTIR) spectroscopy was conducted on a Vertex 70 instrument (Bruker, Ettlingen, Germany).

Scanning electron microscopy (SEM) was recorded with FIB-SEM, Zeiss Auriga. Thermogravimetric analysis (TGA) was performed in air using a TG 209F1 Libra instrument (Netzsch, Selb, Germany).

Ammonium ion concentrations were measured colorimetrically using a Hanna Instruments HI700 Digital Absorption Photometer Checker HC (Ammonia Nitrogen LR, Fresh Water, Smithfield, RI, USA), with a detection range of 0.00–3.00 ppm (mg/L), a resolution of ±0.05 ppm, and an accuracy of ±5% at 25 °C.

Acid Vapor Adsorption via FTIR Gas Cell Tests

The foam's capacity to retain volatile acids was evaluated using quantitative FTIR spectrometry. Acetic acid vapors were introduced into a sealed gas cell (path length: 100 mm; internal diameter: 40 mm; volume: 125 cm^3) fitted with KCl windows. One

https://doi.org/10.3390/buildings16010056

window was modified to accommodate a PTFE-based dosing device that released a defined foam mass directly into the cell.

To prevent water vapor condensation during gas recirculation, the acetic acid solution was cooled in an ice bath, and aerosols were removed using a 0.2 μm polyethersulfone syringe filter (30 mm diameter). This setup generated acetic acid vapors with a calculated partial pressure of 0.58 mmHg.

The gas was recirculated using a diaphragm-type micropump for 20 min. After sealing the system, 80 mg of foam was introduced, and FTIR spectra were recorded every 300 s over a 65 min period. The experimental setup is illustrated in Figure A1 (Appendix A). This ability of the foams to capture volatile acids will be further explored as a functional property, with the perspective of integrating such materials into building applications where they could contribute to improved indoor air quality and enhanced protection of structural elements.

2.3. Mechanical Characterization

Foam specimens (18 ± 0.2 mm diameter, 5 ± 0.3 mm height) were subjected to uniaxial compression testing using a Metrotec universal testing system. All tests were conducted at room temperature under displacement control, with a constant crosshead speed of 2 mm/min.

Force-displacement data were converted into engineering stress–strain curves using the measured specimen dimensions. These curves were used to extract relevant mechanical parameters, including stiffness, strength, and energy absorption (EA) capacity.

Following the initial thermal treatment and subsequent cooling, the resulting carbonaceous foams were extracted. Cylindrical specimens with a diameter of 18 ± 0.2 mm and a height of 5 ± 0.3 mm were cut from the foams using a CNC machine as in Figure 2.

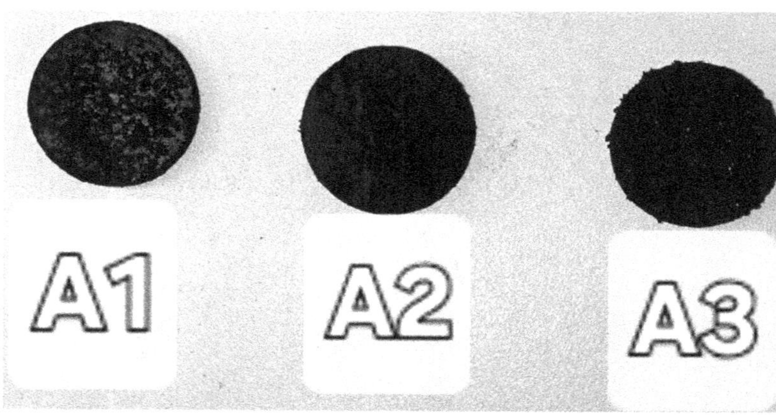

Figure 2. Cylindrical Foam Specimens Prepared for Experimental where A1–A3 are described in Figure 1.

3. Results and Discussions

3.1. X-Ray Diffraction

Upon heating the glucose–$CaCl_2$ mixture to approximately 150 °C, glucose begins to melt and dissolves anhydrous $CaCl_2$, forming eutectic mixtures with reduced viscosity at specific molar ratios. As the temperature increases, reaction (1) is promoted, leading to the formation of 5-hydroxymethylfurfural (HMF) through a three-step dehydration of glucose:

$$C_6H_{12}O_6 \rightarrow C_6H_6O_3 + 3H_2O. \quad (1)$$

 https://doi.org/10.3390/buildings16010056

$CaCl_2$ plays a dual role in this process: it lowers water activity, promoting dehydration, and acts as a Lewis acid catalyst, facilitating glucose isomerization to fructose—a known precursor to HMF. The resulting HMF undergoes polymerization and carbonization, especially under prolonged heating, increasing the viscosity of the reaction medium.

The hygroscopic nature of $CaCl_2$ further reduces water vapor pressure, forming crystal hydrates that melt below 200 °C. This molten phase enhances polymer chain mobility and stabilizes gas bubbles, preventing premature coalescence. When internal pore pressure equilibrates with the vapor pressure above the foam, expansion ceases. Experimental observations show that decreasing the glucose to $CaCl_2$ mass ratio from 4 to 1.5 leads to significantly more expanded structures with increased porosity. In the absence of $CaCl_2$, gas bubbles coalesce and escape, resulting in less uniform pore distribution.

Foams with smaller, closed-cell pores exhibit greater structural integrity and reduced gas diffusion, which is advantageous for achieving high internal surface area and low thermal conductivity. Smaller bubbles withstand higher pressures, and closed-cell foams formed by such bubbles possess more cell walls, limiting long-range gas diffusion. Therefore, controlling reactor pressure is key to regulating foam porosity.

During hydrothermal treatment, glucose undergoes a sequence of dehydration, isomerization, and polymerization steps that are strongly influenced by the presence of $CaCl_2$. Calcium ions can coordinate with hydroxyl groups of glucose and intermediate species, promoting local cross-linking and stabilizing partially dehydrated structures. This coordination lowers the activation energy for dehydration and accelerates the formation of oligomers and carbon-rich nuclei. In parallel, chloride ions do not catalyze the decomposition of glucose into 5-hydroxymethylfurfural (HMF), but they enhance both the selectivity and the rate of fructose conversion to this intermediate compound [41].

The isomerization of glucose to fructose can be readily catalyzed by the mildly acidic environment generated through salt hydrolysis; however, the salt does not act as a specific or selective catalyst. By lowering the water activity in the system and through the complexing effect of the cation, $CaCl_2$ stabilizes the 1,2-enediol intermediate, which plays a decisive role in the isomerization process.

In study Zhi-Xiang Xu and co-authors [42] investigated biocarbon production Via hydrothermal synthesis in the presence of metal salts at moderate concentrations (~10%). They reported that the carbon yield increased in the presence of such salts, which were partially adsorbed onto the carbon surface. Xue-Qin Ma [43] likewise observed that NaCl promoted hydrolysis, dehydration, and decarboxylation. Process yield was also strongly temperature-dependent, with a maximum at 220 °C, followed by a decline at higher temperatures, most likely due to hydrolytic degradation of polymer chains at elevated temperatures.

In our study, the concentration of $CaCl_2$ was substantially higher (20–40%), and thus the expected catalytic and structural effects are correspondingly amplified. This higher salt loading not only accelerates glucose conversion but also enhances the stabilization of intermediate carbonaceous structures, leading to improved foam formation and more pronounced porosity control compared to the moderate salt concentrations reported in the literature.

Following carbonization, the foams consist of compact cellular carbon structures and carbon–$CaCl_2$ composites. Due to $CaCl_2$'s high solubility, water or saturated steam treatment can open pores in the walls containing solid $CaCl_2$. However, in larger samples, water diffusion is limited, and the formation of a saturated $CaCl_2$ solution within pores can inhibit further dissolution. As shown in Figure A2 (Appendix A), pore opening after water washing is clearly visible, confirming that foams with tunable porosity—either closed or open—can be obtained depending on the applied treatment. During subsequent

 https://doi.org/10.3390/buildings16010056

heating of the foams in NH_4HCO_3 vapor, reaction (2) initiates the thermal decomposition of ammonium carbonate, yielding NH_3, H_2O, and CO_2; These gaseous products then react with $CaCl_2$ embedded in the foam matrix, leading to the in situ formation of calcite ($CaCO_3$) and ammonium chloride (NH_4Cl) via reaction (3).

$$(NH_4)_2CO_3 \rightarrow 2\ NH_3 + H_2O + CO_2 \quad (2)$$

$$CaCl_2 + CO_2 + 2\ NH_3 + H_2O \rightarrow CaCO_3 + 2\ NH_4Cl \quad (3)$$

As observed in the XRD patterns shown in Figure 3, samples A1N–A3N contain both calcite and NH_4Cl, as expected according to reaction (3). To remove this phase from the system, the samples were subjected to thermal treatment for 60 h at 200 °C, as described in the experimental section. At this temperature, the vapor pressure of NH_4Cl is approximately 1 mmHg, sufficiently high to enable complete sublimation of this crystalline phase. Consequently, the XRD spectra of samples A1–A3 no longer exhibit the NH_4Cl phase. At 200 °C, the vapor pressure of NH_4Cl is approximately 1.0 mmHg, which is sufficiently high to enable complete sublimation of the crystalline phase from the foam matrix. This value aligns with literature data from Tomida et al. (2008) [44], who measured vapor pressures of NH_4Cl–NH_3 mixtures and reported values consistent with this range at elevated temperatures. Their work provides a reliable basis for interpreting sublimation behavior under your experimental conditions. The XRD analysis confirming the coexistence of carbon phases and in situ formed calcite further supports the potential of these foams to be adapted as building materials, since the mineral reinforcement combined with the porous carbon matrix can provide both mechanical stability and enhanced thermal insulation properties.

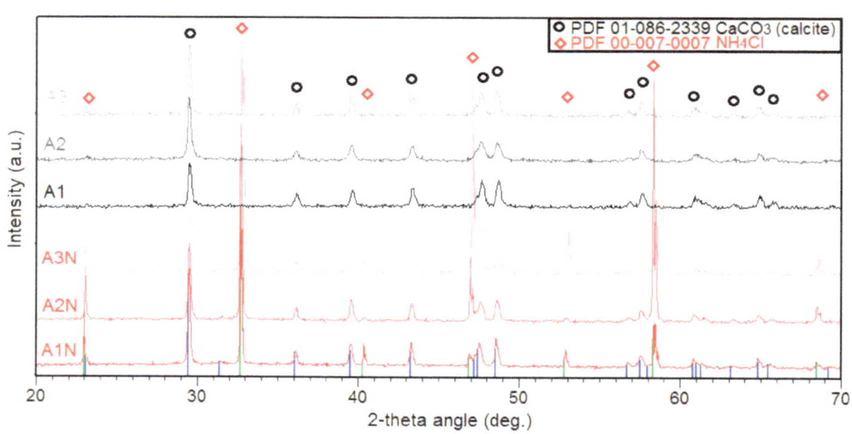

Figure 3. XRD comparison of ammonia-treated (A1N–A3N; the curves are displayed in shades of red) and thermally purified (A1–A3; the curves are plotted in different shades of gray) foams.

This complete sublimation can only occur if the foams exhibit open porosity, which enables pressure equalization between the internal pore network and the external atmosphere, as well as the diffusion of volatile species such as NH_3 and HCl. This structural feature is fundamentally distinct from the closed porosity required during the formation of C/$CaCl_2$-based foams. In those systems, closed-cell architecture is essential to sustain foam expansion; without it, the structure would collapse under the influence of surface tension forces acting on the liquid phase. Thus, the porosity type not only governs the foam's

https://doi.org/10.3390/buildings16010056

mechanical integrity during synthesis but also determines its suitability for post-synthetic treatments such as sublimation or gas-phase conversion.

3.2. Compressive Behavior and Mechanical Performance

Before testing, all samples were dried at 220 °C for 48 h to remove absorbed moisture, given the high hygroscopicity of calcium chloride. This thermal conditioning ensured stable and reproducible mechanical measurements by preventing variations in stiffness associated with residual water. After drying, the specimens were stored in airtight containers until testing to avoid moisture reabsorption.

The compressive stress–strain responses of the investigated carbon-based composite foams are presented in Figure 4a, accompanied by magnified views of the initial elastic region (Figure 4b) and the plateau region (Figure 4c). All samples follow the characteristic deformation sequence of cellular solids, consisting of an initial linear elastic regime, a plateau region governed by the progressive collapse of the cellular structure, and a final densification stage marked by a rapid increase in stress with strain [45–47]. Despite this shared deformation pattern, the detailed shapes of the curves differ markedly among the foams, reflecting variations in composition and processing history.

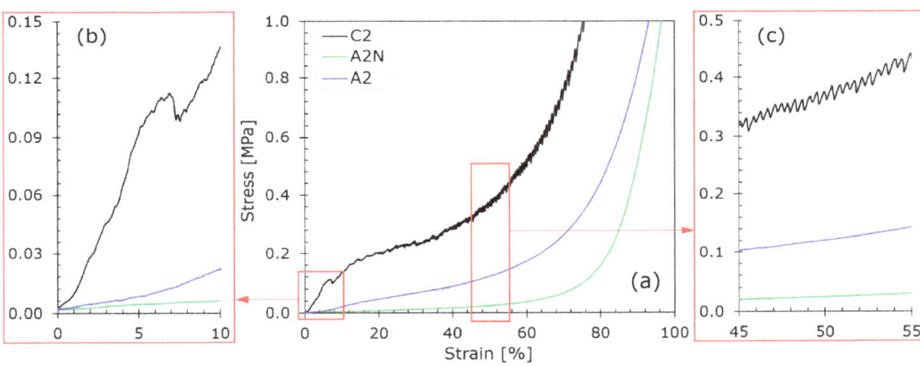

Figure 4. Compressive stress–strain curves (**a**), with magnified views of the elastic (**b**) and plateau (**c**) regimes.

In the elastic regime (Figure 4b), foam C2 exhibits the steepest slope among the three materials, confirming its higher stiffness relative to A2 and A2N. The latter two samples display a more gradual elastic response, consistent with their lower measured moduli. In the plateau region (Figure 4c), foam C2 shows a serrated stress–strain profile with noticeable fluctuations, whereas A2 and A2N present smooth and nearly monotonic plateaus. These differences illustrate the more stable and uniform deformation behavior of the A-series foams during progressive collapse. The densification stage also varies between the formulations. Foam A2 reaches densification at the highest strain level, reflecting its ability to accommodate large deformations before pore compaction. Foam C2 densifies at intermediate strains, whereas A2N exhibits the earliest densification, consistent with its substantially reduced stiffness. Overall, the mechanical responses separate the foams into two behavioral groups: C2, characterized by higher stiffness and fluctuating plateau stress, and A2/A2N, which exhibit lower stiffness and more stable plateau regions. These differences highlight the strong influence of composition and thermal treatment on stiffness, plateau stability, and strain capacity; parameters essential for predictable compressive behavior and reliable deformation performance.

The compressive properties of foams C2, A2, and A2N are summarized in Figure 5, revealing clear distinctions in stiffness, strength, densification strain, and EA. Foam C2

 https://doi.org/10.3390/buildings16010056

exhibits the highest stiffness, with a compressive modulus (E) of 3.08 MPa (Figure 5a), and the highest strength-related parameters, including a compressive strength (CS) of 0.21 MPa and a plateau stress (PS) of 0.15 MPa (Figure 5b). By contrast, A2 and A2N show significantly lower E of 0.13 MPa and 0.06 MPa, respectively. Their PSs remain limited to 0.06 MPa (A2) and 0.01 MPa (A2N), reflecting their reduced load-bearing capacity. The densification strain (ε_d) values (Figure 5c) further differentiate the materials: A2 densifies at 71.5% strain, the highest among the samples; C2 densifies at ~61%; and A2N reaches ε_d at 54.4%. EA (Figure 5d) mirrors these trends. Foam C2 absorbs 0.15 MJ/m^3, outperforming A2 (0.06 MJ/m^3) and A2N (<0.01 MJ/m^3). Despite its broad deformation interval, A2 shows only moderate EA due to its low PS, while A2N exhibits minimal performance across all metrics. Overall, C2 demonstrates superior stiffness, strength, and EA, whereas A2 combines low rigidity with large strain capacity, and A2N shows the weakest mechanical response.

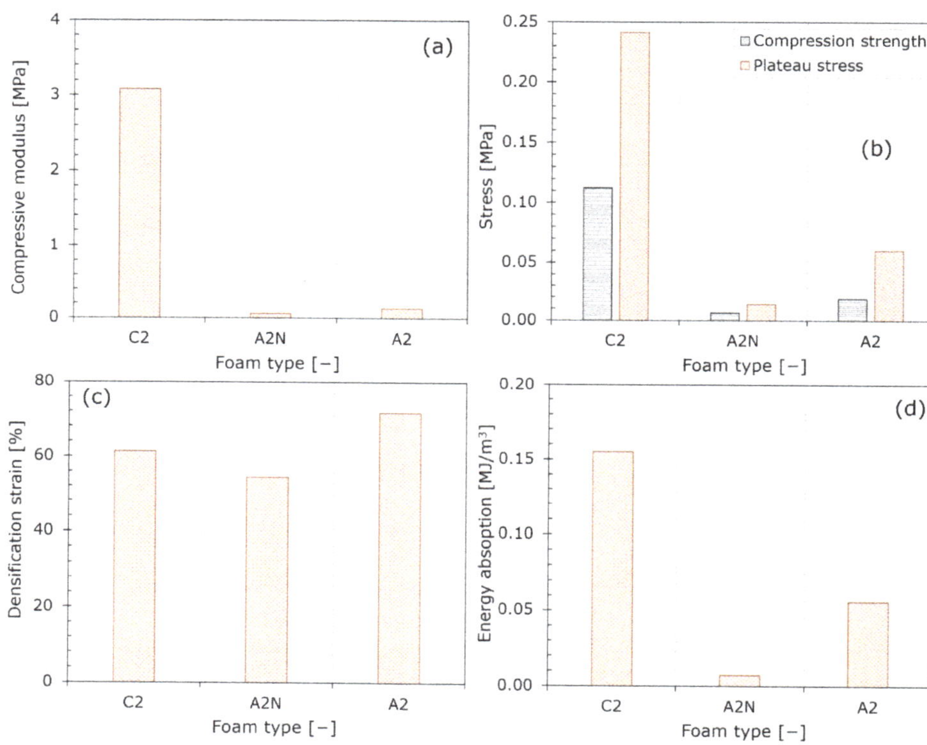

Figure 5. Compression-derived mechanical properties: modulus (**a**), stresses (**b**), densification strain (**c**), and absorbed energy (**d**).

The mechanical behavior of the foams is consistent with the structural changes induced during processing. In the case of A2N, the reduced stiffness and strength can be attributed to the growth of NH_4Cl crystals inside the cells during treatment with NH_3, CO_2, and H_2O vapor. These crystals generate cracks in the cell walls and weaken the carbon framework, as also reflected in the SEM observations reported for treated versus untreated samples. Following the sublimation of NH_4Cl, the larger pores tend to collapse, resulting in a more fragile structure with diminished load-bearing capacity. Preventing such degradation requires heat treatment conditions that inhibit NH_4Cl formation or ensure its removal without compromising the integrity of the pore network. These effects highlight the critical

 https://doi.org/10.3390/buildings16010056

influence of NH_4Cl formation and sublimation on the mechanical resilience of the foams and reinforce the need for controlled thermal treatment strategies when such materials are intended for applications requiring long-term structural stability.

3.3. Optical Microscopy

The SEM images presented in Figure 6 provide valuable insight into the morphological evolution of the foam material synthesized from $CaCl_2$ and glucose under thermal pressure in an autoclave.

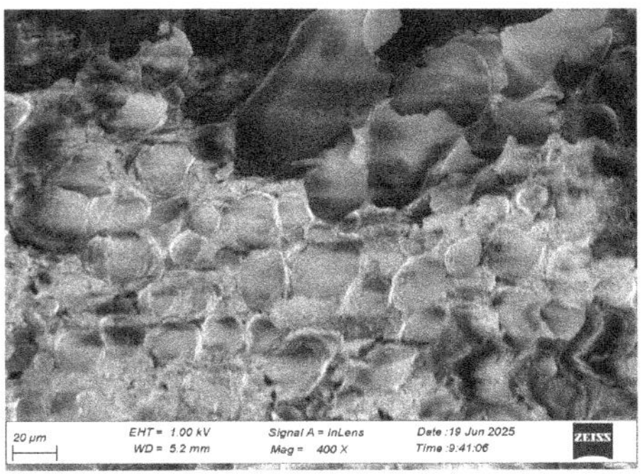

(a)

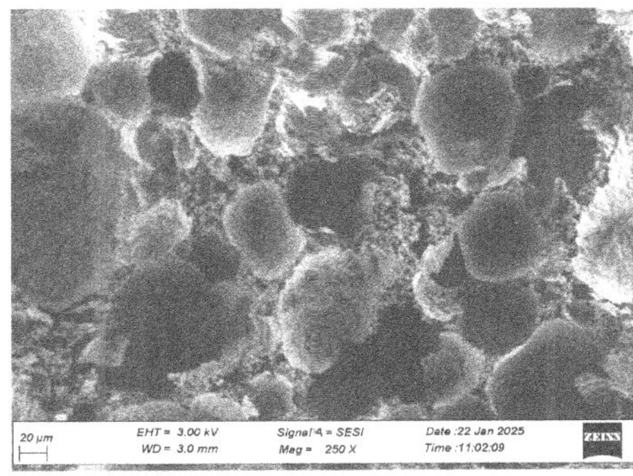

(b)

Figure 6. SEM micrographs showing morphological evolution of $CaCl_2$–glucose-derived foam structure (C1 sample) (**a**) and Water washed sample (**b**).

In Figure 6a the foam exhibits early signs of pore formation and partial breakdown of cell walls. These features are attributed to the formation of gas–solid mixture, which locally disrupt the carbon matrix. Figure 6b demonstrates tunable porosity, which can be modulated by adjusting the water washing and thermal treatment parameters.

The conversion of $CaCl_2$ to $CaCO_3$ during ammonia exposure not only alters the chemical environment but also enhances the mechanical and morphological stability of the final material. This is supported by Guo et al. [48], who observed that NH_4Cl additives influence phase transitions and micromorphology in ceramic systems, reducing transition temperatures and promoting uniform porosity. In our system, the resulting foam—with its open, interconnected pore network—shows promise for filtration, thermal insulation, and as a scaffold for catalytic or adsorption-based applications. The ability to manipulate porosity through controlled chemical pathways offers a versatile platform for designing advanced porous materials with tailored structural and functional properties. C2 foams show smaller, more uniform pores, which limit long-range gas diffusion and enhance barrier properties. Such control over pore architecture, combined with the in situ formation of $CaCO_3$, highlights the potential of these foams to be engineered as building materials, where tunable porosity and enhanced barrier properties can be leveraged to achieve both mechanical stability and efficient thermal insulation.

3.4. Thermogravimetry Recorded in Air

In the TG curves, mass loss occurs in two main stages. The first weight loss, up to approximately 170 °C, is primarily attributed to the decomposition of $CaCl_2 \times XH_2O$. Because the calcium chloride crystal hydrates that crystallize with 6 and 4 water molecules, respectively, according to reactions R4–R7, melt at temperatures of approximately 30 and 45.4 °C, the water formed will be in liquid form and will evaporate along with the water adsorbed on the carbon mass. The similarity between our results and those of Pathak et al. [49] reinforces the interpretation that the early-stage mass loss is governed

https://doi.org/10.3390/buildings16010056

by hydrate decomposition and water release mechanisms. However, our foam system introduces an additional complexity: the porous carbon matrix facilitates vapor migration and may enhance water desorption kinetics compared to bulk salt systems. This distinction highlights the role of pore architecture in modulating thermal behavior under oxidative conditions. In the DTA spectrum, the melting of $CaCl_2 \times 4H_2O$ is manifested by small endothermic peaks around 50 °C (Figure 7) for samples C1–C3). The thermal behavior observed in our TG curves is further supported by the work of Karunadasa [50], who examined the dehydration of calcium chloride using high-temperature X-ray powder diffraction. His study confirms the presence of intermediate hydrates such as $CaCl_2{\cdot}4H_2O$ and $CaCl_2{\cdot}2H_2O$, which remain stable at room temperature and begin to decompose progressively upon heating. In our system, the initial mass loss up to 170 °C corresponds well with this staged dehydration, suggesting that the foam matrix retains both surface-bound and structurally integrated water. The presence of carbon also introduces additional thermal pathways, which may influence the rate and completeness of dehydration. These distinctions highlight the importance of matrix composition and microstructure in modulating thermal decomposition behavior under oxidative conditions. The loss of the last two water molecules occurs at temperatures between 100 and 170 °C, manifested by two other endothermic peaks at temperatures around 120 and 170 °C.

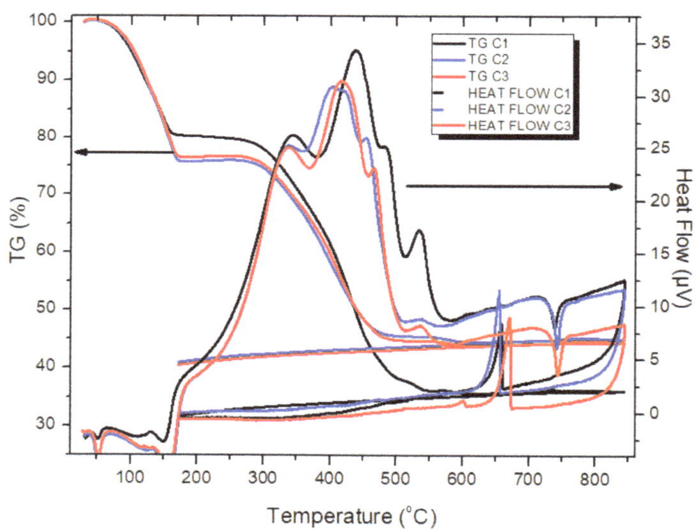

Figure 7. TG/DTA and comparative of C1–C3 samples recorded in air.

The reactions that take place are

$$CaCl_2 \times 6H_2O \rightarrow CaCl_2 \times 4H_2O + 2H_2O, \quad (4)$$

$$CaCl_2 \times 4H_2O \rightarrow CaCl_2 \times 2H_2O + 2H_2O, \quad (5)$$

$$CaCl_2 \times 2H_2O \rightarrow CaCl_2 \times H_2O + H_2O, \quad (6)$$

$$CaCl_2 \times H_2O \rightarrow CaCl_{2\ (anhydrous)} + H_2O. \quad (7)$$

 https://doi.org/10.3390/buildings16010056

At temperatures ranging from 270 °C to 600 °C, the mass loss observed in air is predominantly due to the oxidation of the glucose-derived carbon structure. The exothermic peaks appearing above 270 °C suggest a complex decomposition pathway involving at least four partially superimposed sub-stages, likely reflecting the progressive breakdown of carbon domains with varying degrees of oxidation. The staged dehydration behavior revealed by TG/DTA not only clarifies the thermal stability of the $CaCl_2$–carbon foams but also points to practical opportunities for their use in construction, where controlled water release and endothermic buffering could enhance both moisture regulation and fire resistance in insulation systems.

TG analysis also enabled the quantitative determination of $CaCl_2$ content in the samples, calculated as 45.0% for C1, 59.5% for C2, and 57.38% for C3, consistent with the initial glucose-to-salt ratios used during synthesis. These values confirm the reproducibility of salt incorporation and validate the compositional control achieved during foam preparation.

Upon further heating, an endothermic peak near 750 °C is observed during the heating ramp, followed by an exothermic peak around 650 °C during cooling. These thermal events correspond to the melting and crystallization of $CaCl_2$, respectively. The temperature difference between these transitions—approximately 100 °C is attributed to the supercooling phenomenon, where the salt crystallizes at a lower temperature than its melting point due to delayed nucleation and thermal hysteresis. This behavior is consistent with the findings of Eberbach et al. [51], who reported that $CaCl_2$ exhibits path-dependent hydration and crystallization behavior, with significant sensitivity to vapor pressure, cooling rate, and matrix confinement. In our foam system, the porous carbon architecture likely influences the crystallization kinetics by modulating heat transfer and restricting nucleation sites, thereby amplifying the supercooling effect.

These values align with the initial glucose-to-salt ratios, validating the compositional control during synthesis and supporting the reproducibility of the foam formulation. The TG/DTA data reveal that the thermal behavior of the samples is governed by a combination of hydrate decomposition, carbon oxidation, and salt phase transitions. C1, with the lowest $CaCl_2$ content, shows weaker signals around 50 °C, 120 °C, and 170 °C, indicating fewer hydrate transitions. In contrast, C2 and C3 exhibit more pronounced endothermic peaks, consistent with the presence of $CaCl_2{\cdot}6H_2O$ and $CaCl_2{\cdot}4H_2O$, as previously reported in phase change material studies [52]. This suggests that higher salt loading promotes more extensive hydrate formation during synthesis. All samples exhibit a distinct thermal profile, with an endothermic peak near 750 °C corresponding to the melting of $CaCl_2$, followed by an exothermic peak around 650 °C as crystallization occurs during cooling. The reproducible incorporation of $CaCl_2$ and its well-defined phase transitions, amplified by the porous carbon matrix, highlight the potential of these foams to function as thermally responsive building materials, where controlled salt loading and supercooling behavior can be exploited to enhance insulation efficiency, moisture regulation, and fire-resistant performance

3.5. FT-IR

FTIR spectra of the three samples (C2, A2N, and A2) are shown in Figure 8. All spectra feature a absorption band centered around 3435 cm^{-1}.

This band is due to O–H stretching vibrations from adsorbed water or hydroxyl groups linked to hydrated calcium salts. A band at 3333 cm^{-1} matches with combination or overtone vibrations typical of $CaCO_3$ polymorphs. In the spectrum of sample A2N, there are two additional absorptions at 3176 and 3047 cm^{-1} that correspond to the N–H stretching modes of ammonium species, indicating the presence of NH_4Cl. In the carbonate region, all three spectra show a strong asymmetric stretching band of the $CO_3{}^{2-}$ group at

 https://doi.org/10.3390/buildings16010056

1395 cm^{-1}. This is accompanied by the out-of-plane bending vibration at 871 cm^{-1} and the in-plane bending vibration at 711 cm^{-1}. In the C2 spectrum (black), the absence of carbonate-specific bands indicates that no crystalline $CaCO_3$ phase is present. Instead, the spectrum is dominated by weak O–H stretching and broad background features, characteristic of amorphous carbonaceous frameworks and residual hydroxylated species. This observation aligns with previous reports on hydrothermal carbonization of glucose in the presence of $CaCl_2$, where mineral phases do not form unless post-treatment with CO_2 or alkaline agents is applied [53,54]. In contrast to the amorphous carbon matrix observed in C2, the FTIR spectra of A2 (red) and A2N (green) display the diagnostic triplet of calcite ($CaCO_3$): the asymmetric stretching vibration of the carbonate ion at 1395 cm^{-1}, the out-of-plane bending at 871 cm^{-1}, and the in-plane bending at 711 cm^{-1}. These three bands are widely recognized as fingerprint features of calcite and have been consistently reported in both Raman and infrared studies of calcium carbonate polymorphs [55,56]. Their presence confirms the successful in situ conversion of $CaCl_2$ to calcite during the CO_2–NH_3–H_2O vapor-phase treatment applied to A2N, and the retention of this phase in A2 following NH_4Cl sublimation. In addition to the carbonate bands, A2N exhibits two distinct absorptions at 3176 cm^{-1} and 3047 cm^{-1}, which correspond to N–H stretching vibrations of ammonium ions. These features are characteristic of NH_4Cl, and their positions fall within the 3300–3000 cm^{-1} range typically assigned to NH_4^+ species in crystalline environments [57,58]. Their absence in C2 and only weak presence in A2 strongly suggest that NH_4Cl is introduced during the vapor-phase reaction (reaction (3)) and is subsequently removed by thermal sublimation, as also confirmed by XRD data. This spectral evolution aligns with the known thermal behavior of ammonium chloride–ammonia systems, where NH_4Cl can volatilize or sublimate upon controlled heating [44]. The FTIR data thus provide complementary evidence to the XRD results, confirming both phase identity and treatment-dependent compositional changes across the foam series. The FTIR data thus provide complementary evidence to the XRD results, confirming both phase identity and treatment-dependent compositional changes across the foam series, which are strongly influenced by the initial glucose-to-$CaCl_2$ ratio. The identification of calcite fingerprint bands alongside treatment-dependent NH_4Cl absorptions demonstrates that both phase composition and pore architecture can be chemically engineered, offering a pathway to foams with tailored mechanical stability and barrier properties that are highly relevant for their prospective use as multifunctional building materials with integrated insulation and environmental control capabilities.

3.6. Determination of Ammonium Concentration Using Colorimetric Analysis

This portable device operates within a detection range of 0.00–3.00 ppm (mg/L), with a resolution of ±0.05 ppm and an accuracy of ±5% at 25 °C, ensuring reliable quantification under controlled conditions. Samples were collected after exposure to the NH_3–CO_2–H_2O vapor mixture and 60 h of heat treatment, ensuring that no residual ammonia remained in the system during analysis. The analysis was performed according to the manufacturer's protocol, using pre-measured reagents and a fixed reaction time to ensure consistency. The resulting absorbance values were compared against a calibration curve generated from standard ammonium solutions. The results confirmed that no residual ammonia was present in the system after heat treatment, indicating complete removal. The confirmed absence of residual ammonia after vapor treatment and prolonged heating underscores the chemical stability of the foams, an essential prerequisite for their safe integration into building applications where long-term compositional reliability and environmental compatibility are critical.

https://doi.org/10.3390/buildings16010056

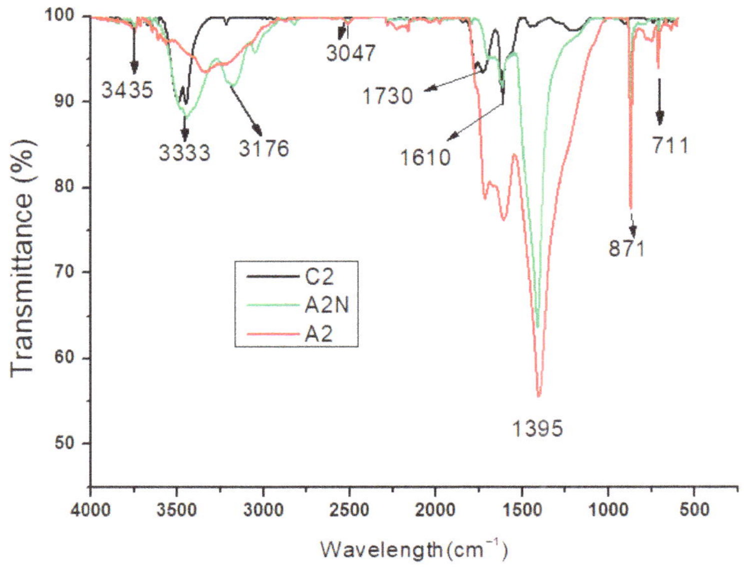

Figure 8. FTIR Spectra of Foam Samples C2, A2N (Ammonia-Treated) and A2 (Post-Sublimatio) corresponding to G/C Ratio of 1.5.

3.7. Determination of Water Absorption Capacity

The water absorption capacity of the foam was evaluated by recirculating sub-saturated water vapor through a bed of absorbent material housed in a Teflon mini-reactor. The vapor flow rate was maintained at 5 mL/s, and the temperature was stabilized at 28 °C. For each test, 390 mg of absorbent material was placed at the reactor's opening (diameter: 12 mm) on a polymer mesh with a mesh size of 600 µm. The sample was compacted under a pressure of 23,719.46 Pa, and a second mesh was placed on top to secure the material. The reactor was then hermetically sealed, and its total mass was recorded at regular time intervals to monitor water uptake. At 25 °C, using a $CaCl_2$ solution with a mass concentration of 25.27%, humidified air was generated with a water vapor concentration of approximately 0.0115 mg H_2O/cm^3. This controlled vapor stream was then passed through the absorbent material bed to evaluate its water uptake capacity.

To generate air with controlled humidity, a bubbling system was employed. Air was passed through a calcium chloride ($CaCl_2$) solution of known concentration, producing sub-saturated vapor. Aerosols were removed using a 0.2 µm pore-size filter, and the humidity of the incoming air was measured in-line using a calibrated hygrometer positioned just before the reactor inlet. It is known that the dependence of the relative vapor pressure π, on the mass fraction, ξ, of dissolved $CaCl_2$ and on temperature for an aqueous $CaCl_2$ solution, defined as the ratio between the vapor pressures of the solution and the saturated water vapor, both determined at the same temperature, is given by an equation that depends on two functions, π_{25} and f(x,q). The function π_{25} provides the equation for the dependence of the relative pressure π determined at a temperature of 25 °C on the mass fraction ξ, while the function f(x,q) is a correction function that extends the validity of the equation over the entire temperature range. The dependence of the relative vapor pressure of the $CaCl_2$ solution on the functions π_{25} and f(x,q) is of the following form [59]:

$$\pi = \frac{p_{solCaCl2}(\xi, T)}{p_{H2O}(T)} = \pi_{25} \cdot f(\xi, \theta),$$

https://doi.org/10.3390/buildings16010056

where

$p_{H_2O}(T)$ is the vapor pressure of liquid water at temperature T;
ξ is the mass fraction of $CaCl_2$ dissolved in the solution.

$$\theta = \frac{T}{T_{c,H_2O}}$$

Here
θ is reduced temperature;
T is absolute temperature, K;
$T_{c,H2O}$ is the critical temperature of water (647.096 K),
where

$$\pi_{25} = \left[1 + \left(\frac{\xi}{\pi_6}\right)^{\pi_7}\right]^{\pi_8} - \pi_9 \cdot e^{-\frac{(\xi - 0.1)^2}{0.005}},$$

and

$$f(\xi, \theta) = A + B \cdot \theta,$$

$$A = 2 - \left[1 + \left(\frac{\xi}{\pi_0}\right)^{\pi_1}\right]^{\pi_2}, B = \left[1 + \left(\frac{\xi}{\pi_3}\right)^{\pi_4}\right]^{\pi_5} - 1$$

This air was then passed through a bed of absorbent material for analysis of water uptake.

The parameters from π_0 to π_9 used to calculate the relative vapor pressure are listed in Table 1.

Table 1. Polynomial coefficients for the calculation of $\pi_{25}\xi$.

i	0	1	2	3	4	5	6	7	8	9
π_i	0.31	3.698	0.6	0.231	4.584	0.49	0.478	−5.2	−0.4	0.018

The temperature dependence of the saturated water vapor pressure, expressed in Pa, is given by the Wexler equation [60]:

$$lnp_{sat\ H_2O}(T) = \sum_{i=0}^{6} g_i \cdot T^{i-2} + g_7 \cdot lnT$$

The values of the coefficients g_0–g_7 can be found in Table 2:

Table 2. Coefficients of the Wexler equation.

i	g_i
0	-0.29912729×10^4
1	-0.60170128×10^4
2	0.1887643854×10^2
3	$-0.28354721 \times 10^{-1}$
4	$0.17838301 \times 10^{-4}$
5	$-0.84150417 \times 10^{-9}$
6	$0.44412543 \times 10^{-12}$
7	0.2858487×10^1

https://doi.org/10.3390/buildings16010056

The concentration of water vapor in moist air, expressed in mg/cm^3, is given by the relation:

$$C_{H_2O} = \frac{\pi \cdot p_{sat\ H_2O} \cdot M_{H_2O}}{1000 \cdot R \cdot T}$$

where:

π—relative vapor pressure of aqueous calcium chloride solutions;

M_{H_2O}—molar mass of water (18.02 g/mol);

R—universal gas constant [Pa·m^3/mol·K]

The graphical representation of the dependence between relative vapor pressure and $CaCl_2$ concentration is shown in Figure 9.

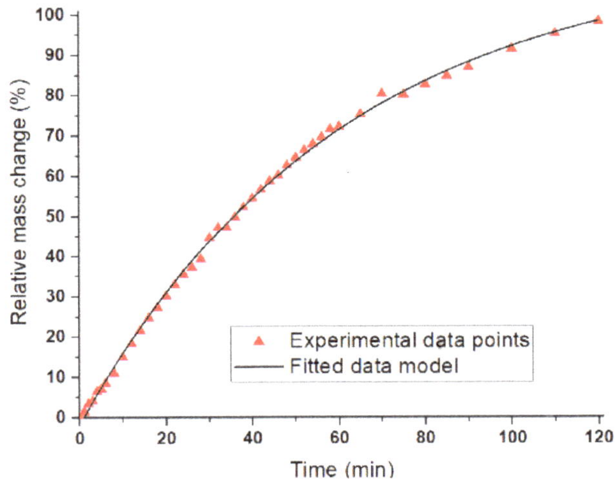

Figure 9. Time-resolved relative mass change (%) of the absorbent material during exposure to sub-saturated water vapor at 25 °C. Experimental data (red triangles) are fitted with an exponential decay model (black curve), indicating rapid initial uptake followed by saturation. Model parameters: A_1 = 114.87 ± 1.46, t_1 = −116.32 ± 1.31, y_0 = 60.37 ± 1.47; R^2 = 0.99871.

A gradual decrease in absorption capacity over time is observed; in the first 30 min, the sample completely absorbs the water vapor from the air, according to graph 8. The fitted exponential decay model describes the kinetics of water vapor uptake, where Δ_1 and t_1 represent the characteristic amplitude and time constant of the absorption process, respectively, and y_0 corresponds to the equilibrium offset. The high coefficient of determination (R^2 = 0.99871) demonstrates that the model accurately captures the experimental behavior, with rapid initial uptake followed by saturation. This statistical validation supports the reproducibility of the observed absorption kinetics.

Table A1 (Appendix A) represents the time-resolved mass measurements of the absorbent foam during exposure to sub-saturated water vapor at 25 °C. The initial dry mass was 56.3510 g, and subsequent increases reflect progressive water uptake. The column "Δ (g)" indicates cumulative mass gain relative to the initial mass, while "Δ Drop (g)" captures the incremental change between consecutive time points.

As shown in Figure 9, the foam absorbed water vapor efficiently during the first 30 min, after which the uptake rate declined, indicating the onset of saturation.

The foam achieved an absorption efficiency exceeding 98% within 2 h. Initially, $CaCl_2$ absorbed water to form sequential hydrates—$CaCl_2 \cdot 2H_2O$, $CaCl_2 \cdot 4H_2O$, and $CaCl_2 \cdot 6H_2O$—before dissolving into the additionally absorbed water. The resulting liquid

https://doi.org/10.3390/buildings16010056

phase wetted the foam's cellular walls, increasing the effective surface area for vapor–liquid exchange and enhancing the absorption rate during intermediate stages.

The foam demonstrated strong hygroscopic behavior, reaching a cumulative uptake of 0.384 g. The absorption profile featured a rapid initial gain (Δ = 0.0845 g at 14 min), followed by a gradual plateau (Δ = 0.3840 g at 120 min), consistent with diffusion-limited transport as internal pores became saturated. The "Δ Drop" metric revealed distinct kinetic phases, with values exceeding 0.01 g in successive intervals, suggesting intensified uptake driven by microstructural rearrangements or thermal feedback. The absorption mechanism follows a known hydrate formation sequence:

- Initial uptake forms $CaCl_2 \cdot 2H_2O$;
- Progresses to $CaCl_2 \cdot 4H_2O$ and $CaCl_2 \cdot 6H_2O$;
- Excess water dissolves $CaCl_2 \cdot 6H_2O$, forming a liquid phase.

This liquid wets the foam's cellular walls, increasing the geometric surface area of the vapor–liquid interface and enhancing uptake. The presence of $CaCl_2$ in the vapor stream further moderates partial pressure, enabling controlled condensation and deeper vapor penetration into the foam matrix. These behaviors align with documented effects in open-cell foams, where capillary condensation improves transport and retention.

Collectively, the results confirm that Foam C3, with its high $CaCl_2$ content and engineered porosity, is highly effective for ambient-temperature dehumidification. Its > 98% absorption efficiency within 2 h, combined with mechanical resilience and scalable synthesis, positions it as a promising candidate for air treatment, moisture control in packaging, and passive climate regulation technologies.

3.8. Acetic Acid Vapor Absorption Tests

As can be seen in Figure 10, during the first 35 min of stabilization, there is a decrease in the absorbance value (recorded absorbance curve) 1. This is probably due to the adsorption of acetic acid on the inner walls of the reactor and its dissolution in the silicone grease used to seal the valves and window gaskets.

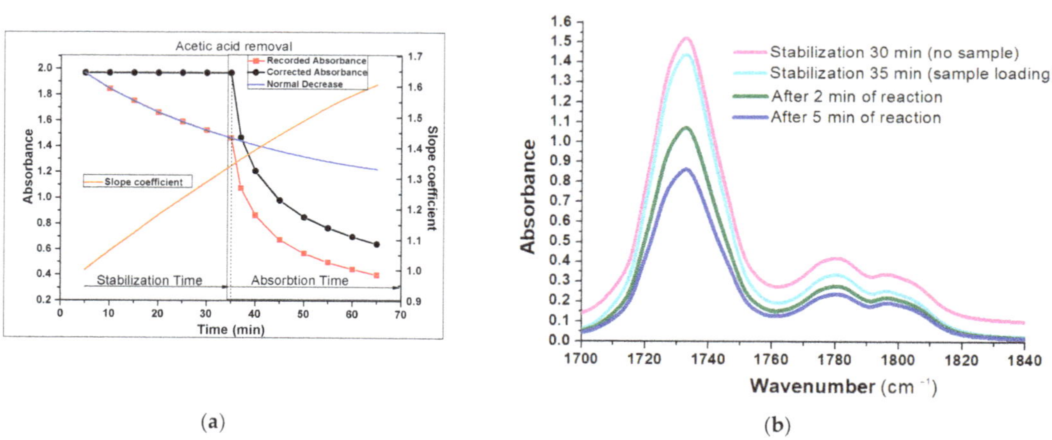

Figure 10. (**a**) Absorbance profile over time during acetic acid vapor exposure. The stabilization phase (0–30 min) shows a gradual decrease in recorded absorbance (red squares); The slope coefficient (orange) quantifies the rate of change in absorbance (**b**) Absorbance Evolution in the carboxyl vibrations.

Subfigure a show the change in absorbance and its slope during stabilization and acetic acid removal, while subfigure b presents the evolution of the carbonyl peak

https://doi.org/10.3390/buildings16010056

(~1730 cm^{-1}) during stabilization and early reaction. Silicone grease is composed of polydimethylsiloxane and amorphous silicon dioxide as a thickening agent. The dissolution process is diffusion-limited. After powder release at 35 min, the absorption phase begins, with corrected absorbance (black circles) and fitted curve (blue line) indicating active uptake and reaction progression; After 35 min, the powder was released into the gas cell. During absorption, the following reaction takes place:

$$CaCO_3 + 2CH_3COOH \rightarrow Ca(CH_3COO)_2 + H_2O + CO_2. \quad (8)$$

Strong gas absorption occurs in the first 5 min, after which absorption is limited by the diffusion of the gas into the solid porous mass. Water vapor and the formed CO_2 are adsorbed onto the porous carbon; for this reason, the characteristic CO_2 band does not appear in the FTIR spectra. Overall, the foam's performance supports its applicability in dehumidification systems, moisture control in packaging, and passive climate regulation technologies, offering a scalable and eco-friendly solution for vapor-phase water capture.

The chemical reaction (R8) is evidenced by the decrease in the absorption bands at wavenumbers 1788, 1430, and 1382 cm^{-1}, corresponding to the C=O stretching, CH_3 deformation, and OH bending vibrations of the carboxyl, methyl, and hydroxyl groups in the gaseous CH_3COOH molecule (Figure 10b). The combined water and acetic acid vapor absorption experiments demonstrate that the $CaCl_2$–carbon foams are not only highly effective hygroscopic materials, capable of rapid and sustained uptake under ambient conditions, but also chemically reactive sorbents that undergo in situ transformations (e.g., $CaCO_3 \rightarrow Ca(CH_3COO)_2$) during acid exposure. This dual functionality—moisture regulation through hydrate formation and pollutant capture Via acid neutralization—underscores the versatility of the foams as scalable, eco-friendly platforms for integrated dehumidification, packaging protection, and passive climate regulation technologies.

Intense gas absorption is observed within the first 5 min, followed by a diffusion-limited regime as acetic acid penetrates the porous solid. Both water vapor and the generated CO_2 are adsorbed onto the carbon matrix, which explains the absence of the characteristic CO_2 band in the FTIR spectra.

These results reinforce the foam's suitability for dehumidification, moisture control in packaging, and passive climate regulation, highlighting its scalability and ecological potential for vapor-phase capture applications.

3.9. Economic Feasibility and Scalability

A key advantage of the one-pot synthesis route is its simplicity. The process relies on inexpensive and widely available precursors—glucose and calcium chloride—whose combined cost is typically below \$2 per kg of foam material [61]. Energy consumption is primarily associated with autoclave heating at 220 °C for 24 h, which translates into an estimated \$5–10 per kg depending on batch size and electricity costs [62]. Additional steps, such as CNC machining and vapor-phase treatment, add only marginal expenses. Overall, the production cost of glucose–$CaCl_2$ foams is estimated at \$10–20 per kg at laboratory scale, with potential reductions at industrial scale through batch processing.

When compared to existing materials, the economic advantage becomes clearer. Silica foams, while offering excellent thermal insulation, remain costly at \$100–300 per kg [63,64] due to complex drying processes and fragile handling requirements. Activated carbon, on the other hand, is inexpensive (\$5–15 per kg) but lacks multifunctionality, offering adsorption capacity without the combined benefits of moisture regulation and pollutant capture [65]. Glucose–$CaCl_2$ foams occupy a middle ground: they deliver balanced mechanical resilience, strong hygroscopic behavior, and chemical reactivity at a fraction of the cost of silica foams, while offering broader functionality than activated carbon.

https://doi.org/10.3390/buildings16010056

Scalability challenges remain, particularly in maintaining uniform pore architecture and consistent salt loading across larger batches. Managing byproducts during vapor-phase treatments also requires optimization. Nevertheless, the simplicity of the synthesis and the affordability of raw materials make these foams promising candidates for building material applications, especially in resource-poor settings where cost efficiency is critical. The trade-off between performance and cost is favorable: while slightly less insulating than silica foams, glucose–$CaCl_2$ foams provide multifunctional benefits at a significantly lower price point, positioning them as scalable, eco-friendly alternatives for passive climate regulation and insulation.

4. Conclusions

This study shows that the composition and processing conditions of glucose–$CaCl_2$ foams directly determine their structural and functional performance. By tuning porosity and microstructure, we established clear links between material architecture and performance. The mechanical tests revealed a tunable response, ranging from higher-stiffness foams with enhanced energy absorption to more compliant structures with extended strain capacity, while independent structural analyses confirmed the features governing vapor transport and chemical reactivity. Thermal analysis confirmed reproducible salt incorporation and staged dehydration of $CaCl_2$ hydrates, with the carbon matrix accelerating vapor release compared to bulk salts. Spectroscopic and microscopic evidence further validated the in situ conversion of $CaCl_2$ to calcite and the removal of NH_4Cl, transformations that control pore opening and stability. Functionally, the foams showed strong hygroscopic behavior, efficiently capturing water vapor under ambient conditions, and demonstrated chemical reactivity toward acetic acid, neutralizing it through carbonate chemistry. These combined properties—including mechanical resilience, thermal buffering, moisture regulation, and pollutant capture—are directly relevant to building applications. Taken together, the results position glucose–$CaCl_2$ foams as scalable, eco-friendly candidates for construction materials, particularly in insulation and passive climate regulation. Their tunable porosity and integrated chemical functionality offer a pathway to building components that combine durability with environmental control. Future work will focus on optimizing synthesis parameters to balance mechanical strength with sorption capacity and on evaluating long-term performance under realistic building conditions. In addition to their functional performance, the foams demonstrate economic feasibility, with low-cost precursors and moderate energy requirements positioning them as competitive alternatives to silica foams and activated carbon for large-scale building applications. Despite the promising results, several limitations must be acknowledged. The scalability of the synthesis process remains to be fully validated, as maintaining uniform pore architecture and consistent salt incorporation in large-scale batches poses technical challenges. Long-term stability under fluctuating humidity, temperature, and pollutant exposure also requires systematic evaluation to ensure durability in real building environments. Furthermore, while preliminary cost estimates suggest competitive production compared to silica aerogels, a comprehensive techno-economic analysis is needed to confirm feasibility at industrial scale. Future research should focus on optimizing synthesis parameters to balance mechanical strength with sorption capacity, while also investigating accelerated aging tests to assess performance over extended lifetimes. Expanding applications beyond building insulation—such as indoor air purification, moisture buffering in heritage conservation, or integration into composite materials—could broaden the impact of these foams. Finally, considering climate change challenges, the development of multifunctional, low-cost materials that simultaneously regulate indoor environments and mitigate pollutant loads represents a critical pathway toward sustainable construction solutions.

 https://doi.org/10.3390/buildings16010056

Author Contributions: Conceptualization, R.B. (Radu Banica); methodology, F.S.R.; software, N.B. and E.L.; validation, R.B. (Raul Bucur) and F.S.R.; formal analysis, F.S.R., C.M. and R.B. (Radu Banica); investigation, F.S.R., R.B. (Radu Banica), D.R., E.L. and R.B. (Radu Banica); resources, F.S.R.; data curation, M.P.M., E.L. and F.S.R.; writing—R.B. (Raul Bucur), F.S.R., E.L. and C.M.; visualization, F.S.R.; supervision, R.B. (Radu Banica); project administration, R.B. (Radu Banica); funding acquisition, F.S.R. All authors have read and agreed to the published version of the manuscript.

Funding: This work was supported by a grant of the Ministry of Research, Innovation, and Digitization, CCCDI-UEFISCDI, project number PN-IV-P8-8.3-PM-RO-BE-2024-0004 within PNCDI IV.

Data Availability Statement: All data used in this study were generated by the authors as part of their research activities. These data may be obtained from the corresponding authors upon request.

Conflicts of Interest: The authors declare no conflicts of interest.

Appendix A

This experimental configuration was selected following a series of preliminary trials, aimed at achieving the highest possible stability of the absorbance baseline prior to the release of the foam into the cell. To investigate vapor-phase interactions, acetic acid vapors were introduced into a sealed gas cell (path length: 100 mm; internal diameter: 40 mm; total volume: 125 cm^3) equipped with KCl windows for infrared transmission. One of the windows was modified to integrate a PTFE-based dosing mechanism, allowing precise delivery of a defined foam mass directly into the cell without disrupting the sealed environment.

To minimize water vapor condensation during gas recirculation, the acetic acid solution was cooled in an ice bath prior to vapor generation. Additionally, aerosols were removed using a 0.2 µm polyethersulfone syringe filter (30 mm diameter), ensuring a clean vapor phase. Under these conditions, the system produced acetic acid vapors with a calculated partial pressure of 0.58 mmHg.

Gas circulation was maintained using a diaphragm-type micropump over a 120 min period. Once the system was sealed, 80 mg of foam was introduced into the cell, and FTIR spectra were collected at 300 s intervals for a total duration of 65 min. A schematic representation of the experimental setup is provided in Figure A1:

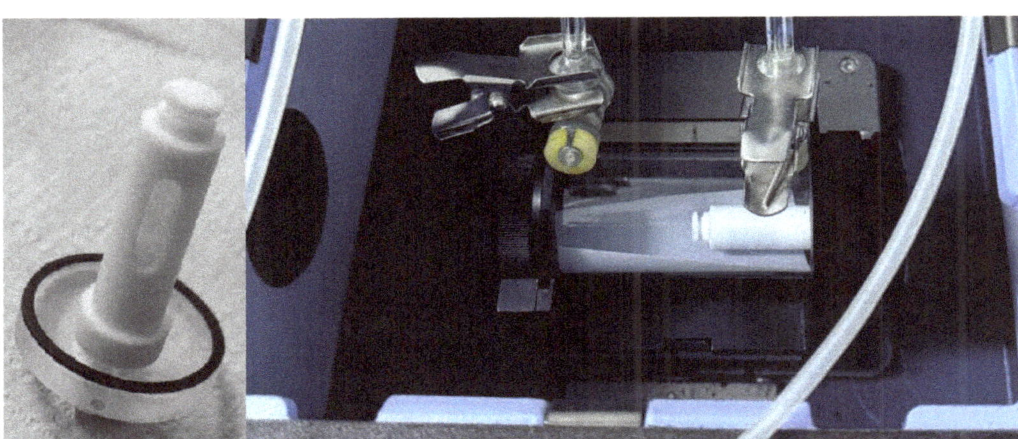

(**a**) (**b**)

Figure A1. Teflon sample container mounted in the KCl window (**a**) and the gas cell installed in the FTIR setup (**b**).

https://doi.org/10.3390/buildings16010056

SEM Figure A2 illustrates the morphological changes in the foams following water washing. After carbonization, the foams exhibit compact cellular carbon structures interspersed with carbon–$CaCl_2$ composites. Given the high solubility of $CaCl_2$, treatment with water or saturated steam facilitates pore opening in regions where solid $CaCl_2$ is embedded within the cell walls. However, in larger samples, limited water diffusion and the localized formation of saturated $CaCl_2$ solutions may hinder complete dissolution.

The SEM image clearly reveals the development of open porosity after washing, confirming that the foam architecture can be tuned to yield either closed or open pores depending on the treatment conditions.

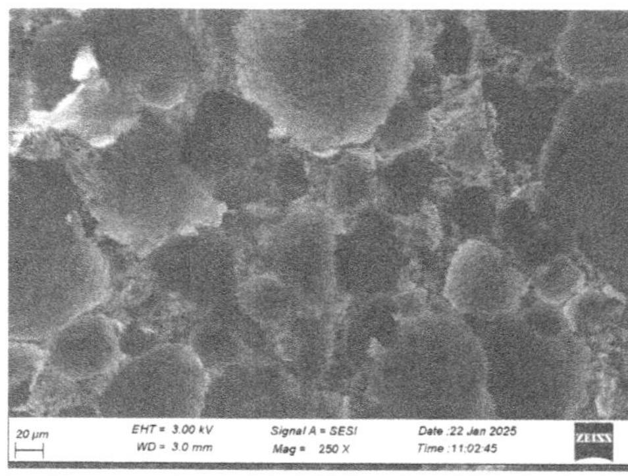

C1

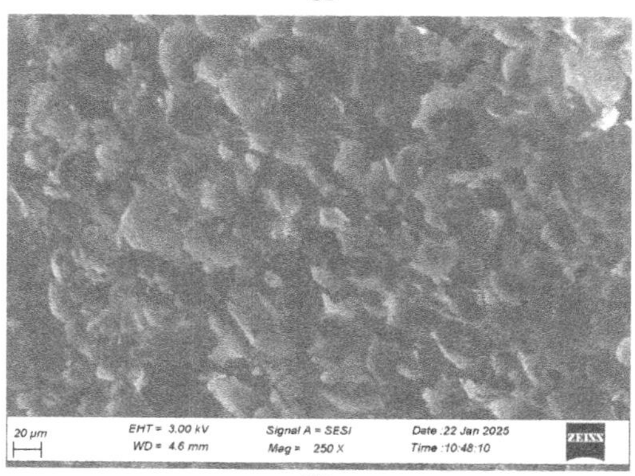

C2

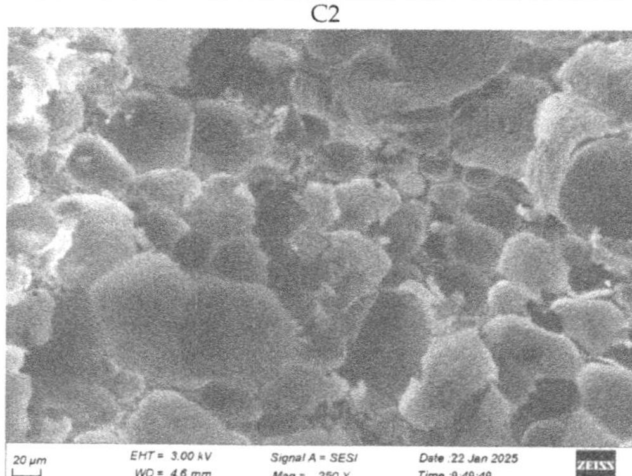

C3

Figure A2. Micrographs of (C1–C3) foams recorded after washing.

https://doi.org/10.3390/buildings16010056

Table A1. Time-resolved mass measurements of the absorbent material during vapor exposure, showing cumulative water uptake (Delta) and incremental changes (Delta Drop). Initial dry mass: 56.3510 g. The initial dry mass was 56.3510 g, and subsequent increases reflect progressive water uptake. The column "Δ (g)" indicates cumulative mass gain relative to the initial mass, while "Δ Drop (g)" captures the incremental change between consecutive time points.

Time (Min)	Mass (g)	Delta (g)	Delta Drop (g)
0	56.3510	0.0000	
1	56.3563	0.0053	0.0053
2	56.3645	0.0135	0.0082
3	56.3669	0.0159	0.0024
4	56.3763	0.0253	0.0094
5	56.3785	0.0275	0.0022
6	56.3842	0.0332	0.0057
8	56.3933	0.0423	0.0091
10	56.4093	0.0583	0.0160
12	56.4223	0.0713	0.0130
14	56.4355	0.0845	0.0132
16	56.4476	0.0966	0.0121
18	56.4575	0.1065	0.0099
20	56.4688	0.1178	0.0113
22	56.4796	0.1286	0.0108
24	56.4897	0.1387	0.0101
26	56.4957	0.1447	0.0060
28	56.5043	0.1533	0.0086
30	56.5253	0.1743	0.0210
32	56.5355	0.1845	0.0102
34	56.5359	0.1849	0.0004
36	56.5458	0.1948	0.0099
38	56.5552	0.2042	0.0094
40	56.5638	0.2128	0.0086
42	56.5721	0.2211	0.0083
44	56.5806	0.2296	0.0085
46	56.5861	0.2351	0.0055
48	56.5956	0.2446	0.0095
50	56.6032	0.2522	0.0076
52	56.6100	0.2590	0.0068
54	56.6161	0.2651	0.0061
56	56.6226	0.2716	0.0065
58	56.6302	0.2792	0.0076
60	56.6326	0.2816	0.0024
65	56.6446	0.2936	0.0120
70	56.6645	0.3135	0.0199
75	56.6639	0.3129	−0.0006
80	56.6737	0.3227	0.0098
85	56.6820	0.3310	0.0083
90	56.6902	0.3392	0.0082
100	56.7082	0.3572	0.0180
110	56.7230	0.3720	0.0148
120	56.7350	0.3840	0.0120

References

1. Kistler, S.S. Coherent Expanded Aerogels and Jellies. *Nature* **1931**, *127*, 741. [CrossRef]
2. Niculescu, A.-G.; Tudorache, D.-I.; Bocioagă, M.; Mihaiescu, D.E.; Hadibarata, T.; Grumezescu, A.M. An Updated Overview of Silica Aerogel-Based Nanomaterials. *Nanomaterials* **2024**, *14*, 469. [CrossRef]
3. Fricke, J.; Emmerling, A. Aerogels. *J. Am. Ceram. Soc.* **1992**, *75*, 2027–2035. [CrossRef]

 https://doi.org/10.3390/buildings16010056

4. Idumah, C.I.; Ezika, A.C.; Okpechi, V.U. Emerging Trends in Polymer Aerogel Nanoarchitectures, Surfaces, Interfaces and Applications. *Surf. Interfaces* **2021**, *25*, 101258. [CrossRef]
5. Mosoarca, C.; Hulka, I.; Şchiopu, P.; Rus, F.S.; Bănică, R. $CaCO_3$-Infused Carbon Fiber Aerogels: Synthesis and Characterization. *Ceramics* **2024**, *7*, 777–795. [CrossRef]
6. Banica, R.; Taranu, B.; Ladasiu, C.; Hulka, I.; Linul, P. Three-Dimensional Porous Electrode Based on Silver Nanowires for Hydrogen Sulfide Detection. *Mater. Lett.* **2021**, *304*, 130720. [CrossRef]
7. Linul, P.; Bănică, R.; Grad, O.; Linul, E.; Vaszilcsin, N. Highly Electroconductive Metal-Polymer Hybrid Foams Based on Silver Nanowires: Manufacturing and Characterization. *Polymers* **2024**, *16*, 608. [CrossRef]
8. Pekala, R.W.; Farmer, J.C.; Alviso, C.T.; Tran, T.D.; Mayer, S.T.; Miller, J.M.; Dunn, B. Carbon Aerogels for Electrochemical Applications. *J. Non-Cryst. Solids* **1998**, *225*, 74–80. [CrossRef]
9. Maleki, H.; Hüsing, N. Current Status, Opportunities and Challenges in Catalytic and Photocatalytic Applications of Aerogels: Environmental Protection Aspects. *Appl. Catal. B Environ.* **2018**, *221*, 530–555. [CrossRef]
10. Gan, G.; Li, X.; Fan, S.; Wang, L.; Qin, M.; Yin, Z.; Chen, G. Carbon Aerogels for Environmental Clean-Up. *Eur. J. Inorg. Chem.* **2019**, *2019*, 3126–3141. [CrossRef]
11. Voiconi, T.; Linul, E.; Marşavina, L.; Sadowski, T.; Kneć, M. Determination of Flexural Properties of Rigid PUR Foams Using Digital Image Correlation. *Solid State Phenom.* **2014**, *216*, 116–121. [CrossRef]
12. Mourshed, M.; Ahmed, R.; Kibria, M.G.; Islam Sarker, M.R. Carbon Foam for Next-Generation Electrochemical Energy Storage: Advances, Challenges, and Outlook. *Energy Convers. Manag. X* **2025**, *28*, 101281. [CrossRef]
13. Job, N.; Pirard, R.; Marien, J.; Pirard, J.-P. Porous Carbon Xerogels with Texture Tailored by pH Control during Sol–Gel Process. *Carbon* **2004**, *42*, 619–628. [CrossRef]
14. Raymundo-Piñero, E.; Azaïs, P.; Cacciaguerra, T.; Cazorla-Amorós, D.; Linares-Solano, A.; Béguin, F. KOH and NaOH Activation Mechanisms of Multiwalled Carbon Nanotubes with Different Structural Organisation. *Carbon* **2005**, *43*, 786–795. [CrossRef]
15. Wang, J.; Kaskel, S. KOH Activation of Carbon-Based Materials for Energy Storage. *J. Mater. Chem.* **2012**, *22*, 23710–23725. [CrossRef]
16. Biener, J.; Stadermann, M.; Suss, M.; Worsley, M.A.; Biener, M.M.; Rose, K.A.; Baumann, T.F. Advanced Carbon Aerogels for Energy Applications. *Energy Environ. Sci.* **2011**, *4*, 656–667. [CrossRef]
17. Hong, J.-Y.; Wie, J.J.; Xu, Y.; Park, H.S. Chemical Modification of Graphene Aerogels for Electrochemical Capacitor Applications. *Phys. Chem. Chem. Phys.* **2015**, *17*, 30946–30962. [CrossRef]
18. Gu, W.; Yushin, G. Review of Nanostructured Carbon Materials for Electrochemical Capacitor Applications: Advantages and Limitations of Activated Carbon, Carbide-Derived Carbon, Zeolite-Templated Carbon, Carbon Aerogels, Carbon Nanotubes, Onion-like Carbon, and Graphene. *WIREs Energy Environ.* **2014**, *3*, 424–473. [CrossRef]
19. Moreno-Castilla, C.; Maldonado-Hódar, F.J. Carbon Aerogels for Catalysis Applications: An Overview. *Carbon* **2005**, *43*, 455–465. [CrossRef]
20. Shen, Y.; Wu, Y. Review on Synthesis of Carbon Aerogels for CO_2 Capture. *Fuel* **2025**, *387*, 134370. [CrossRef]
21. Mohammadi, A.; Moghaddas, J. Synthesis, Adsorption and Regeneration of Nanoporous Silica Aerogel and Silica Aerogel-Activated Carbon Composites. *Chem. Eng. Res. Des.* **2015**, *94*, 475–484. [CrossRef]
22. Cao, Y.; Cheng, Z.; Wang, R.; Liu, X.; Zhang, T.; Fan, F.; Huang, Y. Multifunctional Graphene/Carbon Fiber Aerogels toward Compatible Electromagnetic Wave Absorption and Shielding in Gigahertz and Terahertz Bands with Optimized Radar Cross Section. *Carbon* **2022**, *199*, 333–346. [CrossRef]
23. Long, L.-Y.; Weng, Y.-X.; Wang, Y.-Z. Cellulose Aerogels: Synthesis, Applications, and Prospects. *Polymers* **2018**, *10*, 623. [CrossRef]
24. Payanda Konuk, O.; Alsuhile, A.A.A.M.; Yousefzadeh, H.; Ulker, Z.; Bozbag, S.E.; García-González, C.A.; Smirnova, I.; Erkey, C. The Effect of Synthesis Conditions and Process Parameters on Aerogel Properties. *Front. Chem.* **2023**, *11*, 1294520. [CrossRef]
25. Yao, Z.; Zhang, W.; Yu, X. Fabricating Porous Carbon Materials by One-Step Hydrothermal Carbonization of Glucose. *Processes* **2023**, *11*, 1923. [CrossRef]
26. Chen, L.; Fan, Z.; Mao, W.; Dai, C.; Chen, D.; Zhang, X. Analysis of Formation Mechanisms of Sugar-Derived Dense Carbons via Hydrogel Carbonization Method. *Nanomaterials* **2022**, *12*, 4090. [CrossRef]
27. Khandaker, T.; Islam, T.; Nandi, A.; Anik, M.A.A.M.; Hossain, M.S.; Hasan, M.K.; Hossain, M.S. Biomass-Derived Carbon Materials for Sustainable Energy Applications: A Comprehensive Review. *Sustain. Energy Fuels* **2025**, *9*, 693–723. [CrossRef]
28. Titirici, M.-M.; Antonietti, M. Chemistry and Materials Options of Sustainable Carbon Materials Made by Hydrothermal Carbonization. *Chem. Soc. Rev.* **2009**, *39*, 103–116. [CrossRef]
29. Qiu, J.; Huang, B.; Liu, Y.; Chen, D.; Xie, Z. Glucose-Derived Hydrothermal Carbons as Energy Storage Booster for Vanadium Redox Flow Batteries. *J. Energy Chem.* **2020**, *45*, 31–39. [CrossRef]
30. Sun, J.; Wu, Z.; Ma, C.; Xu, M.; Luo, S.; Li, W.; Liu, S. Biomass-Derived Tubular Carbon Materials: Progress in Synthesis and Applications. *J. Mater. Chem. A* **2021**, *9*, 13822–13850. [CrossRef]

31. Hou, Y.; Sheng, Z.; Fu, C.; Kong, J.; Zhang, X. Hygroscopic Holey Graphene Aerogel Fibers Enable Highly Efficient Moisture Capture, Heat Allocation and Microwave Absorption. *Nat. Commun.* **2022**, *13*, 1227. [CrossRef]
32. Yuan, C.; Wang, J.; Weng, Y.; You, L.; Fang, L.; Luo, X.; Zheng, F. Konjac Glucomannan-Based Aerogels Exhibiting Ultra-Fast Solar-Driven Water Release Rate for Freshwater Harvesting in Low-Humidity Environments. *J. Environ. Chem. Eng.* **2025**, *13*, 119515. [CrossRef]
33. Ello, A.S.; Yapo, J.A.; Trokourey, A. N-Doped Carbon Aerogels for Carbon Dioxide (CO_2) Capture. *Afr. J. Pure Appl. Chem* **2013**, *7*, 61–66. [CrossRef]
34. Franco, P.; Cardea, S.; Tabernero, A.; De Marco, I. Porous Aerogels and Adsorption of Pollutants from Water and Air: A Review. *Molecules* **2021**, *26*, 4440. [CrossRef] [PubMed]
35. You, J.H.; Chiang, H.L.; Chiang, P.C. Comparison of Adsorption Characteristics for VOCs on Activated Carbon and Oxidized Activated Carbon. *Environ. Prog.* **1994**, *13*, 31–36. [CrossRef]
36. Azadi, E.; Dinari, M. Green Synthesis, Characterization, and Properties of Carbon Aerogels. In *Green Carbon Materials for Environmental Analysis: Emerging Research and Future Opportunities*; ACS Symposium Series; American Chemical Society: Washington, DC, USA, 2023; Volume 1441, pp. 1–23.
37. Balci, M.P.; Bayat, R.; Karakurt, C.; Sen, F. Development of Environmentally Friendly Lightweight Aerogel Composites as Sustainable Building Materials: High Insulation Performance and Application Potential. *Int. J. Environ. Sci. Technol.* **2025**, *22*, 14403–14416. [CrossRef]
38. CARBIP—Circular Aerogel Reusable Building Insulation Products. Available online: https://carbip.eu/ (accessed on 11 December 2025).
39. Lu, L.; Wang, H.; Yun, S.; Hu, J.; Wang, M. A State-of-the-Art Review of Novel Aerogel Insulation Materials for Building Exterior Walls. *Energy Sources Part A Recovery Util. Environ. Eff.* **2024**, *46*, 16231–16252. [CrossRef]
40. Aerogel: The Innovative Force in Low-Carbon Architecture, Building a New Line of Defense for Safe and Green Living. Available online: https://insulatewool.com/news/aerogel-the-innovative-force-in-low-carbon-architecture-building-a-new-line-of-defense-for-safe-and-green-living (accessed on 6 December 2025).
41. Wu, X.; Fu, J.; Lu, X. Hydrothermal Decomposition of Glucose and Fructose with Inorganic and Organic Potassium Salts. *Bioresour. Technol.* **2012**, *119*, 48–54. [CrossRef]
42. Xu, Z.-X.; Shan, Y.-Q.; Zhang, Z.; Deng, X.-Q.; Yang, Y.; Luque, R.; Duan, P.-G. Hydrothermal Carbonization of Sewage Sludge: Effect of Inorganic Salts on Hydrochar's Physicochemical Properties. *Green Chem.* **2020**, *22*, 7010–7022. [CrossRef]
43. Ma, X.-Q.; Liao, J.-J.; Chen, D.-B.; Xu, Z.-X. Hydrothermal Carbonization of Sewage Sludge: Catalytic Effect of Cl^- on Hydrochars Physicochemical Properties. *Mol. Catal.* **2021**, *513*, 111789. [CrossRef]
44. Tomida, D.; Suzuki, K.; Katou, Y.; Yokoyama, C. Vapor Pressures and Liquid Densities of Ammonium Chloride + Ammonia Mixtures. *J. Chem. Eng. Data* **2008**, *53*, 1583–1586. [CrossRef]
45. Khezrzadeh, O.; Mirzaee, O.; Emadoddin, E.; Linul, E. Anisotropic Compressive Behavior of Metallic Foams under Extreme Temperature Conditions. *Materials* **2020**, *13*, 2329. [CrossRef]
46. Movahedi, N.; Linul, E. Radial Crushing Response of Ex-Situ Foam-Filled Tubes at Elevated Temperatures. *Compos. Struct.* **2021**, *277*, 114634. [CrossRef]
47. Şerban, D.-A.; Linul, E. Fatigue Behaviour of Closed-Cell Polyurethane Rigid Foams. *Eng. Fail. Anal.* **2023**, *154*, 107728. [CrossRef]
48. Guo, Q. Effect of NH_4Cl Additives on the Phase Transformation and Morphology of α-Al_2O_3. *Ceram. Silik.* **2023**, *67*, 52–57. [CrossRef]
49. Pathak, A.D.; Nedea, S.; Zondag, H.; Rindt, C.; Smeulders, D. A DFT-Based Comparative Equilibrium Study of Thermal Dehydration and Hydrolysis of $CaCl_2$ Hydrates and $MgCl_2$ Hydrates for Seasonal Heat Storage. *Phys. Chem. Chem. Phys.* **2016**, *18*, 10059–10069. [CrossRef] [PubMed]
50. Sujith Karunadasa Dehydration of Calcium Chloride as Examined by High-Temperature X-Ray Powder Diffraction. Available online: https://acrobat.adobe.com/id/urn:aaid:sc:EU:5885d871-ee39-43c2-a7e0-f577959ab8cd (accessed on 10 April 2024).
51. Eberbach, M.C.; Seo, H.; Shkatulov, A.I.; Tinnemans, P.; Huinink, H.P.; Fischer, H.R.; Adan, O.C.G. Path-Dependent Hydration and Dehydration of $CaCl_2$. *Cryst. Growth Des.* **2025**, *25*, 5679–5688. [CrossRef]
52. Xu, X.; Dong, Z.; Memon, S.A.; Bao, X.; Cui, H. Preparation and Supercooling Modification of Salt Hydrate Phase Change Materials Based on $CaCl_2 \cdot 2H_2O/CaCl_2$. *Materials* **2017**, *10*, 691. [CrossRef]
53. Sevilla, M.; Fuertes, A.B. The Production of Carbon Materials by Hydrothermal Carbonization of Cellulose. *Carbon* **2009**, *47*, 2281–2289. [CrossRef]
54. Lin, Y.; Xu, H.; Gao, Y.; Zhang, X. Preparation and Characterization of Hydrochar-Derived Activated Carbon from Glucose by Hydrothermal Carbonization. *Biomass Convers. Biorefinery* **2023**, *13*, 3785–3796. [CrossRef]
55. Gunasekaran, S.; Anbalagan, G.; Pandi, S. Raman and Infrared Spectra of Carbonates of Calcite Structure. *J. Raman Spectrosc.* **2006**, *37*, 892–899. [CrossRef]
56. Socrates, G. *Infrared and Raman Characteristic Group Frequencies: Tables and Charts*, 3rd ed.; Wiley: Hoboken, NJ, USA, 2001.

 https://doi.org/10.3390/buildings16010056

57. Nakamoto, K. *Infrared and Raman Spectra of Inorganic and Coordination Compounds*, 6th ed.; Wiley: Hoboken, NJ, USA, 2009.
58. Coates, J. Interpretation of Infrared Spectra, A Practical Approach. In *Encyclopedia of Analytical Chemistry*; Meyers, R.A., Ed.; Wiley: Hoboken, NJ, USA, 2000; pp. 10815–10837.
59. Conde, M.R. Properties of Aqueous Solutions of Lithium and Calcium Chlorides: Formulations for Use in Air Conditioning Equipment Design. *Int. J. Therm. Sci.* **2004**, *43*, 367–382. [CrossRef]
60. Wexler, A. Vapor Pressure Formulation for Water in Range 0 to 100 °C. A Revision. *J. Res. Natl. Bur. Stand. Sect. A Phys. Chem.* **1976**, *80*, 775. [CrossRef] [PubMed] [PubMed Central]
61. Haines, P.J. *Principles of Thermal Analysis and Calorimetry*; Royal Society of Chemistry: London, UK, 2002; ISBN 978-0-85404-610-2.
62. Available online: https://www.sigmaaldrich.com/RO/en/search/calcium-chloride-anhydrous?focus=products&page=1&perpage=30&sort=relevance&term=calcium%20chloride%20anhydrous&type=producthttps://kindle-tech.com/faqs/do-autoclaves-use-a-lot-of-electricity (accessed on 11 December 2025).
63. Borzova, M.; Lenigk, V.; Gauvin, F.; Schollbach, K. Life Cycle Assessment of Silica Aerogel Produced from Waste Glass via Ambient Pressure Drying Method. *J. Clean. Prod.* **2024**, *477*, 143839. [CrossRef]
64. Group, I. *Silica Aerogel Manufacturing Plant Project Report*; IMARC Group: Noida, India, 2025.
65. Analysts, M. Activated Carbon Market Outlook and Pricing 2024. Available online: https://businessanalytiq.com/procurementanalytics/index/activated-carbon-prices/ (accessed on 11 December 2025).

 https://doi.org/10.3390/buildings16010056

buildings

MDPI

Article

Mechanical Characteristics Based on the Microstructure Analysis of Cementitious Composites Incorporating Polypropylene Powder

Jeonguk Mun, Dongwook Kim, Sunho Kang and Heeyoung Lee *

Department of Civil Engineering, Chosun University, 10 Chosundae 1-gil, Dong-Gu, Gwangju 61452, Republic of Korea; jeonguk0920@chosun.ac.kr (J.M.); dwkim92@chosun.kr (D.K.); lineho@chosun.ac.kr (S.K.)
* Correspondence: heeyoung0908@chosun.ac.kr

Abstract: Incorporating recycled plastics into construction materials offers environmental and economic benefits. This study examined the properties of cementitious composites incorporating recycled polypropylene (PP) powder to evaluate the feasibility of plastics as construction materials. Experimental parameters included PP content and a curing method. Ninety-six specimens were fabricated for compressive strength tests and 48 for flexural strength tests, with six specimens per parameter. The mechanical behavior of the PP cementitious composites was assessed through compressive and flexural strength tests alongside digital image correlation analysis. Field emission scanning electron microscopy (FE-SEM) and mercury intrusion porosimetry (MIP) were used to analyze the pore structure of cementitious composites. Additionally, X-ray diffraction and thermogravimetric analysis examined the thermal and chemical characteristics. Compared with the control specimens, cementitious composites containing 30% PP exhibited approximately 30% reduction in compressive strength but a 28% increase in flexural strength. FE-SEM and MIP results revealed defects that deteriorated the performance of the cementitious composites. However, the compressive strengths exceeded 30 MPa across all the tested parameters, which is satisfactory for construction applications. Furthermore, the addition of PP enhanced flexural strength, providing structural benefits that render it a viable option for sustainable construction materials.

Keywords: recycled polypropylene; cementitious composite; curing method; mechanical characteristics; internal structure

1. Introduction

Plastic is widely used across various fields owing to its excellent productivity and durability, with global annual production exceeding 400 million tons. However, according to the Organization for Economic Co-operation and Development's Global Plastics Outlook report, only approximately 9% of plastic used globally is recycled [1]. The low recycling rates of plastic pose considerable risks, including environmental pollution and challenges in waste management [2]. Most plastic waste (PW) is either discarded in terrestrial and marine environments or ends up in landfills. The resulting micro and nano plastics harm ecosystems and disrupt global carbon cycling [3–5]. Additionally, plastic production contributes substantially to accelerating global warming, accounting for 5% of total carbon emissions [6]. The COVID-19 pandemic exacerbated existing plastic issues by driving up plastic usage through increased demand for personal protective equipment, food packaging/delivery services, and online shopping. This surge led to a substantial rise in plastic packaging use [7,8]. These consumption habits and lifestyles still persist today, with projections indicating that PW could triple by 2060 [9–11]. Therefore, proper treatment and management of the rapidly rising volume of PW is becoming increasingly critical.

Currently, numerous policies and studies on plastic recycling are actively being developed [12–17]. Incorporating plastic into construction materials is a promising approach [18,19]. The incorporation of plastic into construction materials provides environmental and economic benefits [20,21]. This method not only aids in environmental protection by reducing greenhouse gas emissions from plastic production but also creates new value for plastic as a sustainable construction resource [22–25]. Awoyera and Adesina [26] found that PW could be applied in various forms, such as subbase, asphalt, filler, walls, and bricks for road construction. Incorporating PW into cementitious composites offers a highly efficient method for waste utilization [27–30]. Siddique et al. [31] revealed that incorporating plastic into concrete improves resistance to impact and shrinkage cracks. Numerous studies used various types of plastic in cementitious composites, typically incorporating plastic as either aggregate or fiber [32,33]. The effects of plastic on cementitious composites have been analyzed through experiments assessing mechanical properties, chemical properties, and durability [34–38]. These studies show that factors such as mix ratio, type of plastic, and particle shape and size result in significant variability in the outcomes [39–42].

Previous studies have revealed that small particles with rough and irregular shapes enhance the strength of concrete [43,44]. Juki et al. [45] found that smaller polyethylene (PET) particles (1–16 mm) enhanced concrete strength, whereas a higher plastic content reduced density and strength due to the low density of plastic. Ismail and Al-Hashmi [46] indicated that higher sand replacement rates decreased both compressive and flexural strength, with flexural strength decreasing by up to 30.5% compared to that of the control specimen. Mohammed et al. [47] showed that replacing up to 30% of aggregates with PVC (0.49–0.95 mm) caused less than an 8% strength loss while improving toughness, crack resistance, and wear resistance. Saxena et al. [48] reported that PET aggregates (0–20 mm) maintained compressive strength losses below 10% when the fine aggregate replacement was <10%, while significantly enhancing energy absorption (up to 421% and 378% increases at 20% fine and coarse replacements were observed, respectively). Similarly, Gesoglu et al. [49] found that PVC powder replacements of up to 25% of the cement weight maintained compressive strengths above 45 MPa. Overall, plastic with smaller particle sizes and optimized mixing ratios can improve the performance of cementitious composites.

Yin et al. [50] found that incorporating plastic fibers in concrete is effective in controlling crack formation resulting from the dry shrinkage of concrete, thereby preventing sudden failure [51]. In particular, polypropylene (PP) fibers exhibit high resistance to alkaline environments, contributing to long-term durability. Additionally, PP fibers offer both environmental and economic advantages over steel fibers while also significantly reducing work time.

Orouji et al. [52] reported that incorporating 1.5% PP fibers maximized compressive and flexural strength, with specimens showing 7.3-times greater deformation than controls. Fraternali et al. [53] found that PET and PP fibers improved heat resistance, strength, and toughness, with PET fibers increasing compressive strength by 11.07 MPa. Ghernouti et al. [54] showed that waste vinyl fibers increased compressive strength by up to 13% and flexural strength by up to 14% depending on the fiber length, though shorter fibers had little effect. Overall, plastic fibers enhance the strength, ductility, and fracture resistance of concrete.

The hydrophobic nature of plastic interferes with the interaction between plastic and cement, which reduces the strength and performance of cementitious composites [55]. To address this issue, previous studies have investigated the addition of admixtures to plastic-containing concrete or treating the plastic before mixing [56]. Ali et al. [57] showed that adding silica fume mitigated the strength loss in plastic-containing concrete, increasing the compressive strength from 32.78 to 40.34 MPa by incorporating 20% plastic and 20% silica fume. Albano et al. [58] found that the flexural strength improved by 10% at 200 °C but declined at higher temperatures. Naik et al. [59] and Abbas et al. [60] reported that chemical and surface treatments involving high-density polyethylene (HDPE) enhanced matrix–plastic bonding, increasing the mechanical strength by 5–10%. This improvement was attributed to enhanced adhesion between the matrix and the plastic, which became hydrophilic owing to the sand coating.

Despite considerable advancements in incorporating various types of plastics into construction materials, most existing studies have primarily focused on PET, polyethylene (PE), and PVC. High-purity PP can be extracted from PW using solvent extraction methods [61]. PP is known for high toughness, flexibility, and substantial elongation capacity, which enables it to withstand considerable deformation. PP can absorb energy from cracking and inhibit crack propagation, thereby enhancing the overall durability and flexural strength of cementitious composites [62–66]. However, PP remains underexplored despite its high recycling potential [67]. Prior research primarily examined PP fibers and aggregates, focusing on their mechanical reinforcement capabilities in concrete. By contrast, limited research has addressed the use of PP powder derived from PW. PP in fiber form necessitates mechanical or thermal processing before being incorporated in cementitious composites. This processing can lead to changes in the physical and chemical structures of the material, potentially diminishing mechanical performance and altering surface characteristics. Additionally, the production of fibers increases process cost, which can reduce economic productivity in large-scale construction. By contrast, using PP powder enables direct incorporation without extensive processing. Reduced processing minimizes structural degradation of the polymer and lowers overall production costs. Utilizing PP powder derived from PW offers a practical alternative for enhancing both the economic and environmental performance of cementitious composites.

Understanding the influence of PP powder on the hydration process of cementitious composites is crucial for effective application. The effects of PP powder on cement hydration depend on curing conditions and incorporation rate, which influence the extent of hydration, pore structure development, and matrix development. Therefore, this study aimed to investigate the interaction between PP and hydration processes based on the curing method and the role of microstructural changes in increasing or decreasing the strength. Cementitious composites (hereafter referred to as PP cementitious composites) were fabricated with varying PP content and curing methods. Most previous studies have typically incorporated relatively small amounts of plastic into cementitious composites, usually less than 10% by weight of cement. However, in this study, PP powder was added at levels of up to 30% by weight of cement to explore the effects of high-volume incorporation. Admixtures were incorporated to enhance the strength of the PP cementitious composite [68]. Compressive and flexural strength tests evaluated the effects of PP powder on the mechanical characteristics of the cementitious composite. Digital image correlation (DIC) was used to analyze micro-deformation and cracking in the PP cementitious composite. Internal structure analysis was conducted to investigate the factors contributing to changes in the microstructure of the PP cementitious composite, as well as the effects of PP on the hydration reaction of the composite. Field emission scanning electron microscopy (FE-SEM), mercury intrusion porosimetry (MIP), X-ray diffraction (XRD) analysis, and thermogravimetric analysis (TGA) were conducted. Finally, the effects of experimental parameters and materials on the characteristics of the PP cementitious composite were analyzed using the Pearson correlation coefficient.

2. Experimental Programs

2.1. Fabrication of Specimens

The composition ratios of the PP cementitious composite used in this study are presented in Table 1. The materials in the specimens included Type 1 Portland cement, No. 6 silica sand (0.35–0.7 mm), and PP powder. In this study, PP powder with a particle size range of 50–300 μm was used. This range was selected considering that the mechanical and microstructural properties of cementitious composites are sensitive to the particle size of PP powder. This range is readily obtainable through conventional industrial milling and sieving processes, and the particle size is comparable to that of cement and sand. Therefore, PP particles in this range are less likely to interfere with cement hydration and dispersion within the composite matrix. Silica fume was used to compensate for the expected strength reduction in the PP cementitious composite due to the inclusion of plastic. Table 2 lists the chemical properties of silica fume. A superplasticizer was used to ensure flowability and improve the workability of the cementitious composite. Each admixture complied with ASTM C1240-20 and ASTM C494 regulations [69,70].

Table 1. Composition ratios of PP cementitious composites.

W/C	Cement	Sand	Silica Fume	Superplasticizer
0.4	1.0	2.0	0.125	0.005

Table 2. Chemical properties of silica fume.

	Chemical Composition (%)				
	SiO_2	Fe_2O_3	CaO	Al_2O_3	MgO
Silica fume	90.0	3.0	2.0	1.5	0.3

Figure 1 illustrates the particle morphology of PP. The PP particles have thin, elongated structures with smooth surfaces [71] and exhibit a slightly porous structure. The material properties of the PP powder are presented in Table 3 [72–77].

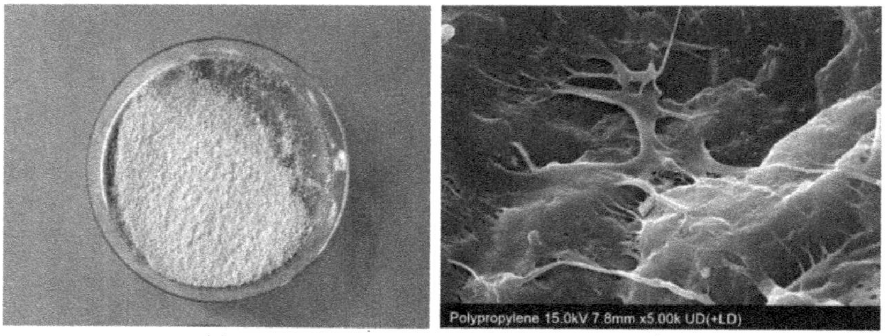

(a) (b)

Figure 1. PP particle: (**a**) PP powder used in the test and (**b**) FE-SEM image of the PP powder.

Table 3. Material properties of PP.

Category	Property	Test Method	Units	Value
Physical properties	Melt index (230 °C, 2.16 kg)	ASTM D1238 [72]	g/10 min	27
	Density	ASTM D792 [73]	kg/m^3	900
Mechanical properties	Yield stress	ASTM D638 [74]	MPa	22.6
	Elongation	ASTM D638 [74]	%	<100
	Flexural modulus	ASTM D790 [75]	MPa	1226
Impact properties	IZOD impact strength (23 °C)	ASTM D256 [76]	J/m	98
	IZOD impact strength (−10 °C)	ASTM D256 [76]	J/m	49
Thermal properties	Heat deflection temperature (0.45 MPa)	ASTM D648 [77]	°C	105

The parameters considered for specimen production were PP incorporation rate, curing method, and curing period. Table 4 lists the corresponding parameter values pertaining to the mechanical experiments. Specimen names were assigned according to the test type, PP content, and curing method. "C" and "F" represent the test type, with "C" indicating a compressive strength test and "F" denoting a flexural strength test. "PP" signifies the incorporation of PP, while the numbers "0," "10," "20," and "30" correspond to the ratio of the PP powder content relative to the cement weight. "R" and "H" indicate the curing methods, with "R" representing the room temperature (RT)-cured specimens and "H" indicating the HT-cured specimens. RT curing was conducted at 23 ± 2 °C, whereas high-temperature (HT) curing was first conducted in a chamber at 200 °C for 2 h and then followed by RT curing. To promote the hydration of cementitious composite and induce physical bonding between the cement matrix and PP powder, HT curing was conducted at 200 °C, exceeding the melting point of PP. A PP cementitious composite (containing 20% PP powder) used for compressive strength testing and cured under HT conditions is designated as "C-PP-20H."

Table 4. Test parameters.

Specimen Name	Test Type	PP Content (%)	Curing Method	Curing Period
C-Control-R	Compressive strength test	0	Room-temperature curing (23 ± 2 °C)	7, 28
C-PP-10R		10		
C-PP-20R		20		
C-PP-30R		30		
C-Control-H		0	High-temperature curing (200 °C)	
C-PP-10H		10		
C-PP-20H		20		
C-PP-30H		30		
F-Control-R	Flexural strength test	0	Room-temperature curing (23 ± 2 °C)	28
F-PP-10R		10		
F-PP-20R		20		
F-PP-30R		30		
F-Control-H		0	High-temperature curing (200 °C)	
F-PP-10H		10		
F-PP-20H		20		
F-PP-30H		30		

The water/cement ratio of the PP cementitious composite was 0.4, and the cement-to-sand ratio was 1:2. Control specimens used for comparison with the PP cementitious composite were made using only cement and sand. The PP powder incorporation ratios were 0%, 10%, 20%, and 30% relative to the cement weight. A total of 144 specimens were fabricated, with six specimens per parameter. Overall, 96 specimens were prepared for compressive strength tests and 48 specimens for flexural strength tests.

The curing periods of specimens for compressive strength were set at 7 and 28 d. The curing period of specimens for flexural strength was 28 d. The specimens for the compressive strength test were prepared with dimensions of 50 mm × 50 mm × 50 mm, in accordance with ASTM C109 regulations [78]. For the flexural test, specimens were fabricated with dimensions of 40 mm × 40 mm × 160 mm, in accordance with ASTM C348 regulations [79].

Figure 2 provides an overview of the specimen manufacturing process for strength tests. Figure 2a illustrates the materials required for specimen production. As illustrated in Figure 2b, cement, sand, PP powder, and silica fume were dry mixed in a mixer bowl for 5 min. Subsequently, water or a mixture of water and superplasticizer was added and mixed for 3 min, as presented in Figure 2c. As shown in Figure 2d, the mixture was compacted into appropriate molds. The compacted specimens were then cured in molds for 1 d, as illustrated in Figure 2e. RT or HT curing was then conducted according to the specified periods. These two curing methods were employed to examine the hardening characteristics of the cementitious composite based on the temperature.

Figure 2. Fabrication process of the PP cementitious composites: (**a**) preparation of materials, (**b**) dry mixing, (**c**) wet mixing, (**d**) compacting, (**e**) curing in a mold, and (**f**) curing in a chamber.

2.2. Mechanical Characteristics of the Specimens

Compressive and flexural strength tests investigated the effects of the curing method and PP powder incorporation on the overall mechanical performance of cementitious composites. Figure 3 presents the longitudinal section of the cementitious composite subjected to the mechanical strength tests, while Figure 4 shows the setup for the mechanical testing.

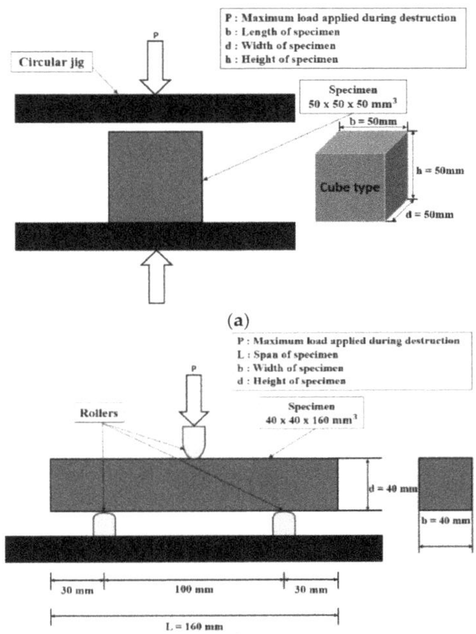

Figure 3. Longitudinal view of mechanical strength testing of cementitious composites: (**a**) compressive strength and (**b**) flexural strength.

The mechanical strength of the cementitious composite was tested using a universal testing machine with a 1000 kN capacity. In accordance with ASTM C109 [78], the compressive strength was measured using a displacement control method at a loading rate of 1 mm/min. The flexural strength of the cementitious composite was measured using a three-point bending test, with a loading rate of 50 N/s in accordance with ASTM C348 [79]. Rollers, placed at 100 mm intervals, were used to support the specimen. The compressive strength and flexural strength of the cementitious composite, measured through the three-point bending test, were calculated using the following equations.

$$f_c = \frac{P}{A} \tag{1}$$

$$f_r = \frac{3PL}{2bd^2} \tag{2}$$

where f_c denotes compressive strength, P denotes the maximum load applied during failure and A represents the cross-sectional area of the material resisting the load. f_r represents flexural strength, L denotes the span of the specimen, b represents the specimen width, and d represents the specimen height.

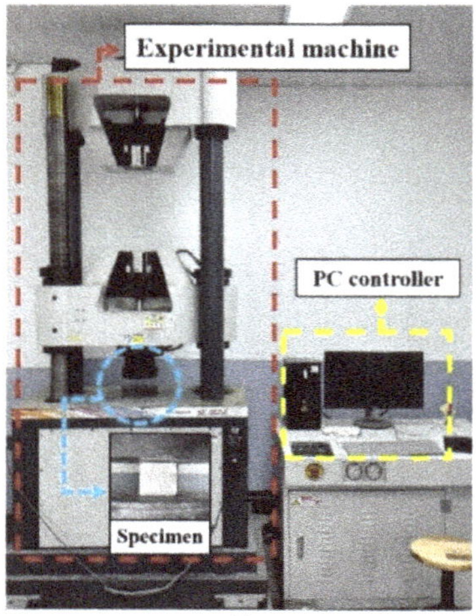

(a)

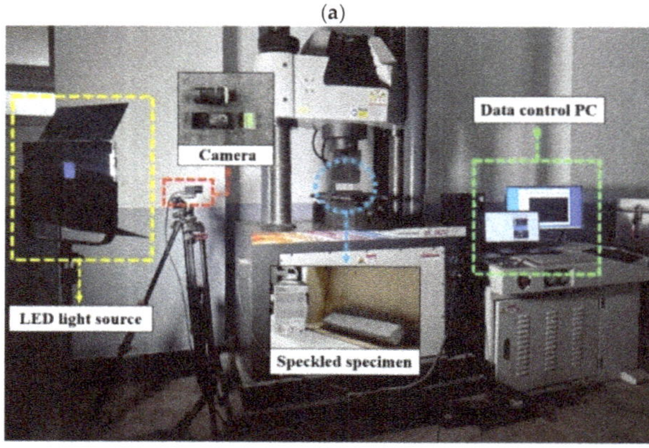

(b)

Figure 4. Set up for mechanical characteristics testing: (**a**) compressive strength and (**b**) flexural strength and DIC system.

Figure 4b presents the setup for the flexural test and DIC analysis. DIC analysis was used to visualize crack behavior under the three-point bending load. The distribution of deformation in the specimen was precisely measured through real-time monitoring. For smooth DIC tracking, a speckled pattern was applied to the surface of the specimen using a spray. An LED light source was used to uniformly illuminate the specimen, enhancing the precision of DIC analysis. A high-speed camera recorded the deformation process of the specimen over time at a rate of 5 fps. Mercury RT software version.3.0.12 was employed to analyze the experimental data [80].

2.3. *Internal Structure Analysis Methods*

Figure 5 shows the equipment used for internal structure analysis of PP cementitious composites. The internal structure was examined across different PP contents and curing methods, focusing on specimens that demonstrated the highest results in the 28 d strength tests. FE-SEM and MIP experiments examined the pore structure of the PP cementitious composites. Additionally, the chemical and thermal properties of the PP cementitious composite were analyzed through XRD and TGA experiments.

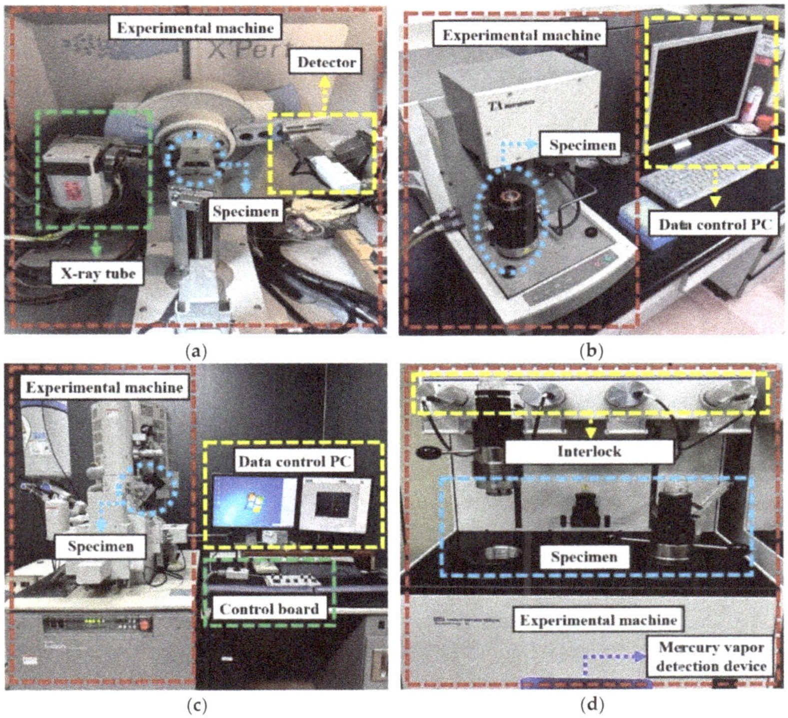

Figure 5. Equipment used for internal structure analysis: (**a**) FE-SEM, (**b**) MIP, (**c**) XRD, and (**d**) TGA.

2.3.1. FE-SEM Test Method

FE-SEM SU8600 observed the structure of PP, while FE-SEM S-4800 observed the microstructure, bonding conditions, and other characteristics of the specimens (Figure 5a). Before testing, dust was removed from the specimens, and specimens were then coated with platinum to improve conductivity. The specimens were viewed at high magnifications in accordance with ISO 19749 [81]. The high-resolution images obtained from FE-SEM provide detailed insights into microdefects and phase distribution within the specimens.

2.3.2. MIP Test Method

AutoPore V 9620 was used to analyze the influences of curing methods and PP on the pore structure of the cementitious composite (Figure 5b). The porosity and pore size distribution of the PP cementitious composite were evaluated by measuring the amount of mercury that intruded into the pores under various pressures. The pressure and pore size ranges were 3.45 kPa–413.69 MPa and 0.003–500 μm, respectively. The contact angle was in the range of 130–140°. The pore diameter was calculated using the Washburn equation [82,83]:

$$d = -\frac{4\gamma cos\theta}{P} \tag{3}$$

where d denotes pore diameter, γ represents the surface tension of mercury, θ denotes the contact angle, and P represents mercury penetration pressure.

2.3.3. XRD Test Method

XRD experiments were performed using the X'Pert PRO system to analyze the effects of PP incorporation and curing methods on hydrate formation in the cementitious composite. X-ray radiation was applied at 40 kV and 40 mA, in accordance with ASTM E915-96, and 2θ scanning was performed in the range of 10–80° (Figure 5c) [84].

2.3.4. TGA Test Method

TGA experiments were conducted in accordance with ASTM C1872 by heating specimens at a rate of 10 °C/min over a temperature range of 25–1000 °C [85]. Mass changes were measured as the specimens thermally decomposed with increasing temperature. The TGA 2050 analyzed the thermal characteristics of the PP cementitious composite by measuring the mass changes in the specimens (Figure 5d).

2.4. Pearson Correlation Analysis

The relationship between the compressive and flexural strength of cementitious composites and the various influencing variables was analyzed by quantifying correlation. The range of the Pearson correlation coefficient was from −1 to 1. The closer the value is to 0, the weaker the correlation, while −1 and 1 indicate strong negative correlation and positive correlation, respectively. The dataset comprised five input variables and two output variables. The input variables affecting the strength of the cementitious composite were curing days, PP content, silica fume, sand, and superplasticizer. The output variables were compressive strength and flexural strength.

3. Experimental Results and Discussion

3.1. Mechanical Characteristics of PP Cementitious Composites

3.1.1. Compressive Strength and Trend Analysis

Figure 6 illustrates how the compressive strength of PP cementitious composites varies with changes in PP content and curing method. The error bars represent the standard deviations calculated from measurements of six specimens tested using each compressive strength test. Generally, irrespective of curing methods, the compressive strength of the cementitious composite tends to decrease as the PP content increases. The trendlines in Figure 6 confirm a consistent decrease in compressive strength with increasing PP content, regardless of the curing method or period. The 28 d compressive strength of C-Control-R and C-Control-H, without PP incorporation, was 45.84 and 48.22 MPa, respectively. The 28 d compressive strength of C-PP-10R and C-PP-10H, with 10% PP incorporation, was 42.47 and 41.58 MPa, respectively, which is 85% of that of the Control specimen. The compressive strength of C-PP-30R and C-PP-30H, with the maximum PP incorporation, was 33.12 and 32.38 MPa, respectively. The decrease in compressive strength in these cementitious composites is attributed to the adhesion problem between plastic and cement paste caused by the hydrophobicity of the plastic. However, the compressive strength of all specimens remained above 30 MPa when appropriate admixtures were used. A compressive strength exceeding 30 MPa is considered sufficient for use in construction applications such as road pavement, indicating that a cementitious composite containing 30% PP can be suitably utilized.

The compressive strength of the RT-cured PP cementitious material increased by up to 20% as the curing period extended. By contrast, the compressive strength of the HT-cured PP cementitious composite showed minimal change with an extended curing period. Compared to the PP cementitious composite cured at room temperature, the composite cured at HT exhibited higher compressive strength at 7 d but lower compressive strength at 28 d. The HT of 200 °C promotes early crystallization, enhancing early strength but reducing the long-term strength of the PP cementitious composite [86]. HT curing enhances early strength but induces microcracking via thermal expansion. These microstructural defects compromise the integrity of the matrix over time, leading to lower compressive strength at later ages compared to RT curing.

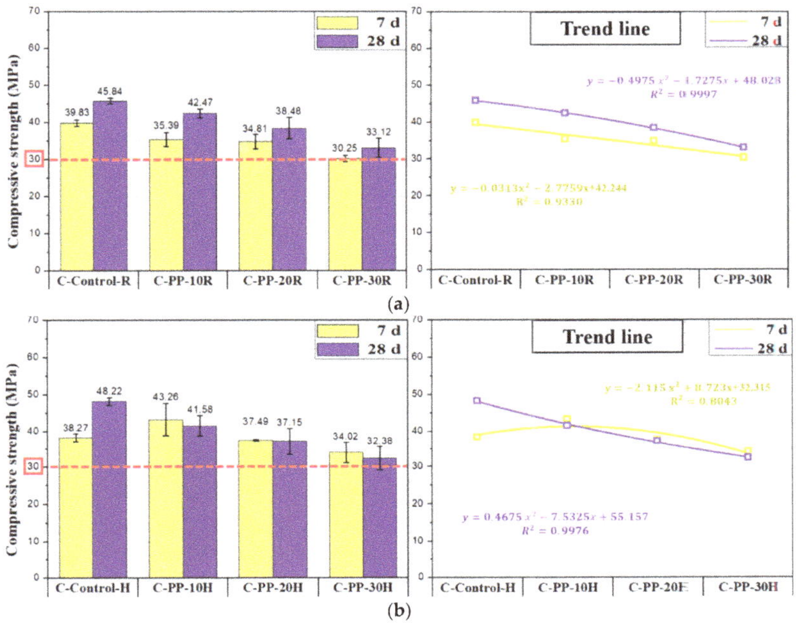

Figure 6. Compressive strengths of the PP cementitious composites under different curing methods: (**a**) RT curing and (**b**) HT curing.

3.1.2. Flexural Strength and DIC Analysis

Figure 7 illustrates the flexural strength of PP cementitious composites cured for 28 d. The error bars represent the standard deviations calculated from measurements of six specimens tested using each flexural strength test. Under RT curing, the incorporation of PP increased the flexural strength of the cementitious composite. The flexural strength of F-PP-10R was 11.81 MPa, which is approximately 10% higher than that of the Control (10.74 MPa), while F-PP-30R exhibited a flexural strength of 13.75 MPa, showing an improvement of approximately 28% compared to the Control. Generally, the flexural strength of cementitious composites tends to be proportional to compressive strength [87–90]. However, the flexural strength of RT-cured PP cementitious composite was inversely proportional to compressive strength. The increase in flexural strength of PP cementitious composites under RT curing can be attributed to the high elongation rate of polypropylene, which allows the plastic to enhance the resistance to bending.

By contrast, the flexural strength of HT-cured PP cementitious composites decreased as the PP content increased. The flexural strengths of F-PP-10H, F-PP-20H, and F-PP-30H were 10.02, 9.83, and 9.78 MPa, respectively, which were up to 40% lower than that of the flexural strength of specimens cured at room temperature. This reduction may be attributed to the thermal degradation of PP at elevated temperatures (200 °C), which diminishes the reinforcing capability of PP within the cementitious composite. During HT curing, thermal expansion and contraction of PP can lead to deformation and non-uniform pore distribution within the matrix. Furthermore, the increase in temperature accelerates thermal degradation, resulting in a decline in the mechanical properties and durability of PP. Although the high elongation rate of PP enhances flexural strength under RT curing, this reinforcing function cannot be maintained effectively under HT curing owing to a combination of the aforementioned reasons.

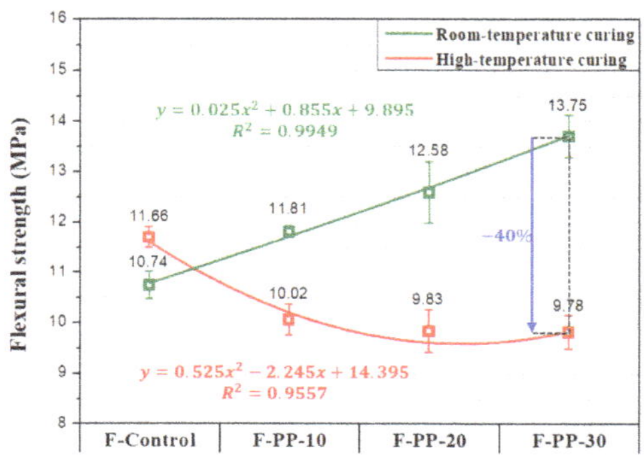

Figure 7. Flexural strengths of the PP cementitious composites.

The trend observed in the flexural strength results is supported by the DIC analysis. DIC analysis was used to monitor the flexural failure deformation of PP cementitious composites. Figure 8 presents the full-field strain distribution images of the composites, showing variations based on the curing method and incorporation rate. During the three-point bending test, compressive forces were applied to the upper part of the neutral axis of the specimen, while tensile forces acted on the lower part. Deformation in all specimens was concentrated around the middle of the span. The greatest deformation was observed close to the crack, which appeared at the lower central part of the specimen, aligning with the point of flexural failure. As shown in Figure 8a, cracks in specimens cured at room temperature propagated vertically in a straight line. Conversely, crack patterns in specimens cured at HTs were characterized by irregular curves. Figure 8b shows that the degree of bending increased with higher PP content. Under HT curing, PP exhibits thermal degradation and disperses irregularly within the cementitious composite. The non-uniformly distributed PP undergoes thermal shrinkage and expansion, resulting in the formation of uneven pores throughout the composite. These uneven pores disrupt the homogeneity of the matrix and contribute to the development of irregular crack patterns, as illustrated in Figure 8b. These internal defects weaken the stress transfer efficiency within the composite and reduce the ability of PP to resist flexural loads. Consequently, these factors are closely associated with the decrease in flexural strength observed in HT-cured specimens.

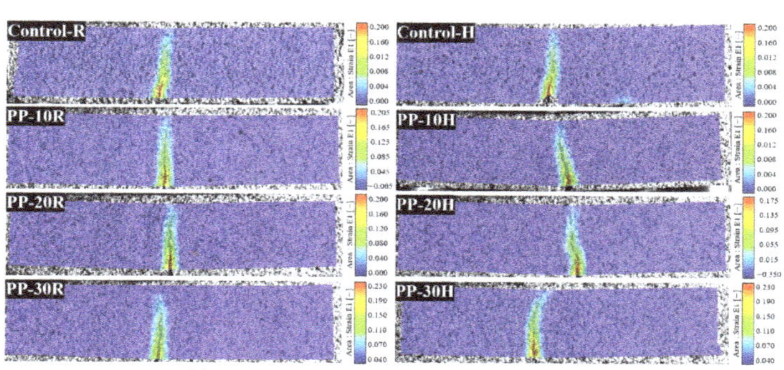

Figure 8. DIC analysis of cracks in the PP cementitious composites: (**a**) RT curing and (**b**) HT curing.

3.2. Internal Structure Analysis

3.2.1. FE-SEM

Using 5000× magnification FE-SEM images, the internal pores and hydrates of PP cementitious composites were examined in relation to PP incorporation and curing temperature. Specimens cured for 28 d were analyzed, and Figure 9 shows the FE-SEM image. Control-R and Control-H, without PP incorporation, exhibited abundant calcium silicate hydrate (C-S-H) and dense internal structures. By contrast, as observed in previous studies, porous structures were present in all specimens containing PP. Belmokaddem et al. [91] analyzed pores resulting from poor adhesion between plastic and the cement matrix using FE-SEM. The interfacial transition zone and pores within PP cementitious composites, caused by the hydrophobic nature of plastic, were observed in several studies [92,93]. FE-SEM observations revealed a weak interfacial transition zone (ITZ) between the PP particles and the cement hydration products, characterized by localized debonding and micropore formation. This poor interface is attributed to the hydrophobic nature of PP, which inhibits effective chemical bonding.

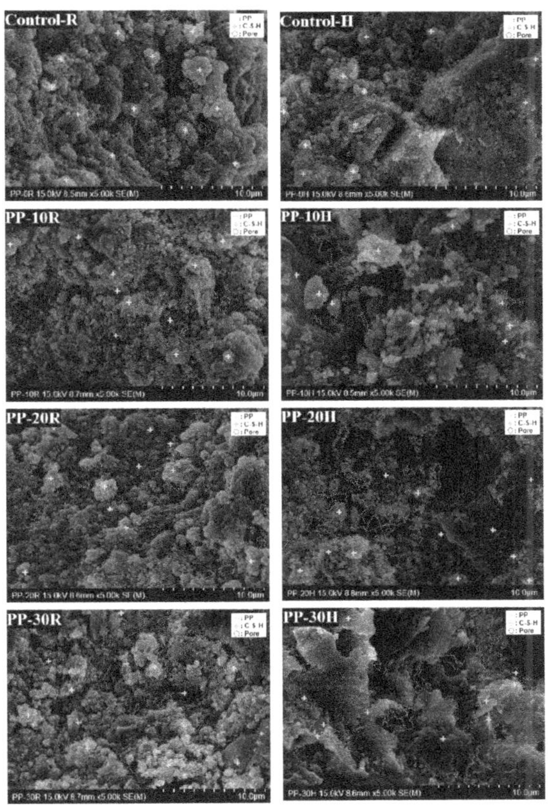

Figure 9. FE-SEM images of the PP cementitious composites: (**a**) RT curing and (**b**) HT curing.

Figure 9a illustrates that the frequency of micropores increases with higher PP content. In PP-30R and PP-30H, pores caused by poor bonding between plastic and cement paste become more prominent as the PP ratio increases. The internal structure of specimens cured at an HT exhibited distinct patterns compared to those cured at room temperature.

While Figure 9a depicts small and uniformly sized pores, Figure 9b reveals a range of pore sizes and a relatively smooth surface. Plastic does not physically or chemically interact with cement paste owing to its hydrophobic nature. However, HT (200 °C), exceeding the melting point of PP (130–171 °C), enables physical bonding between plastic and cement paste by imparting liquid-like properties to the plastic. Consequently, PP-10H, PP-20H, and PP-30H specimens exhibited smoother surfaces. Moreover, at HT curing, PP undergoes shrinkage and expansion during the melting and hardening processes, which facilitates physical bonding with the cement paste. The resulting variation in pore sizes (small and large) within PP cementitious composites is responsible for the diverse-sized pores depicted in Figure 9b, as compared to the more uniform pore sizes illustrated in Figure 9a. While physical bonding between PP and the cement paste can enhance the strength and internal structure of the composite, the detrimental effects of micropores outweigh these benefits, significantly impacting the overall strength of the PP cementitious composite.

3.2.2. MIP

The pore structure of PP cementitious composites observed through FE-SEM was further verified through MIP experiments, which measured specimen porosity under various parameters. Figure 10 illustrates the relationship between the 28 d compressive strength and porosity of the PP cementitious composite. Consistent with the FE-SEM images, the results indicated that the porosity of the PP cementitious composite increased with higher PP content. The porosity of PP-30 increased by approximately 60% compared to that of the Control specimen. This increase in porosity is attributed to poor contact and interference with hydrate formation owing to the hydrophobicity of the plastic. Furthermore, specimens cured at HT exhibited 2–4% higher porosity compared to those cured at room temperature. The growth of these interconnected pores reduces the load-bearing continuity of the cementitious matrix, directly explaining the observed decrease in compressive strength.

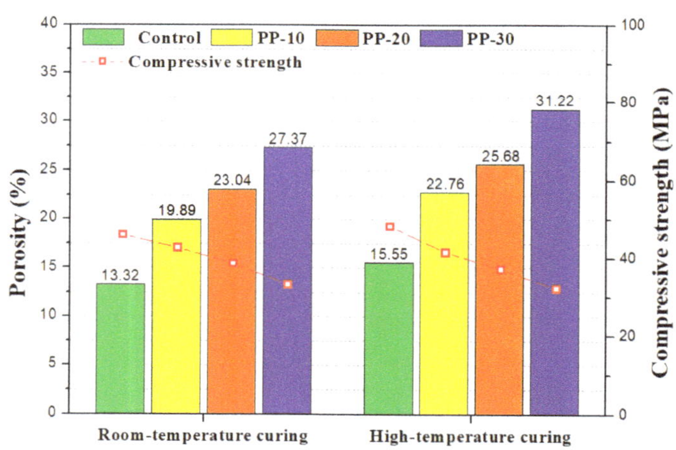

Figure 10. Relationship between the porosity and compressive strength of the PP cementitious composites.

Generally, the porosity of cementitious composite is inversely proportional to compressive strength, as observed in previous studies [94,95]. The porosity of PP cementitious composites in this study followed the same trend. In addition to porosity, the pore size distribution also plays a substantial role in determining the properties of cementitious composites [96]. Figure 11 illustrates the pore size distribution within the total pore volume. In this study, pores were categorized based on size into four types: gel pores (<10 nm), medium capillary pores (10–50 nm), large capillary pores (50–1000 nm), and macro pores (>1000 nm). Among these, large capillary pores and macro pores exceeding 50 nm are considered detrimental to the performance of cementitious composites and are classified as harmful pores [97,98].

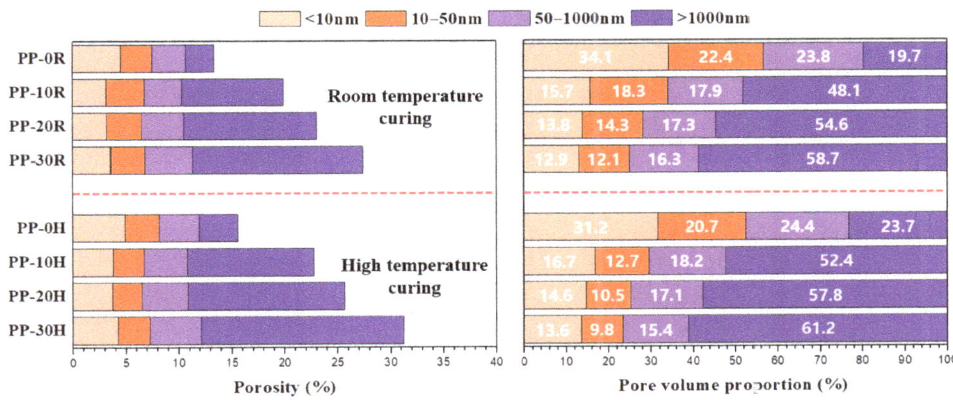

Figure 11. Pore characteristics of the PP cementitious composites.

The pore structure of the Control specimen was dense, with a relatively uniform distribution of pore sizes. Control-R and Control-H, which did not incorporate PP, contained 24–28% gel pores, attributed to the high content of C-S-H gel. By contrast, the pore structure of specimens incorporating PP was predominantly characterized by macro pores. For instance, PP-10R exhibited 15.7% gel pores and 48.1% macro pores, with the gel pore ratio being nearly halved and a significant increase in the macro pore ratio compared to Control-R. While the overall pore structure of cementitious composite was not substantially influenced by the curing method, specimens cured at HT exhibited a 3–4% higher macro pore ratio.

These results clearly demonstrate the effects of PP incorporation on the pore structure of cementitious composites, corroborating the FE-SEM findings. As PP content increased, the porosity of the cementitious composite increased, along with the proportion of harmful pores. This increase in both porosity and the harmful pore ratio contributed to the reduction in compressive strength. In particular, macro pores accounted for the highest proportion of the pore structure of the PP cementitious composite. As a result of analyzing the pore structures of the cementitious composites via the mechanical strength test, macro pores were determined to be a major factor affecting the reduction in the strength of the PP cementitious composites.

3.2.3. XRD Analysis

The XRD analysis verified the amount of hydrate and investigated the effect of hydration on the mechanical properties of PP cementitious composites. To investigate the causes of strength variations in cementitious composites owing to PP incorporation and curing temperature, specimens without PP (Control-R and Control-H) and those with maximum PP content (PP-30R and PP-30H) were analyzed. Figure 12 presents the XRD patterns of specimens. The 2θ angles corresponding to the observed peaks were similar across all specimens. Silica peaks were consistently present in all patterns, with prominent peaks around 27°. Additionally, calcium hydroxide (C-H) and C-S-H peaks appeared at uniform angles, with no evidence of new crystalline phases caused by the addition of PP. The absence of chemical interaction between PP and the cement paste confirms that changes in the mechanical properties of the PP cementitious composite are primarily attributed to physical and microstructural alterations rather than chemical reactions.

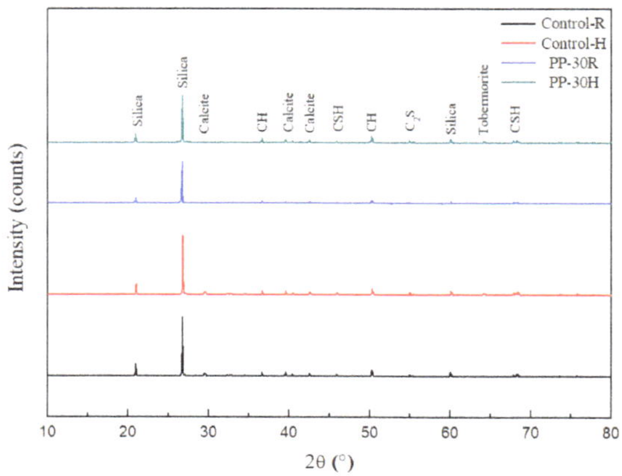

Figure 12. XRD analysis of the PP cementitious composites.

While the XRD patterns of all specimens were generally similar, differences in peak intensity and composition were observed depending on the presence of plastic and curing temperature. The XRD peaks of specimens incorporating PP were reduced, which can be attributed to the hydrophobic nature of the plastic. Weak calcite peaks were identified around 27° in Control-R and Control-H but were absent in PP-30R and PP-30H, indicating that the plastic interfered with the hydration reaction. The HT promoted hydration and increased the amount of hydration products. Tobermorite was observed only in HT-cured specimens (Control-H and PP-30H). In general, an increase in hydration products tends to enhance the strength of cementitious composites; however, factors such as pore structure, cracking, and non-uniform dispersion negatively affect the mechanical strength of the composite.

3.2.4. TGA

TGA was performed to examine the thermal characteristics of PP cementitious composites. Figure 13 presents the weight loss graph as a function of temperature. Each specimen exhibited mass reduction in three distinct stages. The first stage, occurring between 80 and 150 °C, resulted in approximately 3% weight loss due to water evaporation and C-S-H dehydration. The second stage commenced at 400 °C, where rapid mass reduction was noted owing to the decomposition of Portlandite. During this phase, the decomposition of the PP powder between 400 and 480 °C further contributed to weight loss. The final stage began at 750 °C, where the release of CO_2 from the thermal decomposition of $CaCO_3$ caused additional mass reduction, which continued up to 1000 °C. The highest weight loss observed was in PP-30R, which reached 17%.

TGA results indicated that the partial decomposition of PP increased the thermal weight loss of PP cementitious composites. HT curing accelerated dehydration and partial decomposition of hydration products, such as C-S-H and CH, during the curing process. This change reduced the amount of water remaining in the cement matrix. As a result, HT-cured specimens exhibited lower thermal weight loss compared to RT-cured specimens during the TGA test.

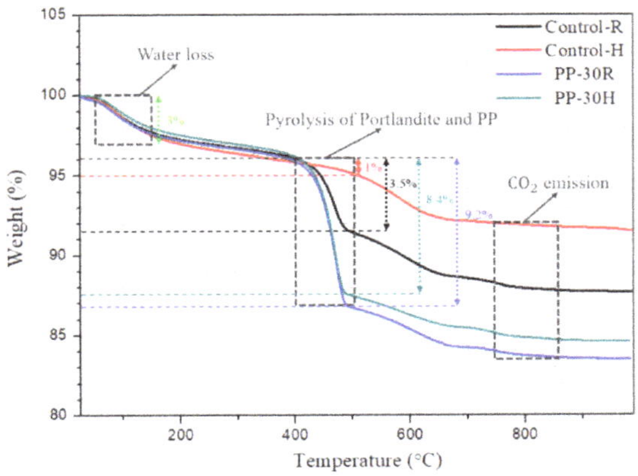

Figure 13. TGA curves of the PP cementitious composites.

3.3. Pearson Correlation Analysis of the PP Cementitious Composites

Figure 14 presents the Pearson correlation coefficients between input variables and the strength properties of cementitious composites. The analysis was conducted without normalization to preserve the original relationships among variables. The correlation between curing duration and flexural strength was excluded because the flexural strength test was performed at a fixed curing age of 28 d.

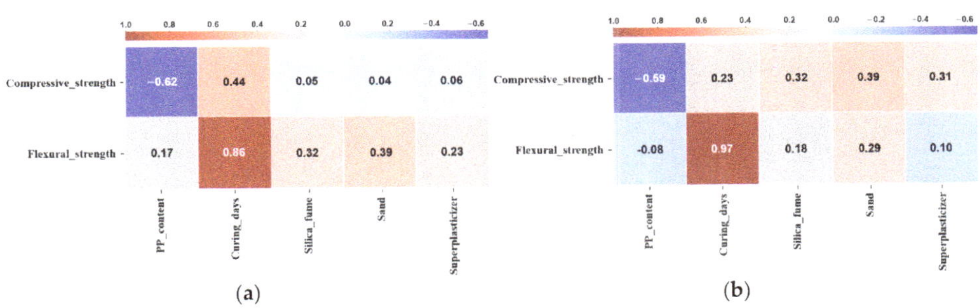

Figure 14. Pearson correlation heatmap: (**a**) RT curing and (**b**) HT curing.

The compressive strength shows a moderately strong negative correlation with PP content (r = −0.62) in Figure 14a, indicating that an increase in PP content reduces compressive performance. The microstructural analysis demonstrated that the presence of PP interferes with cement hydration and induces pore formation, which explains the observed decrease in compressive strength. The correlation between PP content and flexural strength is positive (r = 0.17). While the incorporation of PP impairs hydration and lowers flexural performance, the presence of PP contributes to localized deformation control and supports load transfer during bending. Silica fume shows a weakly positive correlation with compressive strength (r = 0.05), attributed to the pozzolanic reaction that facilitates secondary gel formation and matrix densification. Superplasticizer exhibits a positive correlation with compressive strength (r = 0.06). Increased workability resulting from superplasticizer usage promotes hydration and matrix development. The curing duration demonstrates a positive correlation with the compressive strength (r = 0.44), reflecting continuous hydration under RT curing.

In Figure 14b, the negative correlation between PP content and compressive strength is less pronounced (r = −0.59). Although thermal degradation due to HT curing typically reduces compressive strength, the strength loss appears partially offset by the accelerated hydration and enhanced pozzolanic reaction under elevated temperatures. By comparison, the positive correlation between curing duration and compressive strength is less pronounced (r = 0.44). Exposure to 200 °C causes significant moisture loss in the cementitious composite, which limits hydration owing to insufficient water during the curing period. PP content and flexural strength show a weak negative correlation owing to the deterioration of PP (r = −0.08). HT curing leads to microstructural degradation of the cementitious composite, including increased internal porosity and shrinkage-induced cracking. Silica fume and sand show relatively strong positive correlations with compressive strength. The SiO_2 in silica fume and sand enhances the pozzolanic reaction, which is further activated by elevated temperature, resulting in improved strength of the cementitious composite [99]. Superplasticizer also shows a positive correlation with compressive strength (r = 0.31), which is consistent with enhanced fluidity and reduced void formation in the matrix.

The PP content exhibits a strong negative correlation with compressive strength (r = −0.62), indicating that the compressive strength can be preserved by limiting the amount of incorporated PP. By contrast, under RT curing, the PP content shows a weak positive correlation with flexural strength (r = 0.17), suggesting that the flexural performance can be enhanced by increasing PP incorporation to an appropriate level. Therefore, the PP content in cementitious composites should be strategically adjusted to balance compressive and flexural strengths. Moreover, the correlation analysis reveals that curing temperature has a significant impact on the strength development of PP cementitious composites. Elevated temperatures can enhance strength through accelerated hydration and improved pozzolanic reactions; however, they may also impair mechanical properties owing to matrix degradation and microstructural defects. Therefore, selecting an appropriate curing method is essential to ensure the reliable performance of PP cementitious composites.

4. Conclusions

In this study, compressive and flexural strength tests were conducted to evaluate the mechanical characteristics of PP cementitious composites. Additionally, crack patterns were examined using DIC analysis. FE-SEM, MIP, XRD, and TGA were used to analyze the effects of different parameters on the internal structure of the PP cementitious composites. Finally, overall trends regarding the effects of PP powder on the cementitious composite were analyzed using the Pearson correlation coefficient and machine learning approaches. The main conclusions are as follows:

1. Increased PP content reduced compressive strength under all curing conditions owing to poor bonding and interruption of hydration. HT curing accelerated early strength but decreased long-term strength because of microcracks and porosity.
2. At RT curing, PP enhanced the flexural strength of a cementitious composite by up to 28%. Under HT curing, degradation of the PP and cement matrix reduced flexural strength by up to 40%. DIC analysis revealed irregular cracks resulting from uneven pore distribution and thermal deformation.
3. FE-SEM and MIP confirmed increased porosity and altered pore structure with PP addition. HT curing resulted in the formation of multi-scale pores and defects due to PP shrinkage and expansion, lowering structural integrity. As a result, the compressive strength of the PP cementitious composite decreased by up to 30%.
4. XRD results indicated that PP did not chemically react with the components of the cementitious composite and interfered with hydrate formation. The HT promoted hydration, but the resulting products did not improve mechanical strength owing to microstructural defects in the cementitious composite. TGA revealed greater weight loss with higher PP content owing to thermal decomposition and reduced hydration.

5. Pearson correlation analysis was conducted to investigate the overall effect of parameters on cementitious composites. Curing temperature significantly affected hydration, pore structure, and overall strength development. Ensuring the desired performance of cementitious composites in specific structural or environmental applications requires careful selection of the curing method.
6. With the use of suitable admixtures, the compressive strength of PP cementitious composites exceeded 30 MPa across all tested parameters. These findings highlight the importance of optimizing the mix design to balance mechanical performance and sustainability, ensuring that the incorporation of PP provides environmental benefits while minimizing impact on structural integrity.

Overall, these results suggest that cementitious composites incorporating plastic may be viable construction materials for applications such as road pavements, bridge slabs, and piers. However, the durability of PP cementitious composites, including their resistance to carbonation, freeze–thaw cycles, shrinkage, chloride ion penetration, and life cycle performance, was not evaluated in this study. Without such assessments, the long-term performance and reliability of PP cementitious composites in real-world structural applications remain uncertain. Therefore, future studies will focus on comprehensive durability evaluations to determine the applicability of these composites in practical construction environments.

Author Contributions: J.M.: Conceptualization, Writing and Editing; D.K.: Investigation, Formal analysis, Writing—Original draft preparation; S.K.: Investigation, Formal analysis, Writing—Original draft preparation; H.L.: Writing and Editing, Project administration. All authors have read and agreed to the published version of the manuscript.

Funding: This work was supported by a project for Collabo R&D between the Industry, University, and Research Institute funded by the Korea Ministry of SMEs and Startups in 2025 (RS-2025-02314369), and by the Regional Innovation System & Education (RISE) program through the Gwangju RISE Center, funded by the Ministry of Education (MOE) and the Gwangju Metropolitan City, Republic of Korea (2025-RISE-05-013).

Data Availability Statement: Data will be made available on request.

Conflicts of Interest: The authors declare that they have no competing interests.

References

1. Organization for Economic Co-Operation and Development. *Global Plastics Outlook: Policy Scenarios to 2060*; OECD Publishing: Paris, France, 2022. [CrossRef]
2. MacLeod, M.; Arp, H.P.H.; Tekman, M.B.; Jahnke, A. The global threat from plastic pollution. *Science* **2021**, *373*, 61–65. [CrossRef]
3. Ivar do Sul, J.A.I.I.; Costa, M.F. The present and future of microplastic pollution in the marine environment. *Environ. Pollut.* **2014**, *185*, 352–364. [CrossRef] [PubMed]
4. Thushari, G.G.N.; Senevirathna, J.D.M. Plastic pollution in the marine environment. *Heliyon* **2020**, *6*, e04709. [CrossRef] [PubMed]
5. Shen, M.; Huang, W.; Chen, M.; Song, B.; Zeng, G.; Zhang, Y. plastic crisis: Unignorable contribution to global greenhouse gas emissions and climate change. *J. Clean. Prod.* **2020**, *254*, 120138. [CrossRef]
6. Stegmann, P.; Daioglou, V.; Londo, M.; van Vuuren, D.P.; Junginger, M. Plastic futures and their CO_2 emissions. *Nature* **2022**, *612*, 272–276. [CrossRef]
7. Schlegel, I.; Gibson, C. The Making of an Echo Chamber: How the Plastic Industry Exploited Anxiety About COVID-19 to Attack Reusable Bags. 2020. Available online: https://trashhero.org/wp-content/uploads/2019/11/The-Making-of-an-Echo-Chamber_-How-the-plastic-industry-exploited-anxiety-about-COVID-19-to-attack-reusable-bags-1.pdf (accessed on 26 March 2020).
8. Prata, J.C.; Silva, A.L.P.; Walker, T.R.; Duarte, A.C.; Rocha-Santos, T. COVID-19 pandemic repercussions on the use and management of plastics. *Environ. Sci. Technol.* **2020**, *54*, 7760–7765. [CrossRef]
9. Shams, M.; Alam, I.; Mahbub, M.S. Plastic pollution during COVID-19: Plastic waste directives and its long-term impact on the environment. *Environ. Adv.* **2021**, *5*, 100119. [CrossRef]

10. Patrício Silva, A.L.P.P.; Prata, J.C.; Walker, T.R.; Duarte, A.C.; Ouyang, W.; Barcelò, D.; Rocha-Santos, T. Increased plastic pollution due to COVID-19 pandemic: Challenges and recommendations. *Chem. Eng. J.* **2021**, *405*, 126683. [CrossRef]
11. Lebreton, L.; Andrady, A. Future scenarios of global plastic waste generation and disposal. *Palgrave Commun.* **2019**, *5*, 6. [CrossRef]
12. Yaqub, W.; Aslani, F. Mechanical properties and fire resistance of 3D-printed cementitious composites with plastic waste. *Int. J. Concr. Struct. Mater.* **2024**, *18*, 84. [CrossRef]
13. Da Costa, J.P.; Mouneyrac, C.; Costa, M.; Duarte, A.C.; Rocha-Santos, T. The role of legislation, regulatory initiatives and guidelines on the control of plastic pollution. *Front. Environ. Sci.* **2020**, *8*, 104. [CrossRef]
14. Adam, I.; Walker, T.R.; Bezerra, J.C.; Clayton, A. Policies to reduce single-use plastic marine pollution in West Africa. *Mar. Policy* **2020**, *116*, 103928. [CrossRef]
15. Thiounn, T.; Smith, R.C. Advances and approaches for chemical recycling of plastic waste. *J. Pol. Sci.* **2020**, *58*, 1347–1364. [CrossRef]
16. Babafemi, A.J.; Norval, C.; Kolawole, J.T.; Chandra Paul, S.C.; Ibrahim, K.A. 3D-printed limestone calcined clay cement concrete incorporating recycled plastic waste (RESIN8). *Results Eng.* **2024**, *22*, 102112. [CrossRef]
17. Bharadwaaj, S.K.; Jaudan, M.; Kushwaha, P.; Saxena, A.; Saha, B. Exploring cutting-edge approaches in plastic recycling for a greener future. *Results Eng.* **2024**, *23*, 102704. [CrossRef]
18. Duan, Z.; Deng, Q.; Liang, C.; Ma, Z.; Wu, H. Upcycling of recycled plastic fiber for sustainable cementitious composites: A critical review and new perspective. *Cem. Concr. Compos.* **2023**, *142*, 105192. [CrossRef]
19. Li, L.; Yan, C.; Zhang, N.; Farooqi, M.U.; Xu, S.; Deifalla, A.F. Flexural fracture parameters of polypropylene fiber reinforced geopolymer. *J. Mater. Res. Technol.* **2023**, *24*, 1839–1855. [CrossRef]
20. Ramos, F.J.H.T.V.; Reis, R.H.M.; Grafova, I.; Grafov, A.; Monteiro, S.N. Eco-friendly recycled polypropylene matrix composites incorporated with geopolymer concrete waste particles. *J. Mater. Res. Technol.* **2020**, *9*, 3084–3090. [CrossRef]
21. Almeshal, I.; Tayeh, B.A.; Alyousef, R.; Alabduljabbar, H.; Mohamed, A.M. Eco-friendly concrete containing recycled plastic as partial replacement for sand. *J. Mater. Res. Technol.* **2020**, *9*, 4631–4643. [CrossRef]
22. Adesina, A. Recent advances in the concrete industry to reduce its carbon dioxide emissions. *Environ. Chall.* **2020**, *1*, 100004. [CrossRef]
23. Zulkernain, N.H.; Gani, P.; Chuck Chuan, N.C.C.; Uvarajan, T. Utilisation of plastic waste as aggregate in construction materials: A review. *Constr. Build. Mater.* **2021**, *296*, 123669. [CrossRef]
24. Vijayan, V.; Manthos, E.; Mantalovas, K.; Di Mino, G. Multi-recyclability of asphalt mixtures modified with recycled plastic: Towards a circular economy. *Results Eng.* **2024**, *23*, 102523. [CrossRef]
25. Wetchakornborirak, K.; Prommanee, P.; Supakata, N. Characteristics of aerogel-incorporated concrete with polyethylene terephthalate fiber. *Results Eng.* **2025**, *27*, 107135. [CrossRef]
26. Awoyera, P.O.; Adesina, A. Plastic wastes to construction products: Status, limitations and future perspective. *Case Stud. Constr. Mater.* **2020**, *12*, e00330. [CrossRef]
27. Al-Tayeb, M.M.; Aisheh, Y.I.A.; Qaidi, S.M.A.; Tayeh, B.A. Experimental and simulation study on the impact resistance of concrete to replace high amounts of fine aggregate with plastic waste. *Case Stud. Constr. Mater.* **2022**, *17*, e01324. [CrossRef]
28. Ahdal, A.Q.; Amrani, M.A.; Ghaleb, A.A.A.; Abadel, A.A.; Alghamdi, H.; Alamri, M.; Wasim, M.; Shameeri, M. Mechanical performance and feasibility analysis of green concrete prepared with local natural zeolite and waste PET plastic fibers as cement replacements. *Case Stud. Constr. Mater.* **2022**, *17*, e01256. [CrossRef]
29. Agyeman, S.; Obeng-Ahenkora, N.K.; Assiamah, S.; Twumasi, G. Exploiting recycled plastic waste as an alternative binder for paving blocks production. *Case Stud. Constr. Mater.* **2019**, *11*, e00246. [CrossRef]
30. Blazy, J.; Blazy, R. Polypropylene fiber reinforced concrete and its application in creating architectural forms of public spaces. *Case Stud. Constr. Mater.* **2021**, *14*, e00549. [CrossRef]
31. Siddique, R.; Khatib, J.; Kaur, I. Use of recycled plastic in concrete: A review. *Waste Manag.* **2008**, *28*, 1835–1852. [CrossRef]
32. Mustafa, M.A.T.; Hanafi, I.; Mahmoud, R.; Tayeh, B.A. Effect of partial replacement of sand by plastic waste on impact resistance of concrete: Experiment and simulation. *Structures* **2019**, *20*, 519–526. [CrossRef]
33. Ochi, T.; Okubo, S.; Fukui, K. Development of recycled PET fiber and its application as concrete-reinforcing fiber. *Cem. Concr. Compos.* **2007**, *29*, 448–455. [CrossRef]
34. Silva, R.V.; de Brito, J.; Saikia, N. Influence of curing conditions on the durability-related performance of concrete made with selected plastic waste aggregates. *Cem. Concr. Compos.* **2013**, *35*, 23–31. [CrossRef]
35. Feng, H.; Shao, Q.; Yao, X.; Li, L.; Yuan, C. Investigating the Hybrid Effect of Micro-steel Fibres and Polypropylene Fibre-Reinforced Magnesium Phosphate Cement Mortar. *Int. J. Concr. Struct. Mater.* **2022**, *16*, 35. [CrossRef]
36. Arulrajah, A.; Yaghoubi, E.; Wong, Y.C.; Horpibulsuk, S. Recycled plastic granules and demolition wastes as construction materials: Resilient moduli and strength characteristics. *Constr. Build. Mater.* **2017**, *147*, 639–647. [CrossRef]
37. Marzouk, O.Y.; Dheilly, R.M.; Queneudec, M. Valorization of post-consumer waste plastic in cementitious concrete composites. *Waste Manag.* **2007**, *27*, 310–318. [CrossRef]

38. Aneke, F.I.; Shabangu, C. Green-efficient masonry bricks produced from scrap plastic waste and foundry sand. *Case Stud. Constr. Mater.* **2021**, *14*, e00515. [CrossRef]
39. Kangavar, M.E.; Lokuge, W.; Manalo, A.; Karunasena, W.; Frigione, M. Investigation on the properties of concrete with recycled polyethylene terephthalate (PET) granules as fine aggregate replacement. *Case Stud. Constr. Mater.* **2022**, *16*, e00934. [CrossRef]
40. Akhmetov, D.; Akhazhanov, S.; Jetpisbayeva, A.; Pukharenko, Y.; Root, Y.; Utepov, Y.; Akhmetov, A. Effect of low-modulus polypropylene fiber on physical and mechanical properties of self-compacting concrete. *Case Stud. Constr. Mater.* **2022**, *16*, e00814. [CrossRef]
41. Kim, S.B.; Yi, N.H.; Kim, H.Y.; Kim, J.H.J.; Song, Y.C. Material and structural performance evaluation of recycled PET fiber reinforced concrete. *Cem. Concr. Compos.* **2010**, *32*, 232–240. [CrossRef]
42. Huynh, T.-P.; Ho Minh Le, T.; Vo Chau Ngan, N. An experimental evaluation of the performance of concrete reinforced with recycled fibers made from waste plastic bottles. *Results Eng.* **2023**, *18*, 101205. [CrossRef]
43. Thorneycroft, J.; Orr, J.; Savoikar, P.; Ball, R.J. Performance of structural concrete with recycled plastic waste as a partial replacement for sand. *Constr. Build. Mater.* **2018**, *161*, 63–69. [CrossRef]
44. Saikia, N.; De Brito, J. Mechanical properties and abrasion behaviour of concrete containing shredded PET bottle waste as a partial substitution of natural aggregate. *Constr. Build. Mater.* **2014**, *52*, 236–244. [CrossRef]
45. Juki, M.I.; Awang, M.; Annas, M.M.K.; Boon, K.H.; Othman, N.; Kadir, A.A.; Roslan, M.A.; Khalid, F.S. Relationship between compressive, splitting tensile and flexural strength of concrete containing granulated waste polyethylene terephthalate (PET) bottles as fine aggregate. *Adv. Mater. Res.* **2013**, *795*, 356–359. [CrossRef]
46. Ismail, Z.Z.; Al-Hashmi, E.A. Use of waste plastic in concrete mixture as aggregate replacement. *Waste Manag.* **2008**, *28*, 2041–2047. [CrossRef]
47. Mohammed, A.A.; Mohammed, I.I.; Mohammed, S.A. Some properties of concrete with plastic aggregate derived from shredded PVC sheets. *Constr. Build. Mater.* **2019**, *201*, 232–245. [CrossRef]
48. Saxena, R.; Siddique, S.; Gupta, T.; Sharma, R.K.; Chaudhary, S. Impact resistance and energy absorption capacity of concrete containing plastic waste. *Constr. Build. Mater.* **2018**, *176*, 415–421. [CrossRef]
49. Gesoğlu, M.; Güneyisi, E.; Hansu, O.; Etli, S.; Alhassan, M. Mechanical and fracture characteristics of self-compacting concretes containing different percentage of plastic waste powder. *Constr. Build. Mater.* **2017**, *140*, 562–569. [CrossRef]
50. Yin, S.; Tuladhar, R.; Shi, F.; Combe, M.; Collister, T.; Sivakugan, N. Use of macro plastic fibres in concrete: A review. *Constr. Build. Mater.* **2015**, *93*, 180–188. [CrossRef]
51. Balgourinejad, N.; Haghighifar, M.; Madandoust, R.; Charkhtab, S. Experimental study on mechanical properties, microstructural of lightweight concrete incorporating polypropylene fibers and metakaolin at high temperatures. *J. Mater. Res. Technol.* **2022**, *18*, 5238–5256. [CrossRef]
52. Orouji, M.; Zahrai, S.M.; Najaf, E. Effect of glass powder & polypropylene fibers on compressive and flexural strengths, toughness and ductility of concrete: An environmental approach. *Structures* **2021**, *33*, 4616–4628. [CrossRef]
53. Fraternali, F.; Ciancia, V.; Chechile, R.; Rizzano, G.; Feo, L.; Incarnato, L. Experimental study of the thermo-mechanical properties of recycled PET fiber-reinforced concrete. *Comput. Struct.* **2011**, *93*, 2368–2374. [CrossRef]
54. Ghernouti, Y.; Rabehi, B.; Bouziani, T.; Ghezraoui, H.; Makhloufi, A. Fresh and hardened properties of self-compacting concrete containing plastic bag waste fibers (WFSCC). *Constr. Build. Mater.* **2015**, *82*, 89–100. [CrossRef]
55. Al-Kerttani, O.M.G.; Hilal, N.; Hama, S.M.; Sor, N.H.; Banyhussan, Q.S.; Tawfik, T.A. Durability and hardened characteristics of cement mortar incorporating waste plastic and polypropylene exposed to $MgSO_4$ attack. *Results Eng.* **2024**, *24*, 103310. [CrossRef]
56. Rhee, I.; Lee, J.S.; Kim, Y.A.; Kim, J.M. Mechanical properties of cement mortar with gamma-irradiated recycled plastic powder, pellet, and fiber. *Results Eng.* **2023**, *20*, 101614. [CrossRef]
57. Ali, K.; Qureshi, M.I.; Saleem, S.; Khan, S.U. Effect of waste electronic plastic and silica fume on mechanical properties and thermal performance of concrete. *Constr. Build. Mater.* **2021**, *285*, 122952. [CrossRef]
58. Albano, C.; Camacho, N.; Hernández, M.; Matheus, A.; Gutiérrez, A. Influence of content and particle size of waste PET bottles on concrete behavior at different w/c ratios. *Waste Manag.* **2009**, *29*, 2707–2716. [CrossRef] [PubMed]
59. Naik, T.R.; Singh, S.S.; Huber, C.O.; Brodersen, B.S. Use of post-consumer waste plastics in cement-based composites. *Cem. Concr. Res.* **1996**, *26*, 1489–1492. [CrossRef]
60. Abbas, S.N.; Qureshi, M.I.; Abid, M.M.; Tariq, M.A.U.R.; Ng, A.W.M. An investigation of mechanical properties of concrete by applying sand coating on recycled high-density polyethylene (HDPE) and electronic-wastes (E-Wastes) used as a partial replacement of natural coarse aggregates. *Sustainability* **2022**, *14*, 4087. [CrossRef]
61. Faraca, G.; Astrup, T. Plastic waste from recycling centres: Characterisation and evaluation of plastic recyclability. *Waste Manag.* **2019**, *95*, 388–398. [CrossRef]
62. Zhao, Y.B.; Lv, X.D.; Ni, H.G. Solvent-based separation and recycling of waste plastics: A review. *Chemosphere* **2018**, *209*, 707–720. [CrossRef]

63. Banthia, N.; Gupta, R. Influence of polypropylene fiber geometry on plastic shrinkage cracking in concrete. *Cem. Concr. Res.* **2006**, *36*, 1263–1267. [CrossRef]
64. Yang, G.; Bi, J.; Dong, Z.; Li, Y.; Liu, Y. Experimental study on dynamic tensile properties of macro-polypropylene fiber reinforced cementitious composites. *Int. J. Concr. Struct. Mater.* **2022**, *16*, 66. [CrossRef]
65. Shen, D.; Liu, C.; Luo, Y.; Shao, H.; Zhou, X.; Bai, S. Early-age autogenous shrinkage, tensile creep, and restrained cracking behavior of ultra-high-performance concrete incorporating polypropylene fibers. *Cem. Concr. Compos.* **2023**, *138*, 104948. [CrossRef]
66. Hassan, S.A.; Hanoon, A.N.; Abdulhameed, A.A. Push-out test of eco-friendly steel-concrete–steel composite sections enhanced by polypropylene fibers: An experimental study and statistical analysis. *Results Eng.* **2024**, *23*, 102393. [CrossRef]
67. Rostami, R.; Zarrebini, M.; Mandegari, M.; Sanginabadi, K.; Mostofinejad, D.; Abtahi, S.M. The effect of concrete alkalinity on behavior of reinforcing polyester and polypropylene fibers with similar properties. *Cem. Concr. Compos.* **2019**, *97*, 118–124. [CrossRef]
68. Kang, S.; Pyo, S.; Lee, H. Thermal, electrical, and mechanical performances of ultrahigh-performance cementitious composites with multiwalled carbon nanotubes. *Case Stud. Constr. Mater.* **2024**, *21*, e03691. [CrossRef]
69. *ASTM C1240-20*; Standard Specification for Silica Fume Used in Cementitious Mixtures. ASTM International: West Conshohocken, PA, USA, 2003; Volume 15, pp. 1–6. [CrossRef]
70. *ASTM C494/C494M-19e1*; Standard Specification for Chemical Admixtures for Concrete, Annual Book of ASTM Standards. American Society for Testing and Materials: West Conshohocken, PA, USA, 1999. [CrossRef]
71. López-Buendía, A.M.; Romero-Sánchez, M.D.; Climent, V.; Guillem, C. Surface treated polypropylene (PP) fibres for reinforced concrete. *Cem. Concr. Res.* **2013**, *54*, 29–35. [CrossRef]
72. *ASTM D1238-10*; Standard Test Method for Melt Flow Rates of Thermoplastics by Extrusion Plastometer. American Society for Testing and Materials: West Conshohocken, PA, USA, 2005.
73. *Astm Committee D-20*; Standard Test Methods for Density and Specific Gravity (Relative Density) of Plastics by Displacement. ASTM International: West Conshohocken, PA, USA, 1991.
74. *ASTM D638-03*; Standard Test Method for Tensile Properties of Plastics. ASTM International: West Conshohocken, PA, USA, 2014.
75. *ASTM D790-15e2*; Standard Test Methods for Flexural Properties of Unreinforced and Reinforced Plastics and Electrical Insulating Materials. ASTM International: West Conshohocken, PA, USA, 2015.
76. *ASTM D256-10*; Standard Test Methods for Determining the Izod Pendulum Impact Resistance of Plastics. ASTM International: West Conshohocken, PA, USA, 2010.
77. *ASTM D648*; Standard Test Method for Deflection Temperature of Plastics Under Flexural Load in the Edgewise Position. Annual Book of ASTM Standards; ASTM International: West Conshohocken, PA, USA, 2007.
78. *ASTM C109*; Standard Test Method for Compressive Strength of Hydraulic Cement Mortars (Using 2-in, or [50-mm] cube specimens). Annual Book of ASTM Standards; ASTM International: West Conshohocken, PA, USA, 2016.
79. *ASTM C 348-08*; Standard Test Method for Flexural Strength of Hydraulic-Cement Mortars. ASTM International: West Conshohocken, PA, USA, 2008.
80. Choi, E.; Ostadrahimi, A.; Kim, W.J. Enhancement of compressive strength and strain ductility of SMA fiber reinforced concrete considering fiber's aspect ratios. *Constr. Build. Mater.* **2022**, *345*, 128346. [CrossRef]
81. *ISO 19749:2021*; Nanotechnologies—Measurements of Particle Size and Shape Distributions by Scanning Electron Microscopy. International Organization for Standardization: Geneva, Switzerland, 2021.
82. Washburn, E.W. Note on a method of determining the distribution of pore sizes in a porous material. *Proc. Natl. Acad. Sci. USA* **1921**, *7*, 115–116. [CrossRef]
83. Lee, H.; Jeong, S.; Cho, S.; Chung, W. Enhanced bonding behavior of multi-walled carbon nanotube cement composites and reinforcing bars. *Comput. Struct.* **2020**, *243*, 112201. [CrossRef]
84. *ASTM E915–96*; Test Method for Verifying the Alignment of X-Ray Diffraction Instrumentation for Residual Stress Measurement. ASTM International: West Conshohocken, PA, USA, 2002.
85. *ASTM C1872*; Standard Test Method for Thermogravimetric Analysis of Hydraulic Cement. ASTM International: West Conshohocken, PA, USA, 2018.
86. Ma, Q.; Guo, R.; Zhao, Z.; Lin, Z.; He, K. Mechanical properties of concrete at high temperature—A review. *Constr. Build. Mater.* **2015**, *93*, 371–383. [CrossRef]
87. Köksal, F.; Altun, F.; Yiğit, İ.; Şahin, Y. Combined effect of silica fume and steel fiber on the mechanical properties of high strength concretes. *Constr. Build. Mater.* **2008**, *22*, 1874–1880. [CrossRef]
88. Lee, J.H.; Cho, B.; Choi, E. Flexural capacity of fiber reinforced concrete with a consideration of concrete strength and fiber content. *Constr. Build. Mater.* **2017**, *138*, 222–231. [CrossRef]
89. Abbass, W.; Khan, M.I.; Mourad, S. Evaluation of mechanical properties of steel fiber reinforced concrete with different strengths of concrete. *Constr. Build. Mater.* **2018**, *168*, 556–569. [CrossRef]

90. Xu, B.W.; Shi, H.S. Correlations among mechanical properties of steel fiber reinforced concrete. *Constr. Build. Mater.* **2009**, *23*, 3468–3474. [CrossRef]
91. Belmokaddem, M.; Mahi, A.; Senhadji, Y.; Pekmezci, B.Y. Mechanical and physical properties and morphology of concrete containing plastic waste as aggregate. *Constr. Build. Mater.* **2020**, *257*, 119559. [CrossRef]
92. Abu-Saleem, M.; Zhuge, Y.; Hassanli, R.; Ellis, M.; Rahman, M.M.; Levett, P. Microwave radiation treatment to improve the strength of recycled plastic aggregate concrete. *Case Stud. Constr. Mater.* **2021**, *15*, e00728. [CrossRef]
93. Sáez del Bosque, I.F.; Zhu, W.; Howind, T.; Matías, A.; Sánchez de Rojas, M.I.; Medina, C. Properties of interfacial transition zones (ITZs) in concrete containing recycled mixed aggregate. *Cem. Concr. Compos.* **2017**, *81*, 25–34. [CrossRef]
94. Chan, Y.N.; Luo, X.; Sun, W. Compressive strength and pore structure of high-performance concrete after exposure to high temperature up to 800 °C. *Cem. Concr. Res.* **2000**, *30*, 247–251. [CrossRef]
95. Jin, S.; Zhou, J.; Zhao, X.; Sun, L. Quantitative relationship between pore size distribution and compressive strength of cementitious materials. *Constr. Build. Mater.* **2021**, *273*, 121727. [CrossRef]
96. Mehta, P.K.; Monteiro, P. *Concrete: Microstructure, Properties, and Materials*, 3rd ed.; McGraw-Hill: New York, NY, USA, 2006.
97. Silva, D.A.; John, V.M.; Ribeiro, J.L.D.; Roman, H.R. Pore size distribution of hydrated cement pastes modified with polymers. *Cem. Concr. Res.* **2001**, *31*, 1177–1184. [CrossRef]
98. Zhang, M.H.; Li, H. Pore structure and chloride permeability of concrete containing nano-particles for pavement. *Constr. Build. Mater.* **2011**, *25*, 608–616. [CrossRef]
99. Seleem, H.E.H.; Rashad, A.M.; Elsokary, T. Effect of elevated temperature on physico-mechanical properties of blended cement concrete. *Constr. Build. Mater.* **2011**, *25*, 1009–1017. [CrossRef]

buildings

MDPI

Article

Constitutive Modeling of Coal Gangue Concrete with Integrated Global–Local Explainable AI and Finite Element Validation

Xuehong Dong *, Guanghong Xiong *, Xiao Guan and Chenghua Zhang

School of Civil and Architecture Engineering, Xi'an University of Science and Technology, Xi'an 710054, China; guanxiao@xust.edu.cn (X.G.); zch0819@xust.edu.cn (C.Z.)

* Correspondence: 22404120302@stu.xust.edu.cn (X.D.); xiongguanghong@xust.edu.cn (G.X.)

Abstract: Coal gangue concrete (CGC), a recycled cementitious material derived from industrial solid waste, presents both opportunities and challenges for structural applications due to its heterogeneous composition and variable mechanical behavior. This study develops an ensemble learning framework—incorporating XGBoost, LightGBM, and CatBoost—to predict four key constitutive parameters based on experimental data. The predicted parameters are subsequently incorporated into an ABAQUS finite element model to simulate the compressive–bending response of CGC columns, with simulation results aligning well with experimental observations in terms of failure mode, load development, and deformation characteristics. To enhance model interpretability, a hybrid approach is adopted, combining permutation-based global feature importance analysis with SHAP (SHapley Additive exPlanations)-derived local explanations. This joint framework captures both the overall influence of each feature and its context-dependent effects, revealing a three-stage stiffness evolution pattern—brittle, quasi-ductile, and re-brittle—governed by gangue replacement levels and consistent with micromechanical mechanisms and numerical responses. Coupled feature interactions, such as between gangue content and crush index, are shown to exacerbate stiffness loss through interfacial weakening and pore development. This integrated approach delivers both predictive accuracy and mechanistic transparency, providing a reference for developing physically interpretable, data-driven constitutive models and offering guidance for tailoring CGC toward ductile, energy-absorbing structural materials in seismic and sustainability-focused engineering.

Keywords: coal gangue concrete; uniaxial compressive constitutive model; gradient boosting machine learning; finite element validation; SHAP interpretability; three-stage stiffness evolution

1. Introduction

Coal remains a principal energy source worldwide, with a particularly strong reliance observed across the Asia-Pacific region, where its share has consistently remained elevated. Among the primary by-products of coal mining, coal gangue (CG) represents roughly 10% to 25% of annual coal production, resulting in accumulated stockpiles surpassing 4.5 billion tons—constituting over 40% of China's total industrial solid waste [1]. At present, open-air dumping remains the predominant disposal method, causing extensive land occupation and triggering secondary geological hazards, such as landslides and debris flows. Moreover, prolonged exposure leads to the leaching of heavy metals (e.g., Pb^{2+}, Zn^{2+}, and Cu^{2+}) into the surrounding ecosystems, contaminating soil and water bodies. The spontaneous combustion of residual coal and pyrite further aggravates environmental risks by releasing harmful gases (SO_2, NO_x, CO), intensifying regional air quality concerns [2].

Existing research on CGC has coalesced into three principal strands. First, controlled experiments establish a pronounced non-monotonic dependence of mechanical performance on the gangue-aggregate replacement ratio: moderate substitution (30% coarse aggregate) tends to maximize strength, whereas excessive incorporation induces marked reductions in both elastic modulus and compressive strength [3]. Second, mixture-design optimization—via proportioning strategies and targeted material modifications, such as partial substitution of quartz powder with activated gangue or the inclusion of discrete fibers—has yielded improvements in strength, durability, and fatigue resistance [4,5]. Third, thermal mechanical activation protocols demonstrate that calcination at approximately 800 °C followed by fine grinding produces pozzolanic gangue powders that promote secondary hydration, densify the cementitious matrix, and refine pore structure, thereby enhancing strength development in alkali-activated binders. Collectively, this literature delineates the coupled influences of replacement level, mix design, and activation treatment, while elucidating the microstructural mechanisms governing performance enhancement in CGC [6].

The lack of robust mechanical modeling remains a central impediment to the widespread structural application of CGC. Although numerous constitutive formulations have been advanced across disparate experimental regimes, including uniaxial compression, varying replacement ratios, and freeze–thaw cycles [5,7], persistent deficiencies in generalizability, adaptability, and physical transparency are evident. A critical appraisal isolates three limitations: (1) minimal adoption of machine learning, only a small subset applies ML to CGC [8], with a prevailing reliance on linear regression to calibrate isolated parameters [6,9], a paradigm unable to capture the nonlinear, hierarchical interactions of heterogeneous cementitious systems; (2) reliance on restrictive statistical premises, normality, homoscedasticity, and feature independence—misaligned with CGC's variability and skewed distributions, degrading robustness across loading and curing regimes [9,10]; and (3) weak interpretability, models rarely identify dominant predictors or coupled mechanisms, limiting engineering deployment and mechanistic insight into CGC's behavior [10,11].

In contrast, recent advances in machine learning, particularly in geotechnics, fracture mechanics, and structural health monitoring, have addressed the limitations of conventional constitutive models [12–15]. Building on this, we develop a Gradient Boosted Tree (GBT) framework for predicting CGC constitutive parameters. By iteratively fitting the residuals, the model captures nonlinear interactions without prespecified functional forms and operates effectively in high-dimensional, heterogeneous feature spaces. To reconcile accuracy with physical interpretability, we couple permutation-based global importance with SHAP to quantify both aggregate influence and context-dependent effects. The framework yields three advances: (1) a unified, scalable pipeline linking feature attribution, multi-parameter prediction, stress–strain reconstruction, and finite element validation; (2) a data-driven yet interpretable alternative to regression formulations that identifies critical mix and curing variables—gangue replacement ratio, water–binder ratio, and age governing peak stress and fracture toughness; and (3) identification of a three-stage stiffness evolution—brittle, quasi-ductile, re-brittle—modulated by gangue content and consistent with micromechanical observations and numerical simulations. Collectively, these results establish a robust basis for embedding explainable AI into constitutive modeling and the engineering application of sustainable concrete.

2. Explainable Modeling of CGC Constitutive Parameters

2.1. Constitutive Foundations and Explainable-AI Problem Framing

Rigorous characterization of the compressive stress–strain response is fundamental to evaluating the CGC strength and deformability [16,17]. Convergent evidence shows that increasing gangue-aggregate replacement precipitates substantial capacity attenuation—peak stress reductions of 12.1 to 49.2%—while the macroscopic stress–strain morphology remains broadly consonant with conventional concrete [18,19]. However, CGC exhibits heightened parametric susceptibility of constitutive descriptors to gangue content, as corroborated by normative modeling [20]. Accordingly, we adopt the uniaxial compressive damage model codified in the Code for Design of Concrete Structures (GB50010-2010) as the baseline constitutive schema [21].

This stress–strain curve (Figure 1) is controlled by four key parameters: the elastic modulus E_c dictates the initial stiffness; the uniaxial compressive strength $f_{c,r}$ and peak strain $\varepsilon_{c,r}$ define the location of the curve's apex; and the post-peak parameter α_c characterizes the steepness of the descending branch. These parameters collectively influence the damage evolution function d_c, which in turn governs the slope transitions and geometric shape of the stress–strain trajectory. Accordingly, this study constructs a predictive modeling framework based on three machine learning algorithms, with these four constitutive parameters as the target variables. By learning the complex mapping between input features (material composition and curing conditions) and the key model parameters, this approach aims to provide a robust quantitative basis for developing accurate constitutive models of CGC.

Within this modeling context, XAI has materially advanced the transparency of civil engineering by integrating local, global, and deep learning interpretability. Locally, SHAP and LIME delineate salient predictors and material idiosyncrasies in structural retrofitting [22] and subgrade-strength prediction [23]; system-level frameworks fuse FEM-generated datasets with SHAP and a coherence model to robustly infer failure mechanisms and critical accelerations of flexible retaining walls under seismic loading [24]; globally, permutation importance and partial-dependence (PDP) analyses coupled with Auto ML prognosticate the capacity of ultra-high-performance concrete members [25]; for strengthening, experimental bond tests, literature synthesis, and ANN modeling yield a physically grounded analytical formula for SRG–concrete bond capacity, bridging data-driven inference with mechanics [26]; in deep learning, LIME and Grad-CAM render CNN decision pathways legible for ground-penetrating-radar inversion, exposing hierarchical subsurface features [27]. Collectively, these exemplars consolidate XAI's remit as an interpretable, domain-consonant paradigm across civil engineering applications.

2.2. Dataset Assembly and Feature Characterization

2.2.1. Experimental Data Integration and Predictor Selection

This study leverages a curated corpus of 262 representative CGC samples synthesized from diverse experimental programs spanning gangue typologies, replacement ratios, and curing regimens (Figure 2) [7,9,21,29–34]. The dataset comprises ten predictors and four response variables indexing salient mechanical properties. To foreground feature effects on model outputs, the six most prognostic variables were selected for statistical visualization in Figure 2.

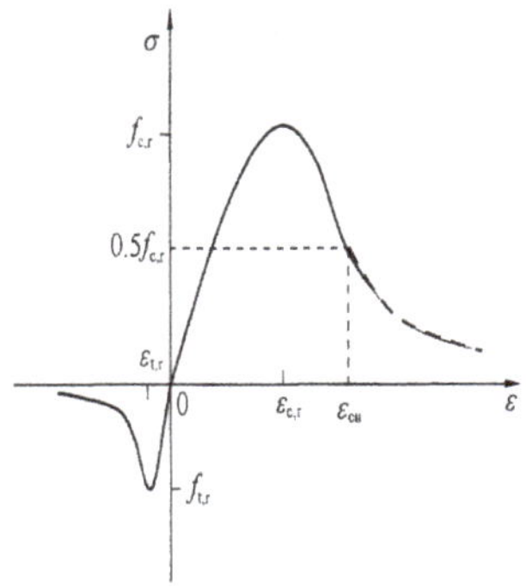

Figure 1. Schematic representation of the uniaxial stress–strain curve in concrete. Adapted from the Chinese Code for Design of Concrete Structures (GB50010-2015) [28].

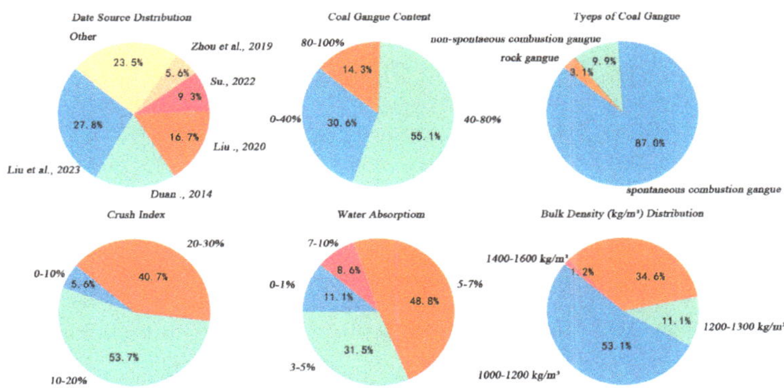

Figure 2. Distribution profiles of selected input features in the CGC dataset, including data sources, coal gangue content, types of coal gangue, crush index, water absorption, and bulk density. Data sources are compiled from previous studies [7,9,21,29–34].

To comprehensively delineate the structural characteristics of the dataset, descriptive statistics, including the minimum, maximum, mean, standard deviation, and number of unique values, were computed for each variable. The mean and interquartile range (Qi) capture the central tendency of the data distribution, whereas the standard deviation and extrema reflect the degree of variability and potential outliers. The summary statistics are listed in Table 1. Notably, variables such as elastic modulus and bulk density exhibit broad value ranges and high standard deviations, indicating pronounced variability in material composition and mechanical performance under diverse experimental conditions.

The final dataset was constructed through a rigorous selection and normalization process, consolidating approximately 500 raw data entries extracted from 16 core publications. Given the inconsistencies in testing standards, variable nomenclature, and measurement units across sources, all samples were manually reviewed, cleaned, and harmonized to ensure coherence. After careful screening, 262 high-quality entries were retained, most of which originated from the same laboratory and followed a consistent testing protocol. This standardized dataset effectively captures the representative parameter space of CGC and provides a robust foundation for subsequent machine learning modeling, particularly in validating model stability and generalizability across varied input domains.

Table 1. Descriptive statistics of key features and target parameters in the CGC dataset.

Parameter	Mean	Std	Min	Q1	Q2	Q3	Max
content (%)	66.3	35.11	0	40.0	70.0	100.0	100
age(day)	27.87	8.5	7	28.0	28.0	28.0	60
Bulk density (kg/m^3)	1205.78	124.29	1044	1100.0	1160.0	1350.0	1590
Cement kg×m^{-3}	395.19	104.19	210.0	325.0	361.0	457.0	640.0
crush index%	17.12	4.73	9.8	12.6	18.6	21.0	24.38
ratio	0.46	0.13	0.25	0.35	0.45	0.55	0.75
water kg×m^{-3}	173.36	23.32	136.5	160.0	160.0	195.0	268.0
water absorption	4.81	1.78	0.7	4.4	5.1	5.7	8.4
water reducer kg×m^{-3}	2.52	1.56	0.0	1.46	2.37	3.41	8.64
E_c	17,562.57	9881.03	1730	10,520.0	13,566.5	26,650.0	40,630
$\varepsilon_{c,r}$	2992.76	830.22	1564.67	2439.5	2808.85	3361.75	5153.32
$f_{c,r}$	29.14	8.86	10.97	22.58	29.04	34.98	56.2
α_c	1.29	1.42	0.67	0.35	0.99	1.69	7.01

2.2.2. Stratified Augmentation for Replacement-Level Balancing

To mitigate distributional sparsity while ensuring physical plausibility, we employed a physics-guided covariance-aware data augmentation strategy [35,36] applied exclusively to the training set. For each stratification tier of coal gangue replacement (0–40%, 50–80%, and 100%), synthetic samples were generated by drawing multivariate Gaussian perturbations as follows:

$$\Delta x \sim \mathcal{N}\left(0, \eta^2 \text{diag}(\sigma) R \text{diag}(\sigma)\right) \tag{1}$$

where R is the empirical correlation matrix of the training features and σ is the vector of their standard deviations (heteroscedastic scaling), perturbed samples were projected back into feasible domains by clipping each feature to its 1st–99th percentile range, enforcing the mix-design constraint ratio, and bounding gangue replacement between 0% and 100%. This procedure preserves observed inter-feature dependencies and mix-design constraints while increasing the effective sample density in each tier without contaminating the validation or test sets.

2.3. Predictive Performance Evaluation and Model Optimization

2.3.1. Metrics for Regression Accuracy and Robustness

To ensure a rigorous, generalizable evaluation across all constitutive targets, we conjointly employed five canonical regression metrics—R^2, MAE, MSE, RMSE, and MAPE (Equations (2)–(7)), per established practice [37,38]. This battery spans complementary error modalities, including absolute deviation, dispersion, relative bias, and explanatory power. Given scale and variance heterogeneity across outputs (elastic modulus and post-peak parameter α), single-metric appraisal is intrinsically susceptible to bias. Accordingly, we define a Composite Performance Index (CPI) that min–max normalizes and aggregates the five indicators, rendering them commensurate and enabling equitable comparison across multiscale responses. A lower CPI denotes a uniformly strong accuracy, whereas higher values indicate metric-specific weaknesses or imbalances, permitting a direct, objective ranking of predictive efficacy across heterogeneous constitutive parameters.

$$R^2 = 1 - \frac{\sum_{i=1}^{m} (\hat{y}_i - y_i)^2}{\sum_{i=1}^{m} (\overline{y} - y_i)^2} \tag{2}$$

$$MAPE = \frac{100\%}{m} \sum_{i=1}^{m} \frac{|(y_i - \hat{y}_i)|}{y_i} \tag{3}$$

$$MAE = \frac{1}{m}\sum_{i=1}^{m} \left|(y_i - \hat{y}_i)\right| \quad (4)$$

$$MSE = \frac{1}{m}\sum_{i=1}^{m} (y_i - \hat{y}_i)^2 \quad (5)$$

$$\mathrm{RMSE} = \sqrt{\frac{1}{m}\sum_{i=1}^{m} (y_i - \hat{y}_i)^2} \quad (6)$$

$$\mathrm{CPI} = \frac{1}{\mathrm{N}}\sum_{\mathrm{j}=1}^{\mathrm{N}} \frac{\mathrm{P_j} - \mathrm{P_{j,min}}}{\mathrm{P_{j,max}} - \mathrm{P_{j,min}}} \quad (7)$$

Let y_i, $\hat{y}_i$ and $\overline{y}$ represent the observed value, predicted value, and mean of the observed values, respectively, across a dataset comprising m samples. In Equation (7), P_j denotes the j-th evaluation indicator used to quantify model performance, with $P_{j,\min}$ and $P_{j,\max}$ corresponding to its lower and upper bounds. The total number of metrics employed, denoted as N, is predefined as 5 in this study.

2.3.2. Cross-Model Comparison Under Unified Conditions

To ensure the rigor and comparability of algorithmic performance evaluation, the following methodological safeguards were implemented:

(1) A suite of complementary performance metrics was employed to capture predictive error from multiple perspectives, including absolute deviation, outlier sensitivity, and relative accuracy, thereby enabling a comprehensive and balanced assessment across models.

(2) To eliminate randomness-induced variance, a fixed random seed was used to partition the dataset into training (64%), validation (16%), and test (20%) subsets [39]. This ensured that all models were trained and evaluated on identical data splits, enhancing the reliability of inter-model comparisons.

(3) Hyperparameter optimization was conducted uniformly across algorithms using the Optuna framework, with the population size and iteration count held constant. A total of 40 independent trials were performed within a unified search space encompassing five key hyperparameters (Table 2). Model performance in each trial was evaluated on the validation set using multiple indicators, and the optimal configuration was selected based on the lowest Composite Performance Index (CPI), ensuring convergence toward high-performing, generalizable solutions.

Table 2. Hyperparameter search space and ranges for ensemble models.

Hyperparameter	Type	Range
n_estimators	Integer	50–300
max_depth	Integer	2–6
learning_rate	Float	0.01–0.2
subsample	Float	0.6–1.0
colsample_bytree	Float	0.6–1.0

2.3.3. Hyperparameter Tuning Guided by Composite Performance Index

To identify the optimal hyperparameter configurations for each ensemble model, CPI-based heatmap visualizations were constructed using a unified Optuna optimization framework (Figure 3). In the heatmaps, darker color indicates lower CPI and thus better overall model performance. These visual gradients highlight sensitive regions in the hyperparameter space, where predictive accuracy is maximized. As shown in Table 3, the search space spanned tree depth and learning rate, with the Composite Performance Index (CPI) serving as the unified evaluation metric. These heatmaps not only facilitated cross-model performance comparison but also guided the selection of final model parameters used in subsequent predictive tasks. For XGBoost, CPI steadily decreased with increasing depth, reaching 0.003 for elasticity modulus at a learning rate of 0.078. LightGBM exhibited similar trends, with optimal performance on UCS and descent section parameters at learning rates of 0.129 and 0.075, and CPI values of 0.040 and 0.008, respectively. CatBoost showed a more scattered behavior but still achieved its best CPI (0.068) for the peak strain at a depth of 6 and a learning rate of 0.138 (Table 3).

Table 3. Optimal hyperparameters and CPI values across models and targets.

Model	Target	Max Depth	Learning Rate	CPI
XGBoost	E_c	6	0.0781	0.003
LightGBM	$f_{c,r}$	6	0.1294	0.04
LightGBM	α_c	6	0.0751	0.008
CatBoost	$\varepsilon_{c,r}$	6	0.1376	0.068

These CPI heatmaps directly informed the selection of final model configurations, ensuring both methodological consistency and improved generalization in the subsequent prediction tasks. As summarized in Figure 4, after database establishment and sample stratification, the dataset was partitioned into training (64%), validation (16%), and testing (20%) datasets. Augmentation was restricted to the training split; no synthetic samples were entered for validation or testing. Standardization was estimated on the augmented training data and transported to the other splits to prevent leakage. XGBoost, LightGBM, and CatBoost were fitted to the training split, with hyperparameters optimized via Optuna on validation (selection landscapes in Figure 3). With selection fixed, the held-out test split alone underpinned downstream reporting: (1) predictive accuracy and stress–strain reconstructions (Figure 5); (2) global attribution via permutation importance (Figure 6); and (3) interaction diagnostics via SHAP dependence plots for key parameters (Figures 7 and 8). Interpretability was strictly out-of-sample; training and validation data were not used.

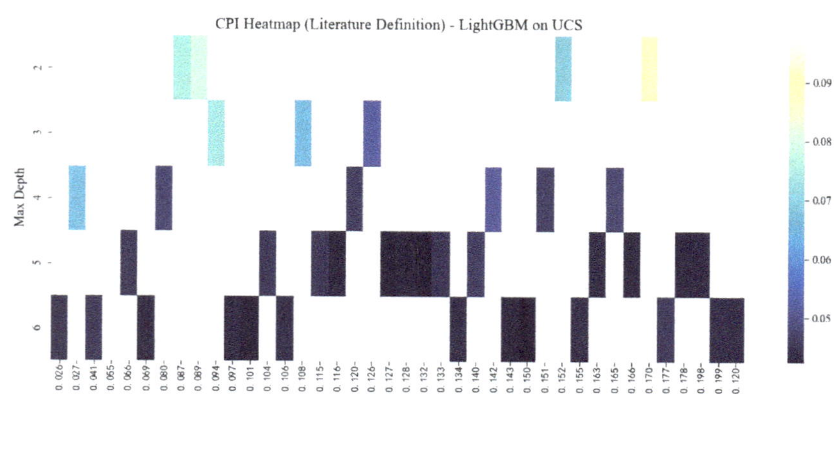

(**a**) CPI heatmap for LightGBM on uniaxial compressive strength prediction

Figure 3. *Cont.*

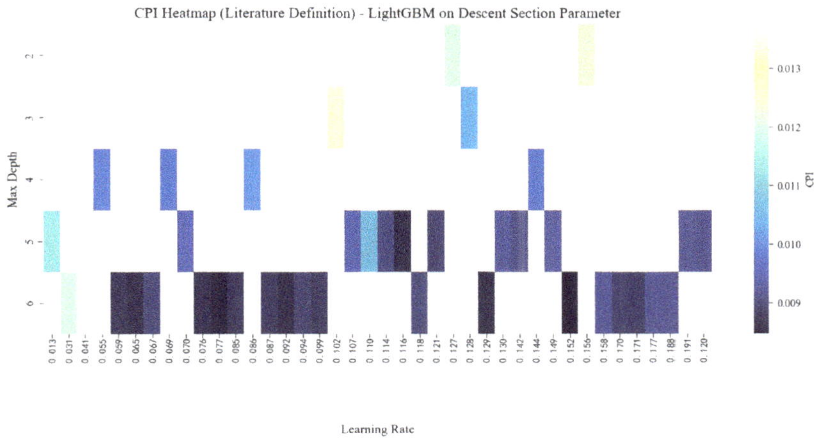

(**b**) CPI heatmap for LightGBM on descent section parameter prediction

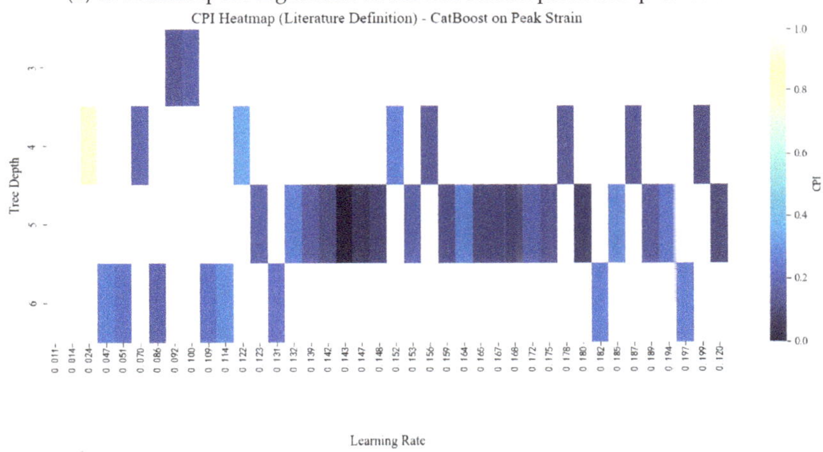

(**c**) CPI heatmap for CatBoost on peak strain prediction

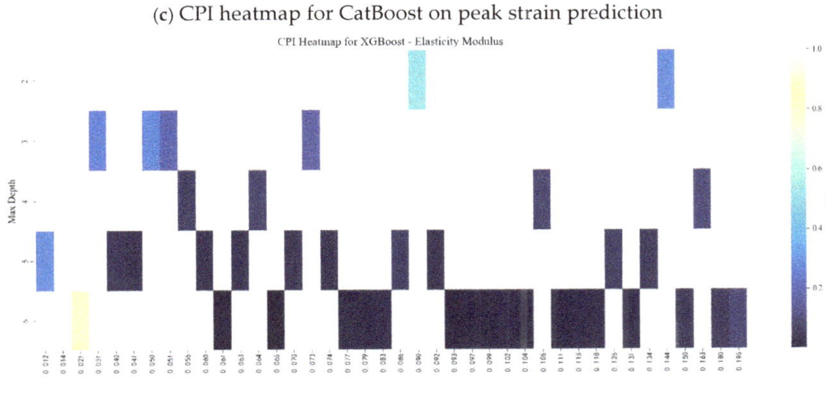

(**d**) CPI heatmap for XGBoost on elasticity modulus prediction

Figure 3. CPI-based hyperparameter performance landscapes across models and prediction targets.

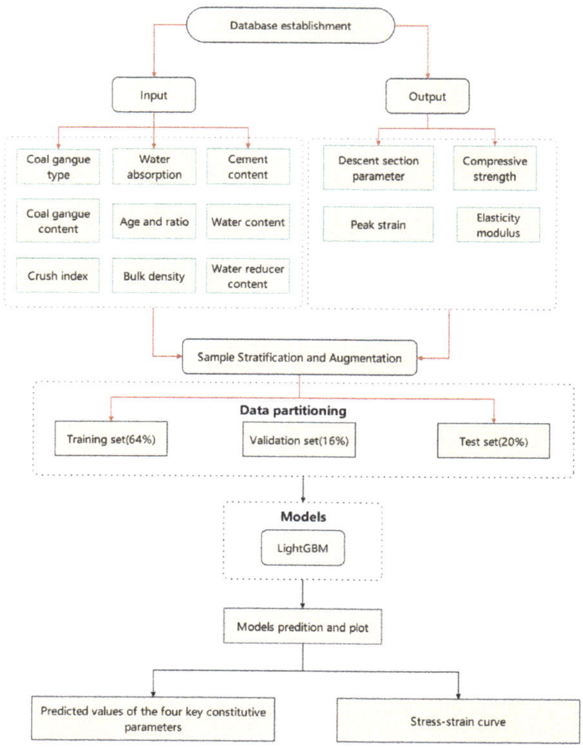

Figure 4. Flowchart of machine learning models.

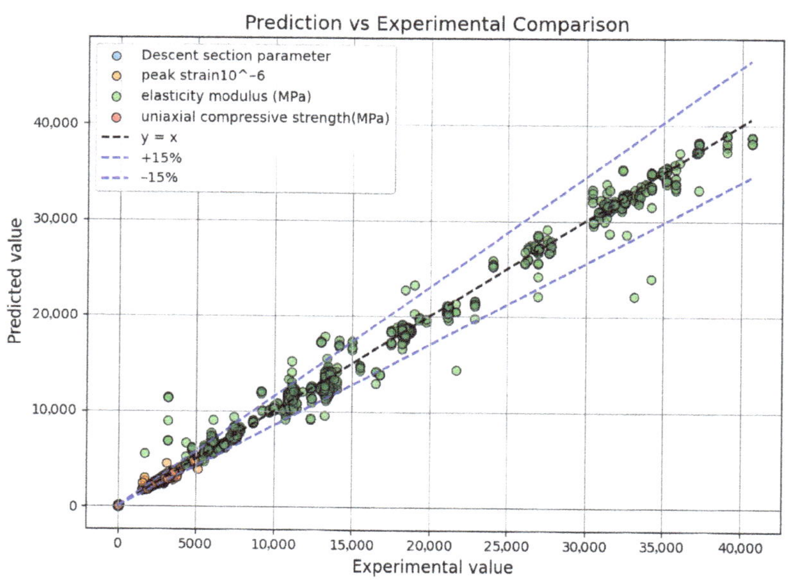

(**a**) XGBoost

Figure 5. *Cont.*

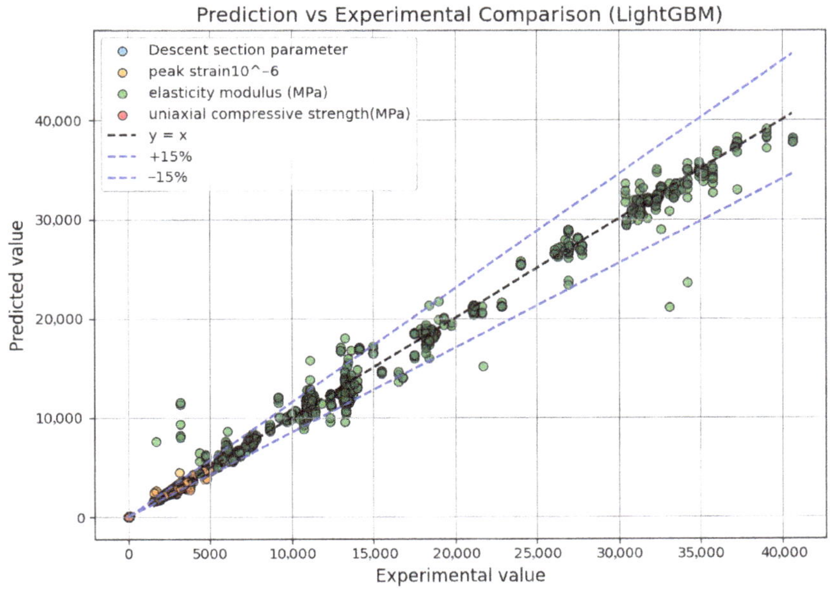

(**b**) LightGBM

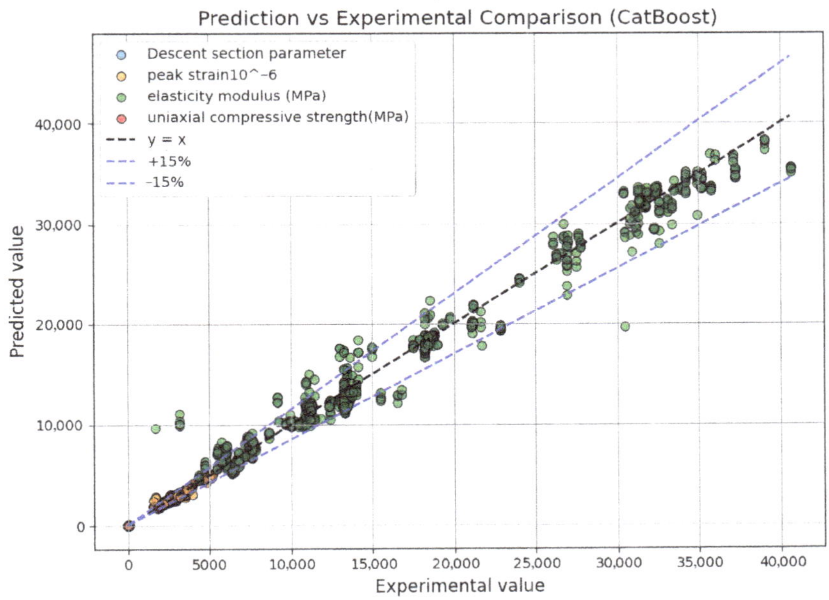

(**c**) CatBoost

Figure 5. Scatter plots of predicted vs actual values for constitutive parameters.

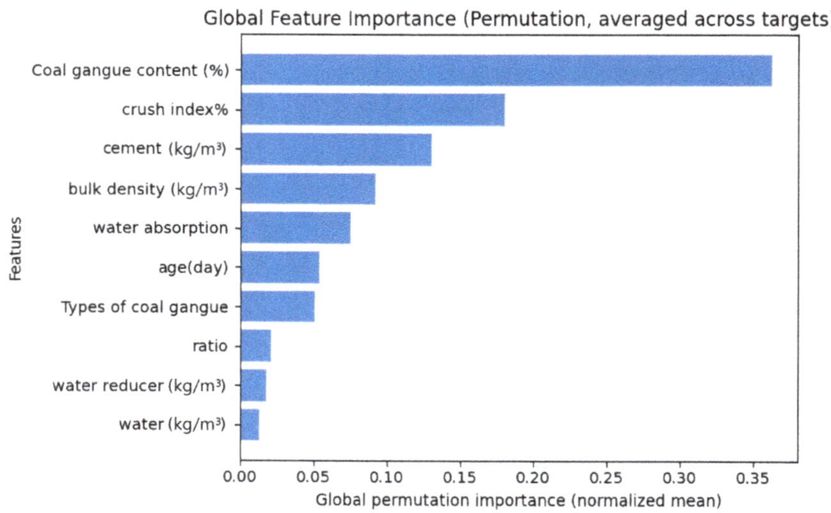

Figure 6. Global feature importance based on the permutation method.

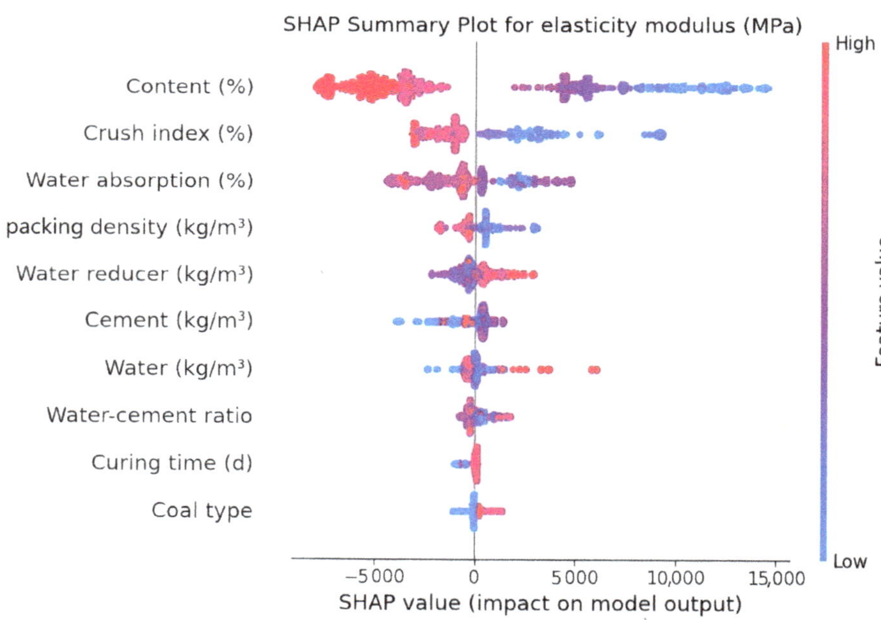

(**a**) Elasticity Modulus

Figure 7. *Cont.*

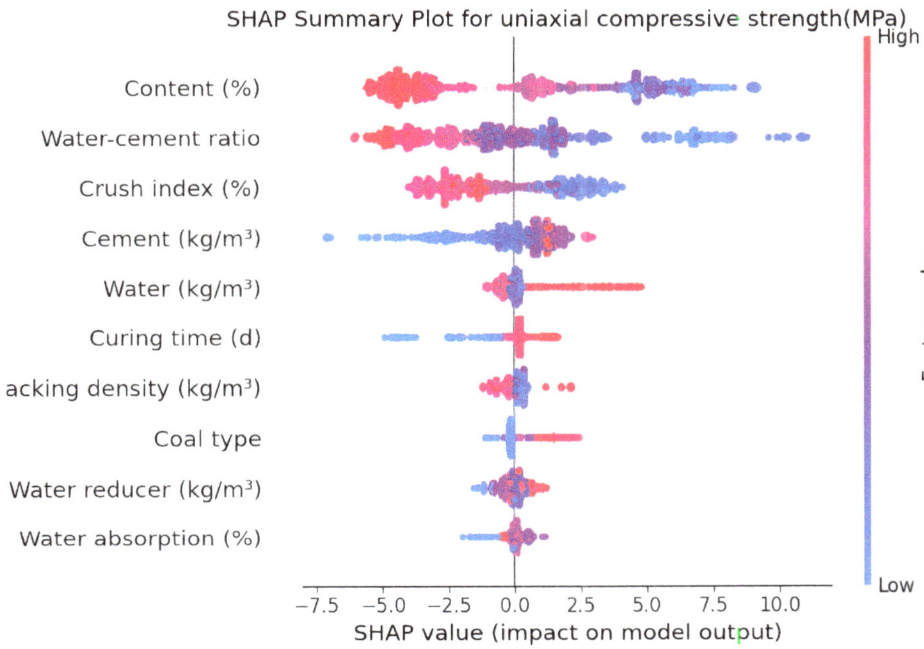

(**b**) Uniaxial compressive strength

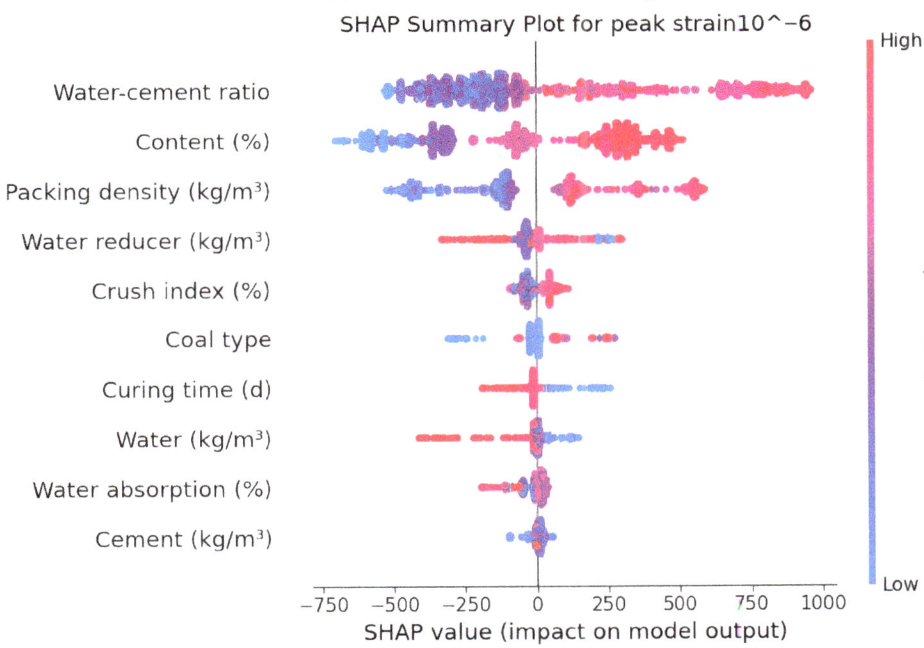

(**c**) Peak strain

Figure 7. *Cont.*

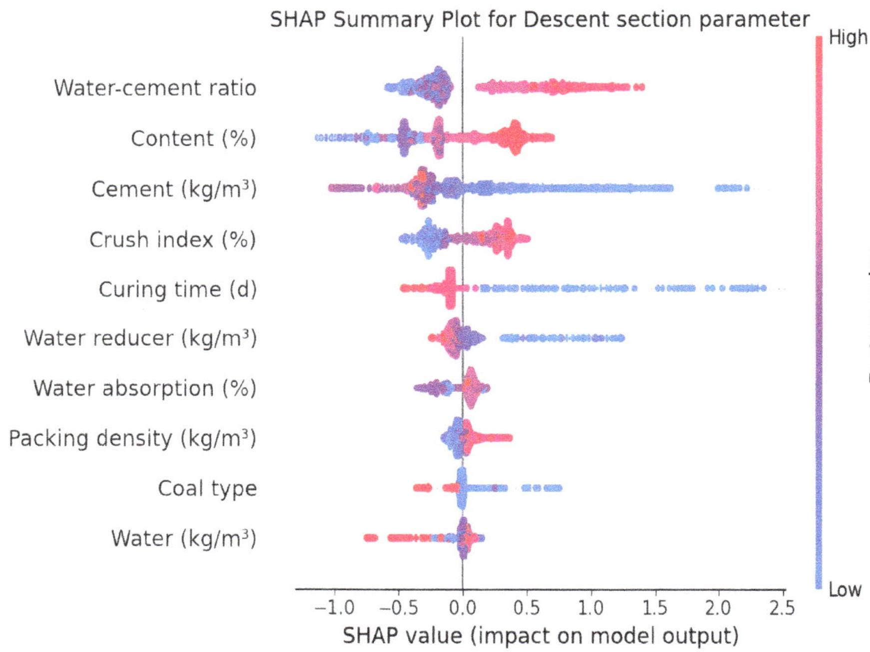

(**d**) Descending Parameter

Figure 7. Comparison of SHAP summary plots for key parameters.

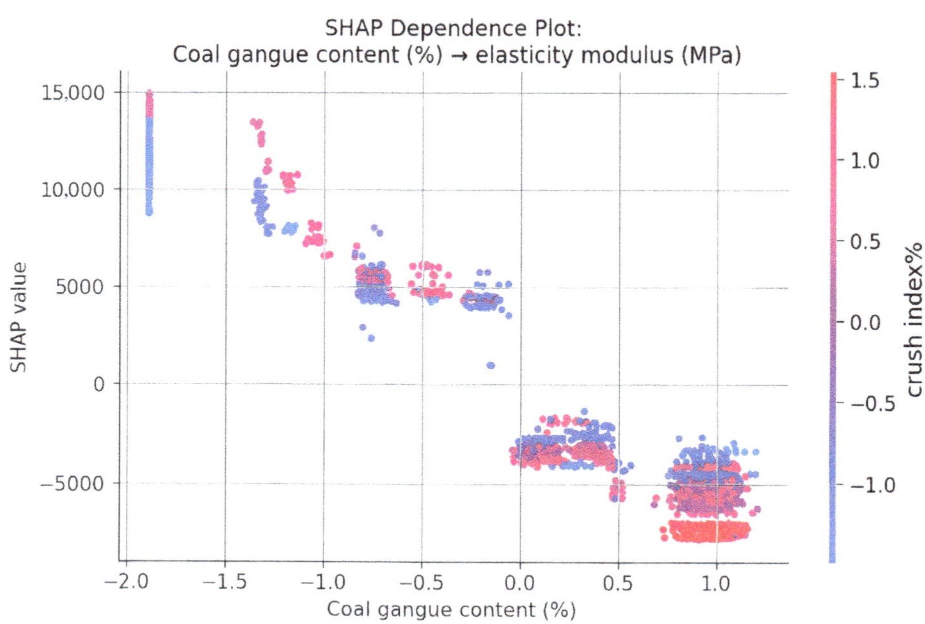

(**a**) Elasticity modulus

Figure 8. *Cont.*

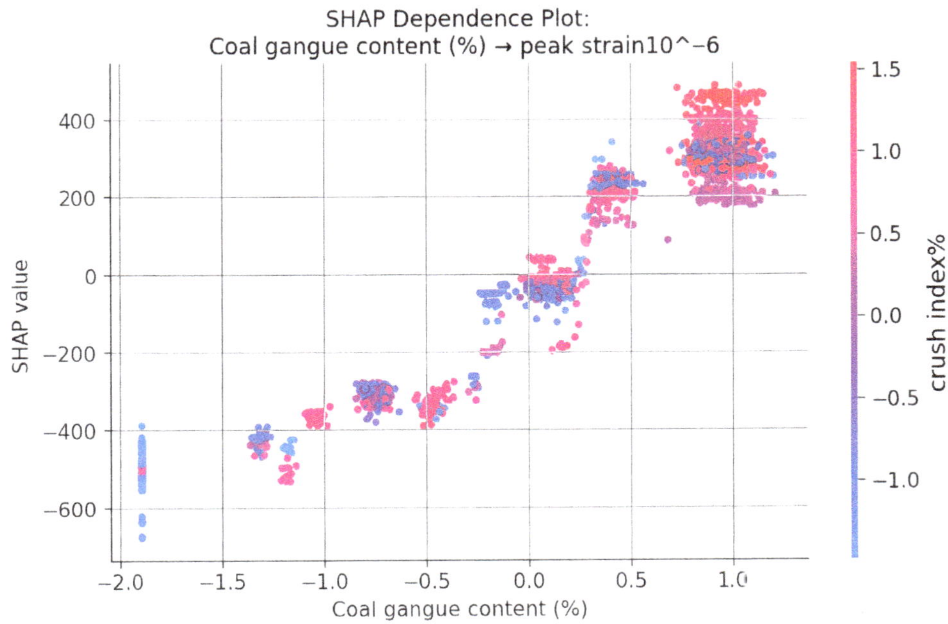

(**b**) Uniaxial compressive strength

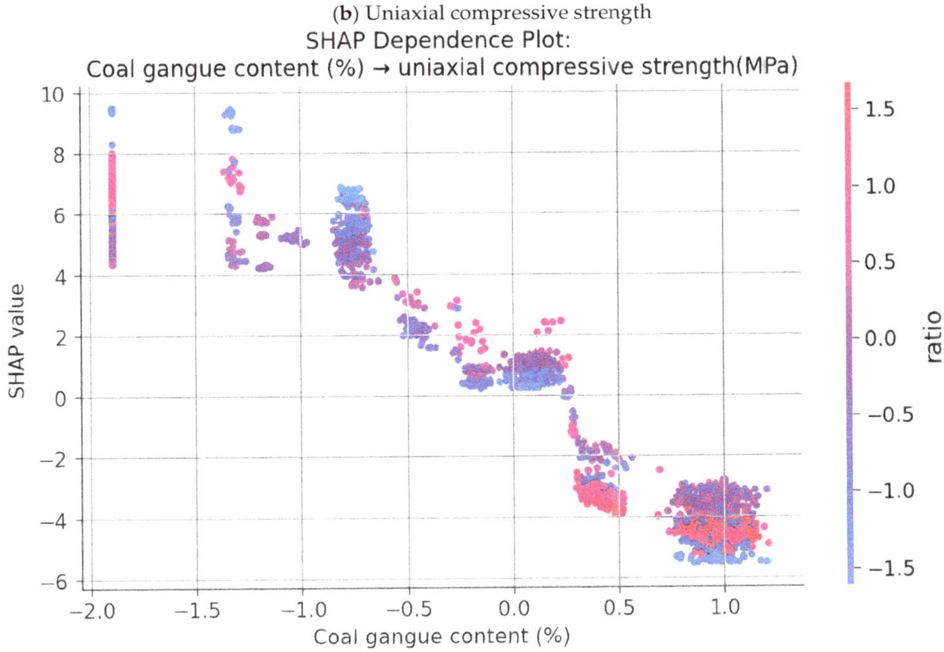

(**c**) Peak strain

Figure 8. *Cont.*

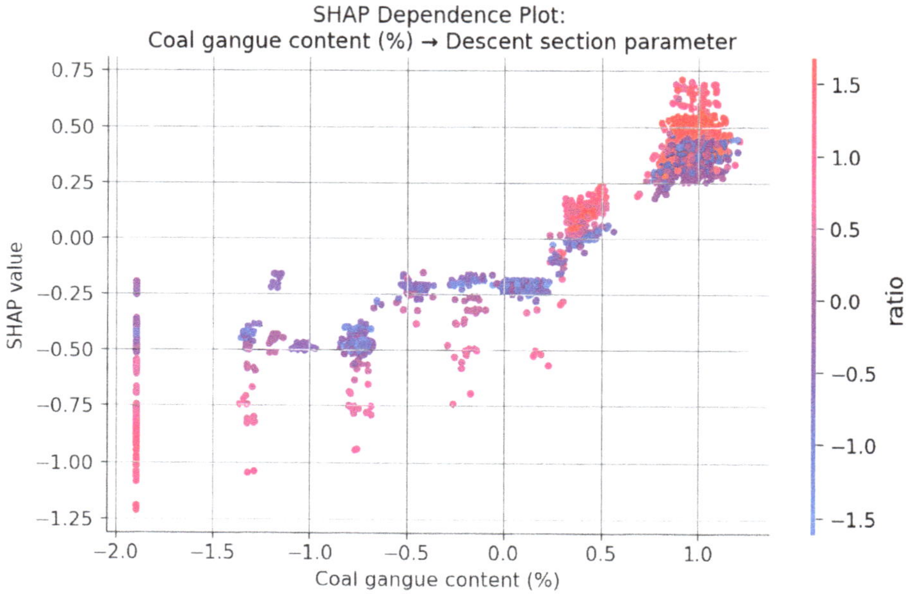

(**d**) Descending Parameter

Figure 8. Comparison of SHAP dependence plots for key parameters.

3. Interpretability and Validation of Predictive Models for CGC Constitutive Behavior

3.1. Predictive Accuracy and Multi-Model Performance Comparison

To fairly assess the predictive performance of the three ensemble models—XGBoost, LightGBM, and CatBoost [40], this setup was designed to reflect realistic prediction scenarios and avoid bias from repeated data exposure. Performance was measured using a set of standard regression metrics, including co and the Composite Performance Index (CPI). Visual validation was also carried out using ±15% error-band scatter plots.

As summarized in Tables 4–7, all models showed reasonable predictive accuracy across the four target parameters, with R^2 values above 0.91 in all instances. Notably, LightGBM maintained low CPI values on the test set—consistently below 0.04—which closely matched its performance on the validation set (Table 3). This level of consistency suggests that the model generalized well without clear signs of overfitting. For XGBoost and CatBoost, the gap between the validation and test CPI was more variable, especially for targets like the peak strain and elastic modulus, indicating some sensitivity to data distribution shifts.

Table 4. Statistics of training model information for descending segment parameters.

Model	R^2	MAPE	MAE	MSE	RMSE	CPI
XGBoost	0.96	3.21	90.74	26,645.40	163.29	0.11
LightGBM	0.96	3.06	88.39	26,108.22	161.58	0.01
CatBoost	0.96	3.62	98.99	22,967.33	151.55	0.03

Table 5. Statistics of peak strain training model information.

Model	R^2	MAPE	MAE	MSE	RMSE	CPI
XGBoost	0.98	7.34	850.48	1,997,496.00	1413.33	0.58
LightGBM	0.98	7.75	880.75	2,099,099.00	1448.83	0.02
CatBoost	0.97	9.78	1141.07	2,792,556.00	1671.09	0.03

Table 6. Information statistics of the uniaxial compressive strength training model.

Model	R^2	MAPE	MAE	MSE	RMSE	CPI
XGBoost	0.96	4.72	1.26	2.95	1.72	0.34
LightGBM	0.96	4.62	1.23	2.86	1.69	0.03
CatBoost	0.95	5.80	1.57	4.21	2.05	0.03

Table 7. Statistics of the elastic modulus training model information.

Model	R^2	MAPE	MAE	MSE	RMSE	CPI
XGBoost	0.93	77.80	0.25	0.14	0.37	0.57
LightGBM	0.93	75.39	0.26	0.15	0.38	0.03
CatBoost	0.91	87.43	0.30	0.18	0.43	0.04

Each model exhibited distinct strengths. LightGBM performed reliably across all targets, particularly for high-variance outputs like the elasticity modulus and compressive strength (Figure 5b). XGBoost effectively captured the structural patterns, which is reflected in the close clustering of the prediction points along the diagonal in Figure 5a. Although slightly less stable overall, CatBoost handled sample dispersion well; its predictions remained within the ±15% error band, even when the data became sparse or unevenly distributed.

3.2. Global–Local Feature Attribution and Mechanistic Interpretation

To further elucidate the feature response mechanisms within the predictive models for different mechanical parameters, this section presents a systematic interpretability analysis using the SHAP method based on the best-performing LightGBM model. As a representative of gradient-boosted tree models, LightGBM offers a refined structure, high training stability, and superior error control, making its interpretability results a reliable reflection of the true feature contribution mechanisms.

A permutation-based global importance audit (Figure 6) quantifies the aggregate predictor influence across all targets. The salience hierarchy isolates coal gangue content, crush index, and cement content as dominant—governing, respectively, for replacement level, particle integrity, and binder proportion within the composite matrix. This system-level attribution provides a synoptic map of the feature–response landscape and a principled sieve for variables prioritized for deeper diagnostics.

3.2.1. Feature Attribution and Mechanistic Interpretation for Elasticity Modulus

As shown in the SHAP summary plot (Figure 7a), coal gangue content and crush index are the most influential features for predicting elasticity modulus, with both the highest mean SHAP values and the broadest dispersion. Figure 8a shows a nonlinear inverse relationship between gangue content and its marginal effect, partitioning three mechanical regimes (brittle → quasi-ductile → re-brittle). Below 28% replacement, the SHAP values are predominantly positive, indicating stiffness preservation via enhanced particle interlock, matrix continuity, and intact ITZ; angular gangue elevates internal friction consistent with a brittle-dominated response [41,42]. Between 28% and 56%, SHAP precipitously crosses into negative territory at an inflection point where porosity-driven weakening supersedes the initial benefits. Microcracking, ITZ debonding, and flaw-controlled inclusions impede stress transmission, with fractography corroborating shear-slip and defect accrual [43,44]. Beyond ~56%, the SHAP stabilizes in the negative band (−4000 to −6000 MPa), indicating a re-brittle regime governed by a friable, low-strength aggregate skeleton, rapid crack coalescence, and localized interfacial failure with minimal strain accommodation [45,46]. A pronounced interaction with the crush index is also observed: jointly high values produce the SHAP nadir and compounded stiffness attrition, consistent with the premature fracture of friable particles, local densification, microvoid nucleation, and accelerated interfacial cracking that nonlinearly amplifies stiffness loss [47].

3.2.2. Feature Attribution and Mechanistic Interpretation for Uniaxial Compressive Strength (UCS)

As shown in the SHAP summary plot for UCS (Figure 7b), coal gangue content and water–cement ratio are the dominant features, followed by crush index and cement dosage. These variables display wide SHAP value dispersion, reflecting pronounced nonlinear and coupled effects. For gangue content, the SHAP values ranged from approximately +10 MPa to –7.5 MPa, indicating a bidirectional influence dependent on the replacement level.

Figure 8b shows the tripartite regime. For replacement < 26%, SHAP values are predominantly positive, indicating maintained or slightly augmented UCS via improved packing and interparticle friction from moderately angular gangue [48]. Across 26–54%, SHAP descends precipitously, an inflection where deleterious effects supersede initial gains, driven by proliferating weak ITZs, heightened capillary porosity, and matrix discontinuities that vitiate stress transmission and precipitate incipient cracking, in concordance with microstructural deterioration in prior studies [41,43]. Beyond 60%, the SHAP plateaus in a negative asymptote, evidencing saturation: additional gangue scarcely worsens UCS as failure is governed by a friable, low-strength aggregate skeleton, consistent with granular framework–collapse models [41]. The color gradient further evinces a strong interaction with the water–cement ratio: elevated ratios exacerbate the negative SHAP effect between 40% and 70% replacement via dual interfacial debilitation, matrix cohesion loss, and ITZ degradation from porous gangue (water uptake, shrinkage mismatch), accelerating microcrack coalescence and attenuating effective stress transfer, in line with observations on porosity- and moisture-induced interfacial weakening [49].

3.2.3. Prediction Mechanism and Feature Attribution for Peak Strain

As revealed in the SHAP summary plots (Figure 7c,d), water–cement ratio and coal gangue content are consistently identified as the principal drivers of both peak strain (ε_p) and the descending parameter α, reflecting their pivotal roles in regulating inelastic deformation and post-peak dissipation characteristics. These variables exhibit strongly monotonic SHAP trends, suggesting that their effects are scale-consistent across the composition spectrum. Additional contributors—such as packing density for ε_p and cement dosage for α—act as localized modulators, fine-tuning fracture progression and residual ductility.

The corresponding SHAP dependence plots (Figure 8c,d) indicate a clear positive shift in SHAP values as gangue content increases beyond approximately 47–60% (standardized values of −0.2 to +0.4), quantitatively supporting the ductility enhancement trend previously observed in the stiffness-related response (see Section 3.2.1). This upward transition reflects the cumulative effects of interfacial degradation, increased pore connectivity, and disrupted aggregate interlocks, which together promote greater strain accommodation capacity and retard stress collapse during the post-peak phase.

Taken together, these results suggest that elevated gangue incorporation induces progressive plasticization and energy dissipation, thereby reshaping the post-peak failure envelope through the coupled evolution of stiffness degradation and strain delocalization mechanisms. These insights reinforce the view that gangue not only alters strength capacity but also fundamentally transforms failure morphology in hybrid cementitious systems.

3.3. Constitutive Curve Prediction and Validation Based on Machine Learning

Building upon the SHAP-based interpretation of individual constitutive parameters, this section reconstructs the full stress–strain trajectories of coal gangue concrete under varying material conditions using LightGBM-predicted inputs(Figures 9 and 10). The generated curves not only replicate the declining trends in stiffness and strength associated with increased gangue content but also reveal the progressive enhancement of ductility, as reflected in elevated peak strains and smoother post-peak descent. These emergent behaviors are consistent with the feature-wise mechanisms previously identified—particularly the interplay between coal gangue content, water–cement ratio, and aggregate integrity.

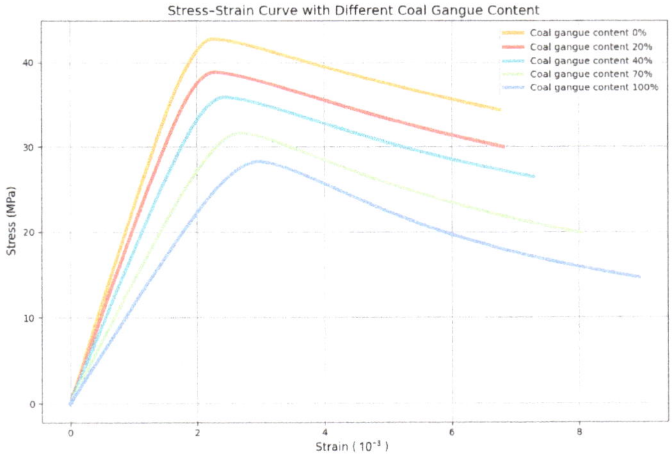

Figure 9. Effect of coal gangue content on stress–strain behavior.

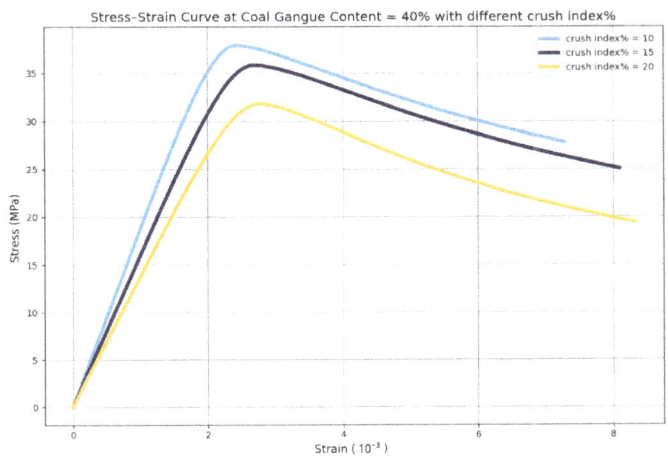

Figure 10. Effect of crushing index on stress–strain behavior at 40% coal gangue content.

Notably, the predicted curves capture both the degradation and toughening effects induced by gangue replacement, with the former driven by microstructural weakening of the interfacial zones and the latter associated with crack deflection and energy absorption mechanisms. This model-driven constitutive reconstruction validates the internal consistency of the proposed framework by bridging parameter-level learning with holistic mechanical behavior. These predictive curves further serve as transferable inputs for the finite element simulations that follow, enabling structural-scale validation under coupled loading conditions.

4. Finite Element Simulation of Coal Gangue Concrete Columns Under Combined Loading in ABAQUS

4.1. Finite Element Modeling Strategy and Material Parameter Input

We employed the ABAQUS Concrete Damage Plasticity (CDP) model to simulate the nonlinear response of CGC columns under combined axial–bending, calibrating and validating against the small-eccentric compression tests of [30]. Specimens were 300 × 300 × 1800 mm RC columns with gangue-coarse-aggregate replacement of 40, 70, and 100% (w/c = 0.40; cement = 475 kg·m^{-3}; mix proportions adjusted accordingly). Axial loading with $e_0/h = 0.25$ was applied under displacement control on a hydraulic frame, and LVDTs and strain gauges were instrumented on both the compression and tension faces. Failures manifested as flexural–compression: incipient horizontal cracking on the tension face, progressive crack coalescence, and compression face spalling near ultimate. Four key material parameters were subsequently rectified via Gradient-Boosted-Tree predictions in accordance with the experimental conditions. (Table 8). Reinforcement was modeled as ideal elastic–plastic using T3D2 truss elements, while concrete was discretized with C3D8R solid elements. The steel–concrete system was assembled into a complete reinforced model, and the overall mesh sizes of the concrete member and the reinforced cage were set as 30 mm and 20 mm, respectively, to consider the calculation efficiency and accuracy, as illustrated in Figure 11a,b.

Table 8. Constitutive parameters of coal gangue concrete predicted by the LightGBM model.

Content of Coal Gangue	$f_{c,r}$(MPa)	$\varepsilon_{c,r}(10^{-6})$	E_c(MPa)	α_c
40	29.67	2489.80	21,653	1.36
70	24.80	2794.01	19,412	4.15
100	20.53	2894.05	13,368	7.21

Table 9 demonstrates the values of dilation angle (φ), eccentricity (ε), flow stress ratio (fb0/fc0), deviatoric shape factor (K), and viscosity coefficient were determined through a combination of literature references [50,51] and extensive numerical calibration. The final configuration was selected based on its ability to ensure numerical stability and accurately reproduce the post-peak softening behavior of coal-gangue concrete under combined loading conditions..

Table 9. Key parameter settings for the Concrete Damage Plasticity (CDP) model in ABAQUS.

Expansion Angle (°)	Eccentricity	fb0/fc0	K	Viscosity Parameter
40	0.1	1.16	0.6667	0.011

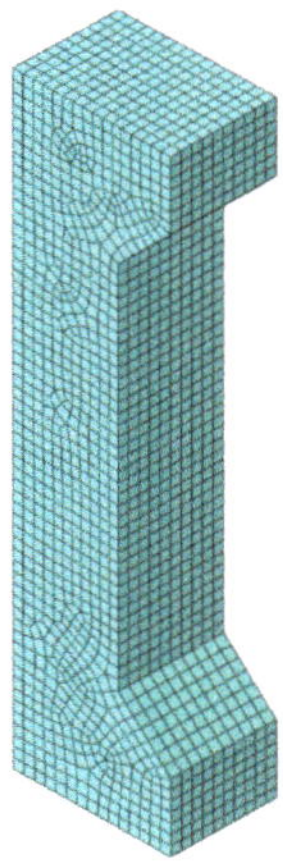

(**a**) Concrete component meshing diagram

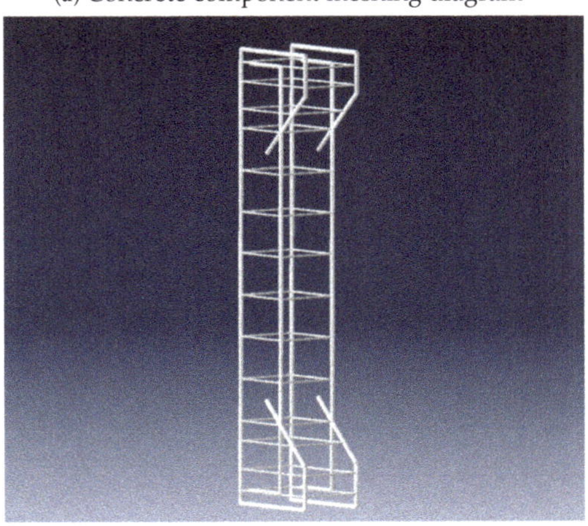

(**b**) Reinforcement cage schematic diagram

Figure 11. Finite element modeling schematic diagram of bent members.

4.2. Numerical Simulation Analysis and Experimental Validation

Due to the inherent high porosity and crush value of coal gangue concrete, early-stage testing (displacement < 2.8 mm) tends to induce localized contact nonlinearity, which may compromise the accuracy of the constitutive response in the primary control zone. Therefore, initial instability data were excluded during numerical validation to ensure the focus remained on evaluating intrinsic constitutive behavior [7].

4.2.1. Validation of Failure Patterns and Crack Propagation Pathways

For the 40% coal gangue replacement specimen, Figure 12 illustrates the simulated tensile and compressive damage distributions at the final loading stage. The simulation indicates that tensile cracks initiate at the tension face and propagate vertically, while compressive crushing develops near the mid-low of the compression face. Figure 13 presents the corresponding experimental failure patterns. A direct comparison shows that the simulated crack initiation locations and damage severity are generally consistent with the experimental observations, indicating that the finite element model provides a reasonable representation of the failure process of the specimen under combined loading.

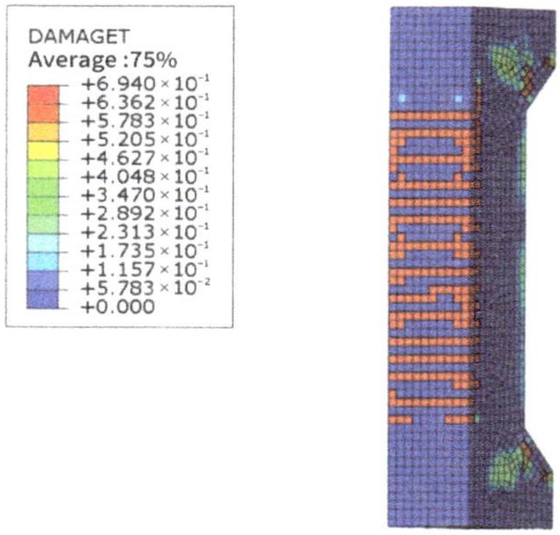

(a)

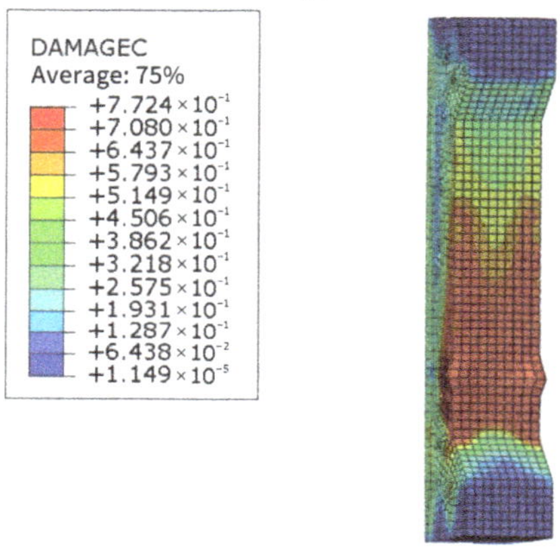

(b)

Figure 12. Numerical simulation of damage evolution in coal gangue concrete columns. (**a**) Crack distribution diagram in the tension zone (C Side). DAMAGET: tensile damage factor. (**b**) Local damage map of compression zone in the numerical simulation (A Side). DAMAGEC: compressive damage factor.

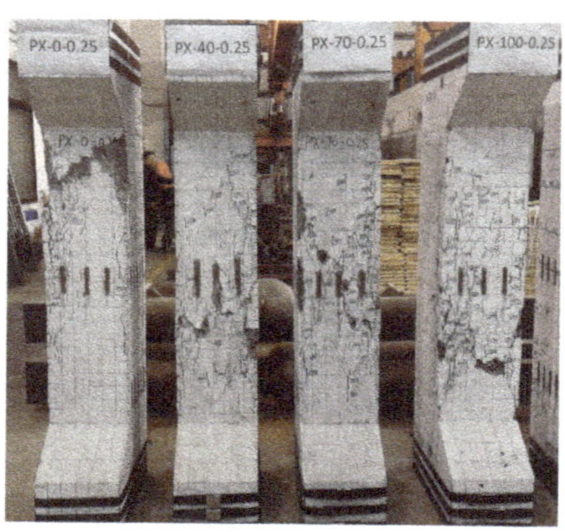

(a) A side

(b) C side

Figure 13. Experimental results. Source: Adapted/reproduced from Liu et al. (2022), with permission from Elsevier.

4.2.2. Comparative Analysis of Characteristic Loads

Table 10 summarizes the comparison between predicted and measured characteristic loads. Overall, the simulation-to-test ratios fall within 0.98–1.05, encompassing the expected variability due to material heterogeneity and experimental error. Although the model slightly overpredicts peak loads, this deviation is within acceptable bounds given the intrinsic differences between the heterogeneous nature of coal gangue aggregates and the isotropic assumptions in the simulation. These findings confirm that the machine learning–enhanced constitutive framework effectively supports strength development simulations under uniaxial stress, meeting practical engineering accuracy requirements.

Table 10. Comparison of numerical simulation and test load parameters.

Content of Coal Gangue%	Cracking Load (KN)			Ultimate Load (KN)		
	TEST	NUM	NUM/TEST	TEST	NUM	NUM/TEST
40%	670	692	1.03	1438	1445	1.01
70%	540	549	1.02	1397	1372	0.98
100%	460	482	1.05	1254	1319	1.05

TEST: the test record data. NUM: shows the numerical simulation data.

4.2.3. Load–Midspan Deflection Response Evolution

Figure 14a–c illustrates the comparative load–midspan deflection curves for specimens with 40%, 70%, and 100% replacement ratios. The finite element model successfully captures the full response trajectory, including the elastic phase, crack initiation, stiffness degradation, reinforcement yielding, and ultimate failure. The simulations showed high fidelity to the experiments at both inflection points and overall deformation trends.

The evolution of load–displacement responses under varying coal gangue replacement levels offers direct mechanical corroboration for the structural transition mechanisms previously inferred through SHAP-based feature attribution. As illustrated in Figure 10, specimens incorporating 40% and 70% gangue demonstrate a progressive attenuation of post-peak load, manifested in extended softening branches and delayed force degradation. These experimental and simulated responses reflect a notable increase in energy dissipation capacity and deformation tolerance, indicative of quasi-ductile behavior. The underlying mechanisms—such as frictional interlock among angular gangue particles, crack deflection, and matrix–aggregate interfacial plasticity—promote strain redistribution and delay catastrophic failure. In contrast, specimens with 100% gangue replacement exhibit a distinctly brittle response profile, characterized by a precipitous post-peak decline in load-bearing capacity and minimal inelastic accommodation. This abrupt structural collapse aligns with the SHAP-derived identification of a terminal degradation regime, wherein excessive porosity, weak interfacial bonding, and skeletal discontinuity coalesce to suppress any residual ductility.

Notably, these observed mechanical trajectories conform closely to the three-stage structural response model (brittle → quasi-ductile → re-brittle) proposed earlier in Section 3.2.1. The finite element results thus serve as an independent validation of the model sensitivity in capturing nonlinear mechanical transitions within coal gangue concrete. More importantly, the pronounced energy absorption behavior identified within the intermediate (saturation) regime suggests a tangible potential for engineering applications requiring seismic resilience and impact mitigation. Given the growing interest in sustainable and high-performance damping materials [52], the SHAP-enhanced interpretability framework established in this study may offer a new paradigm for the data-driven development of energy-dissipative concrete systems based on physical realism and verifiable design logic.

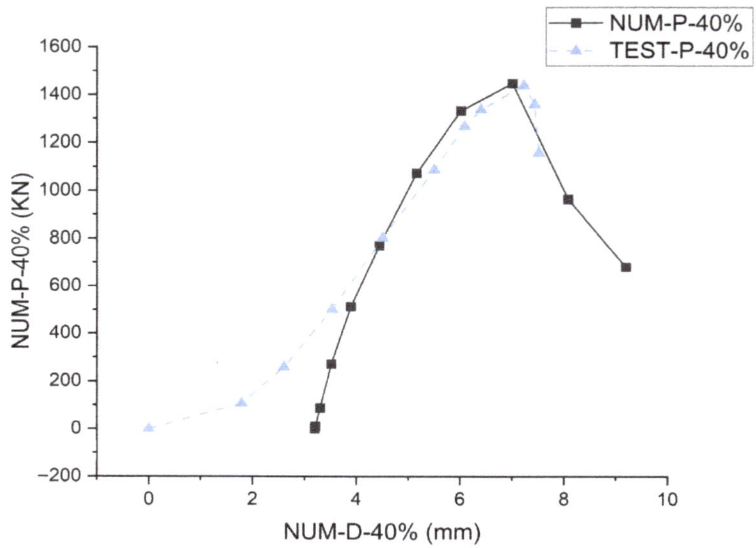

(**a**) Load–deflection curve comparison for 40% coal gangue replacement

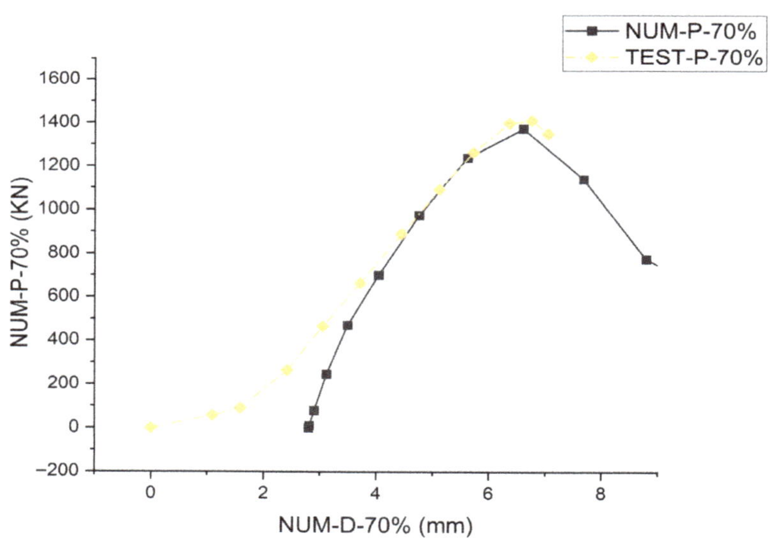

(**b**) Load–deflection curve comparison for 70% coal gangue replacement

Figure 14. *Cont.*

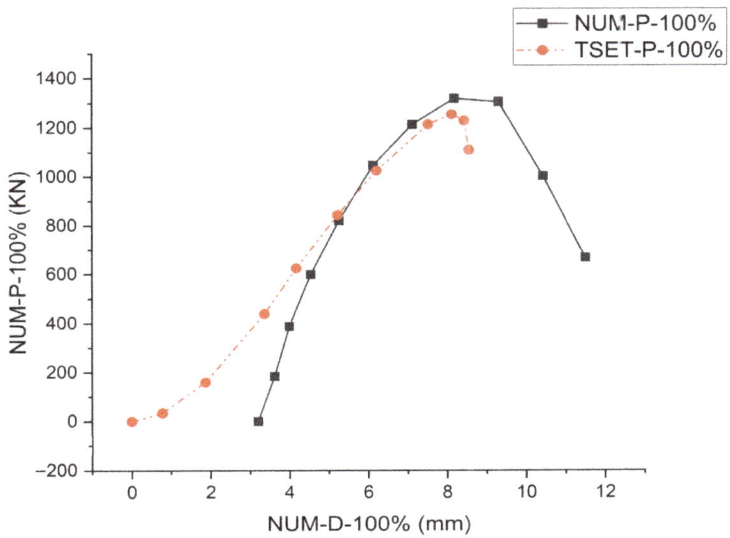

(c) Load–deflection curve comparison for 100% coal gangue replacement

Figure 14. Load–deflection curve comparison (numerical vs. experimental validation).

5. Conclusions and Future Directions

(1) The CPI-based heatmap analysis suggests that ensemble models perform best on coal gangue concrete data when trained with moderately low learning rates (around 0.07–0.13) and relatively shallow tree depths (no more than 6). This combination offers a good trade-off between capturing nonlinear patterns and avoiding overfitting, making it well-suited for modeling the complex and heterogeneous behavior of CGC materials.

(2) Among the evaluated models, LightGBM outperforms in terms of accuracy, robustness, and error mitigation, particularly excelling in predicting high-dimensional targets such as the elasticity modulus and uniaxial compressive strength. CatBoost demonstrates superior adaptability in small-sample, discrete datasets, while XGBoost offers strong boundary detection and local fitting capabilities, making it ideal for scenarios with well-defined structural patterns.

(3) SHAP-based interpretability analysis elucidated a three-stage mechanical response trajectory—brittle, quasi-ductile, and re-brittle—governing the stiffness evolution of CGC with increasing gangue incorporation. This phase-structured framework, grounded in feature–response attribution patterns, was corroborated through micromechanical evidence and deformation morphologies, thereby affirming the model's capacity to capture nonlinear degradation thresholds and interfacial instability regimes intrinsic to hybrid aggregate systems.

(4) Hybrid explainability analysis revealed compounded degradation effects arising from the coupled influence of coal gangue content with both crush index and water–cement ratio. The permutation-based global ranking established these variables as jointly dominant across all targets, while SHAP-based local visualizations elucidated nonlinear synergistic weakening mechanisms—wherein friable gangue particles and elevated moisture levels collectively destabilize the matrix–aggregate interface—thereby intensifying stiffness loss and precipitating compressive failure. These findings align with prior micromechanical observations linking interfacial debonding, pore coalescence, and moisture-induced damage to hybrid aggregate deterioration.

(5) Finite element validation, calibrated through fracture energy–based softening curves, independently substantiated the reliability of the SHAP-informed constitutive model in reproducing key structural responses across failure modes, load trajectories, and post-peak behavior. Notably, the model accurately captured the brittle → quasi-ductile → re-brittle evolution path of CGC, aligning simulation outcomes with SHAP-predicted transitions. The pronounced energy dissipation observed in the intermediate regime highlights the potential of coal gangue concrete as a ductile and damage-tolerant material for seismic applications. These findings establish a mechanistically interpretable and numerically verifiable framework for guiding the development of performance-based, data-driven structural materials. Building on the generalizability of the proposed modeling framework, there is significant potential for cross-domain extension. Future research should focus on the following key directions:

(1) Enhancement of dataset scale and representativeness for advanced interaction modeling.

The present study, while based on a systematically curated dataset, remains constrained by the limited number and diversity of available samples. Future investigations should incorporate more extensive and heterogeneous datasets encompassing broader variations in material composition, curing regimes, and testing protocols. Such expansions would not only strengthen the statistical robustness of machine learning models but also enable a more rigorous identification of nonlinear feature interactions and latent coupling mechanisms that govern the mechanical behavior of coal gangue concrete.

(2) Integration of inter-material coupling mechanisms into numerical simulations.

Although the current finite element modeling framework successfully captures the macroscopic response of CGC components, it inherently simplifies the material system by neglecting the meso-structural interactions among distinct phases, such as aggregates, cement paste, and interfacial transition zones. To enhance the physical fidelity of simulation outcomes, future work should aim to incorporate multi-phase constitutive relationships or cohesive interface models that account for inter-material bonding, slip, and damage evolution. This would facilitate a more comprehensive representation of the hybrid nature of CGC and improve the generalizability of the simulation results for complex structural scenarios.

(3) Integration of advanced machine learning algorithms.

The accelerating convergence of artificial intelligence and materials science is ushering in a new era of modeling paradigms that transcend traditional input–output mappings. Future research may leverage architecture-aware frameworks—such as graph-based learning or physics-informed networks—to capture spatially distributed interactions, multiscale heterogeneity, and path-dependent responses under complex loading regimes. Such approaches hold promise for embedding structural priors and mechanistic constraints directly into learning processes, thereby enabling more robust and generalizable constitutive formulations for sustainable concrete systems.

Author Contributions: Conceptualization, G.X.; methodology, validation, formal analysis, investigation, resources, data curation, writing original draft preparation, X.D.; writing review and editing, X.G.; project administration, G.X.; funding acquisition, X.G. and C.Z. All authors have read and agreed to the published version of the manuscript.

Funding: This research was funded by the National Natural Science Youth Science Foundation Project, grant number 51808443, and Shaanxi Provincial Key Research Project Research on Seismic Performance and Application Technology of Prefabricated Composite Wall Systems with Non-Removable Steel Mesh Frameworks, Grant No. 2024SF-ZDCYL-05-12, funded by the Shaanxi Provincial Department of Science and Technology.

Data Availability Statement: The dataset used in this study has been integrated into a proprietary software titled "Intelligent Prediction System for Constitutive Parameters of Coal Gangue Concrete V1.0", which has been officially registered with the National Copyright Administration of China (Registration No. 2025SR1233551). As the software is protected by copyright and incorporates proprietary algorithms and data structures, the complete dataset cannot be made publicly available. However, data supporting the findings of this study may be provided by the corresponding authors upon reasonable request, strictly for academic and non-commercial use.

Acknowledgments: The financial support received from the funding organizations is gratefully acknowledged. The authors would like to express their sincere gratitude to Yanchang Chen for his valuable guidance during the preparation of this manuscript. Special thanks are also extended to Yu Luo and Zihuan Cao for their kind technical assistance throughout the research process. The authors have reviewed and edited the output and take full responsibility for the content of this publication.

Conflicts of Interest: The authors declare no conflicts of interest.

Abbreviations

The following abbreviations are used in this manuscript:

CGC	Coal gangue concrete
MAE	Mean Absolute Error
MSE	Mean Squared Error
RMSE	Root Mean Squared Error
MAPE	Mean Absolute Percentage Error
SHAP	SHapley Additive exPlanations
LightGBM	Light Gradient Boosting Machine
CatBoost	Categorical Boosting
XGBoost	Extreme Gradient Boosting
ANN	Artificial Neural Network
SVM	Support Vector Machine
GP	Gaussian Process
SVR	Support Vector Regression

References

1. Wang, A. Research Progress of Resourceful and Efficient Utilization of Coal Gangue in the Field of Building Materials. *J. Build. Eng.* **2025**, *99*, 111526. [CrossRef]
2. Gao, S.; Zhang, S.; Guo, L. Application of Coal Gangue as a Coarse Aggregate in Green Concrete Production: A Review. *Materials* **2021**, *14*, 6803. [CrossRef]
3. Qin, X.; Xu, Z.; Liu, M.; Zhang, Y.; Wang, Y.; Yang, Z.; Ling, X. Mechanical Properties and Elastic Modulus Prediction of Mixed Coal Gangue Concrete. *Materials* **2025**, *18*, 1240. [CrossRef]
4. Li, S.; Wang, X.; He, R.; Zhang, Y.; Kang, S. Experimental Investigation into the Mechanical Performance of Roadway Shotcrete with the Partial Replacement of Cement with Recycled Gangue Powder. *Appl. Sci.* **2025**, *15*, 3180. [CrossRef]
5. Li, J.; Chen, L.; Luo, J.; Zhu, W.; Fan, X.; Zhu, Y.; Zhang, Z. Mechanical Properties and Microscopic Characteristics of Steel Fiber Coal Gangue Concrete. *Front. Mater.* **2023**, *10*, 1222481. [CrossRef]
6. Li, C.; Wang, Y.; Qian, L.; Ba, T.; Dong, W.; Yuan, F.; Sun, Z. Mechanical Assisted Thermal and Chemical Composite Activation of Coal Gangue and Its Hydration Characteristics. *J. Build. Eng.* **2025**, *112*, 113687. [CrossRef]
7. Guan, X.; Qiu, J.; Song, H.; Qin, Q.; Zhang, C. Stress–Strain Behaviour and Acoustic Emission Characteristic of Gangue Concrete under Axial Compression in Frost Environment. *Constr. Build. Mater.* **2024**, *395*, 132112. [CrossRef]
8. Zhou, M.; Dou, Y.; Zhang, Y.; Zhang, Y.; Zhang, B. Effects of the Variety and Content of Coal Gangue Coarse Aggregate on the Mechanical Properties of Concrete. *Constr. Build. Mater.* **2019**, *220*, 386–395. [CrossRef]
9. Dong, W.; Ren, Z.; Liu, X. Experimental Study on Constitutive Relation of Coal Gangue Coarse Aggregate Concrete under Uniaxial Compression. *Mater. Today Sustain.* **2024**, *28*, 100987. [CrossRef]
10. Liu, Z.; Zhao, Y.-G.; Ma, L.; Lin, S. Review on High-Strength Recycled Aggregate Concrete: Mix Design, Properties, Models and Structural Behaviour. *Structures* **2024**, *64*, 106598. [CrossRef]

11. Li, Q.; Xu, M.; Pan, C.; Zhang, L.; Zhong, Y.; Li, W.; Gu, S.; Yu, M. Study on Axial Compressive Behavior and Mesoscopic Structural Evolution of FRP Confined Coal Gangue Concrete. *PLoS ONE* **2025**, *20*, e0322647. [CrossRef]
12. Cheng, L.; Zhao, L.; Cheng, L.; Gao, Y.; Guo, H.; Che, Y.; Fu, H. Research Progress of Concrete Preparation Based on Solid Waste Properties of Coal Gangue. *Sustainability* **2025**, *17*, 2007. [CrossRef]
13. Aldwaik, M.; Adeli, H. Advances in Optimization of Highrise Building Structures. *Struct. Multidisc. Optim.* **2014**, *50*, 899–919. [CrossRef]
14. Amezquita-Sanchez, J.P.; Valtierra-Rodriguez, M.; Aldwaik, M.; Adeli, H. Neurocomputing in Civil Infrastructure. *Sci. Iran.* **2016**, *23*, 2417–2428. [CrossRef]
15. Liao, T.W.; Egbelu, P.J.; Sarker, B.R.; Leu, S.S. Metaheuristics for Project and Construction Management—A State-of-the-Art Review. *Autom. Constr.* **2011**, *20*, 491–505. [CrossRef]
16. Shahin, M.A. State-of-the-Art Review of Some Artificial Intelligence Applications in Pile Foundations. *Geosci. Front.* **2016**, *7*, 33–44. [CrossRef]
17. Li, B. Strength Criterion of Recycled Aggregate Concrete under Triaxial Compression: Model Calibration. *Constr. Build. Mater.* **2022**, *322*, 126376. [CrossRef]
18. Lin, J.-X. Analysis of Stress–Strain Behavior in Engineered Geopolymer Composites Reinforced with Hybrid PE–PP Fibers: A Focus on Cracking Characteristics. *Compos. Struct.* **2023**, *311*, 116803. [CrossRef]
19. Al Khazaleh, M.; Kumar, P.K.; Mohamed, M.J.S.; Kandasamy, A. Influence of Coarse Coal Gangue Aggregates on Properties of Structural Concrete with Nano Silica. *Mater. Today Proc.* **2023**, *72*, 2089–2095. [CrossRef]
20. Hao, Y.; Guo, X.; Yao, X.; Han, R.; Li, L.; Zhang, M. Using Chinese Coal Gangue as an Ecological Aggregate and Its Modification: A Review. *Materials* **2022**, *15*, 4495. [CrossRef]
21. Liu, H.; Bai, G.; Wang, J.; Dai, M.; Zhu, K. Experimental Study on Uniaxial Compressive Stress–Strain Curve of Coal Gangue Concrete. *J. Build. Struct.* **2023**, *44*, 236–245, 254. (In Chinese)
22. Shabbir, K.; Noureldin, M.; Sim, S.-H. Data-Driven Model for Seismic Assessment, Design, and Retrofit of Structures Using Explainable Artificial Intelligence. *Sensors* **2024**, *24*, 843. [CrossRef]
23. Thapa, I.; Ghani, S. Explainable AI-Enhanced Machine Learning for CBR Prediction in Stabilized and Un-Stabilized Subgrade Soils. *Appl. Soft Comput.* **2025**, *178*, 113275. [CrossRef]
24. Pistolesi, F.; Baldassini, M.; Volpe, E.; Focacci, F.; Cattoni, E. Fast and Interpretable Prediction of Seismic Kinematics of Flexible Retaining Walls in Sand through Explainable Artificial Intelligence. *Comput. Geotech.* **2025**, *179*, 107007. [CrossRef]
25. Khan, M.S.; Peng, T.; Akhlaq, H.; Khan, M.A. Comparative Analysis of Automated Machine Learning for Hyperparameter Optimization and Explainable Artificial Intelligence Models. *IEEE Access* **2025**, *13*, 84966–84991. [CrossRef]
26. Ombres, L.; Aiello, M.A.; Cascardi, A.; Verre, S. Modeling of Steel-Reinforced Grout Composite System-to-Concrete Bond Capacity Using Artificial Neural Networks. *J. Compos. Constr.* **2024**, *28*, 04024034. [CrossRef]
27. Kumar, A.; Singh, U.K.; Pradhan, B. Enhancing Interpretability in Deep Learning-Based Inversion of 2-D Ground Penetrating Radar Data: An Explainable AI (XAI) Strategy. *IEEE Geosci. Remote Sens. Lett.* **2024**, *21*, 3506505. [CrossRef]
28. *GB50010-2015*; Code for Design of Concrete Structures. Ministry of Housing and Urban–Rural Development of the People's Republic of China. China Architecture & Building Press: Beijing, China, 2015.
29. Dai, M. Experimental Study on the Service Performance of Coal Gangue Concrete Flexural Members. Master's Thesis, Xi'an University of Architecture and Technology, Xi'an, China, 2022. (In Chinese)
30. Liu, H.; Bai, G.; Gu, Y.; Yan, F. The Influence of Coal Gangue Coarse Aggregate on the Mechanical Properties of Concrete Columns. *Case Stud. Constr. Mater.* **2022**, *17*, e01315. [CrossRef]
31. Sun, Q. Study on Mechanical Properties of Coal Gangue Concrete. Master's Thesis, Shenyang Jianzhu University, Shenyang, China, 2019. (In Chinese)
32. Liu, H. Experimental Study on the Bearing Capacity of Coal Gangue Concrete Columns. Master's Thesis, Xi'an University of Architecture and Technology, Xi'an, China, 2020. (In Chinese)
33. Duan, X. Study on Microstructure and Physicomechanical Properties of Coal Gangue Aggregate Concrete. Ph.D. Thesis, China University of Mining and Technology, Xuzhou, China, 2014. (In Chinese)
34. Su, Y. Experimental Study on Basic Mechanical Properties of Coal Gangue Concrete. Master's Thesis, Xi'an University of Architecture and Technology, Xi'an, China, 2022. (In Chinese)
35. Karniadakis, G.E.; Kevrekidis, I.G.; Lu, L.; Perdikaris, P.; Wang, S.; Yang, L. Physics-Informed Machine Learning. *Nat. Rev. Phys.* **2021**, *3*, 422–440. [CrossRef]
36. Shorten, C.; Khoshgoftaar, T.M. A Survey on Image Data Augmentation for Deep Learning. *J. Big Data* **2019**, *6*, 60. [CrossRef]
37. Liu, Y.; Mohammed, Z.M.A.; Ma, J.; Xia, R.; Fan, D.; Tang, J.; Yuan, Q. Machine Learning Driven Fluidity and Rheological Properties Prediction of Fresh Cement-Based Materials. *Materials* **2024**, *17*, 5400. [CrossRef]

38. Mathur, P.; Shaikh, H.; Sheth, F.; Kumar, D.; Gupta, A.K. A Computational Intelligence Framework Integrating Data Augmentation and Meta-Heuristic Optimization Algorithms for Enhanced Hybrid Nanofluid Density Prediction through Machine and Deep Learning Paradigms. *IEEE Access* **2025**, *13*, 35750–35779. [CrossRef]
39. Luo, D.; Qiao, X.; Niu, D. A Predictive Model for the Freeze–Thaw Concrete Durability Index Utilizing the Deeplabv3+ Model with Machine Learning. *Constr. Build. Mater.* **2025**, *459*, 139788. [CrossRef]
40. Raudys, S.J.; Jain, A.K. Small Sample Size Effects in Statistical Pattern Recognition: Recommendations for Practitioners. *IEEE Trans. Pattern Anal. Mach. Intell.* **1991**, *13*, 252–264. [CrossRef]
41. Wang, X.; Zhuang, B.; Smyl, D.; Zhou, H.; Naser, M.Z. Machine Learning for Design, Optimization and Assessment of Steel–Concrete Composite Structures: A Review. *Eng. Struct.* **2025**, *328*, 119652. [CrossRef]
42. Zhang, L.; Lai, X.; Pan, J.; Shan, P.; Zhang, Y.; Zhang, Y.; Xu, H.; Cai, M.; Xi, X. Experimental Investigation on the Mixture Optimization and Failure Mechanism of Cemented Backfill with Coal Gangue and Fly Ash. *Powder Technol.* **2024**, *440*, 119751. [CrossRef]
43. Li, W.; Xiao, J.; Sun, Z.; Kawashima, S.; Shah, S.P. Interfacial Transition Zones in Recycled Aggregate Concrete with Different Mixing Approaches. *Constr. Build. Mater.* **2012**, *35*, 1045–1055. [CrossRef]
44. Chang, T.; Wang, P.; Zhao, Y. Strength and Deformation Characteristics of Gangue–Coal Composite with Arched Interface under Confined Uniaxial Compression: Laboratory Test and Numerical Simulation. *Rock Mech. Rock Eng.* 2025, *in press*. [CrossRef]
45. Nunez, I.; Marani, A.; Flah, M.; Nehdi, M.L. Estimating Compressive Strength of Modern Concrete Mixtures Using Computational Intelligence: A Systematic Review. *Constr. Build. Mater.* **2021**, *310*, 125279. [CrossRef]
46. Wu, A.; Wang, L.; Ruan, Z.; Wang, J.; Wang, S.; Guo, R.; Xu, J.; Bai, L. Impact of Aggregate Segregation on the Mechanical Property and Failure Mechanism of Cemented Coarse Aggregate Backfill. *J. Miner. Met. Mater.* **2025**, *in press*. [CrossRef]
47. Sun, T.; Zhang, Y.; Wang, K.; Yu, Z.; Wang, Y. Effect of Waste Clay Bricks on the Performance of Cemented Tailings Backfill and Its Damage Constitutive Model. *Minerals* **2023**, *13*, 987. [CrossRef]
48. Li, J.; Qin, Y.; Zhang, J.; Zhang, A.; Zhang, X. Compaction and Shear Characteristics of Recycled Construction & Demolition Aggregates in Subgrade: Exploring Particle Breakage and Shape Effects. *J. Clean. Prod.* **2024**, *465*, 142776. [CrossRef]
49. Zhang, T.; Zhang, C.; Song, F.; Zou, J.; Gao, Y.; Yang, W. Breakage Behavior of Silica Sands during High-Pressure Triaxial Loading Using X-ray Microtomography. *Acta Geotech.* **2023**, *18*, 5195–5211. [CrossRef]
50. Zhang, T.; Hou, Z.; Li, X.; Chen, Q.; Zhang, S.; Sun, W.; Zhang, Y. Study on the Calculation Method of Parameters in Concrete Plastic Damage Model. *Chin. J. Comput. Mech.* **2024**, *41*, 1130–1137. (In Chinese)
51. Nie, J.; Wang, Y. Comparative Study on Concrete Constitutive Models for Structural Static Simulation in ABAQUS. *Eng. Mech.* **2013**, *30*, 59–67, 82. (In Chinese)
52. Mao, N.; Wu, D.; Chen, K.; Cao, K.; Huang, J. Combining Experiments and Molecular Dynamics Simulations to Investigate the Effects of Water on the Structure and Mechanical Properties of a Coal Gangue-Based Geopolymer. *Constr. Build. Mater.* **2023**, *389*, 131556. [CrossRef]

buildings

MDPI

Review

Sustainability of Recycling Waste Ceramic Tiles in the Green Concrete Industry: A Comprehensive Review

Ghasan Fahim Huseien [1,2,3,*], Zahraa Hussein Joudah [4,5], Mohammad Hajmohammadian Baghban [6,*], Nur Hafizah A. Khalid [4,*], Iman Faridmehr [7], Kaijun Dong [1,2], Yuping Li [1,2] and Xiaobin Gu [1,2,8]

1 Guangzhou Institute of Energy Conversion, Chinese Academy of Sciences, Guangzhou 510640, China; dongkj@ms.giec.ac.cn (K.D.); liyp@ms.giec.ac.cn (Y.L.); guxb@ms.giec.ac.cn (X.G.)
2 Guangdong Provincial Key Laboratory of New and Renewable Energy Research and Development, Guangzhou 510640, China
3 Department of the Built Environment, School of Design and Environment, National University of Singapore, Singapore 117566, Singapore
4 Faculty of Civil Engineering, Universiti Teknologi Malaysia, Johor Bahru 81310, Johor, Malaysia; zahraa.alsafy85@uomisan.edu.iq
5 Department of Civil Engineering, Faculty of Engineering, University of Misan, Misan 62001, Iraq
6 Department of Manufacturing and Civil Engineering, Norwegian University of Science and Technology (NTNU), 2815 Gjøvik, Norway
7 Civil Engineering Department, Faculty of Engineering, Girne American University, N. Cyprus Via Mersin 10, Kyrenia 99428, Turkey; imanfaridmehr@gau.edu.tr
8 State Key Laboratory of Green Building Materials, China Building Materials Academy, Beijing 100024, China
* Correspondence: ghasan@ms.giec.ac.cn (G.F.H.); mohammad.baghban@ntnu.no (M.H.B.); nur_hafizah@utm.my (N.H.A.K.); Tel.: +6583057143 (G.F.H.)

Abstract: Ceramic tiles classified as non-biodegradable are made from fired clay, silica, and other natural materials for several construction applications. Waste ceramic tiles (WCTs) are produced from several sources, including manufacturing defects; surplus, broken, or damaged tiles resulting from handling; and construction and demolition debris. WCTs do not decompose easily, leading to long-term accumulation in landfills and occupying a significant amount of landfill space, which has substantial environmental impacts. Recycling WCTs offers several critical ecological benefits, including reducing landfill waste and pollution, conserving natural resources, lowering energy consumption, and supporting the circular economy, which in turn contributes to sustainable construction and waste management practices. In green concrete manufacturing, WCTs are widely utilized as replacements for cement, fine, and coarse aggregates, and the recycling level in the concrete industry is an increasingly explored practice aimed at promoting sustainability and reducing construction waste. From this view, this paper reports the innovative technologies, advancements in green concrete performance, and development trends in the reuse of WCTs in the production of systems. The effects of WCTs on fresh, engineering, microstructural, and durable properties, as well as their environmental performance, are reviewed. In conclusion, the use of technologies for recycling WCTs has demonstrated potential in promoting sustainability and supporting the transition toward a more environmentally friendly construction industry. This approach offers a practical contribution to sustainable development and represents significant progress in closing the recycling loop within the construction sector.

Keywords: sustainable concrete; waste tile ceramic; sustainability; low-carbon cement; recycled aggregates; resource efficiency; waste valorization

1. Introduction

Waste from ceramic tiles, primarily generated during manufacturing, construction, renovation, and demolition activities, includes broken tiles, rejected batches, and unused off-cuts [1,2]. These wastes are non-biodegradable and chemically stable, leading to long-term accumulation in landfills. Their disposal poses significant environmental concerns, including land degradation, increased landfill burden, and the unnecessary consumption of natural resources due to the lack of recycling practices [3]. Additionally, the energy-intensive production of ceramic tiles contributes to carbon emissions, and improper disposal can lead to dust generation and the leaching of trace elements, further impacting soil and water quality. Annually, over 10 million square meters of ceramic tiles are produced worldwide [4]. An accurate survey revealed that nearly 15 to 30% of the material goes to waste in the process of ceramic manufacturing [5]. Therefore, addressing ceramic tile waste through sustainable management and recycling is essential to reduce environmental impacts and promote resource efficiency in the construction industry. Several estimates indicate that ceramic tiles-related sectors around the world generate large quantities of waste each year, most of which ends up in landfills. Both the recycling and discarding procedures of ceramic waste are expensive, leading to environmental concerns. Therefore, the concrete industry could gain significant advantages by properly recycling this waste as a substitute for OPC or/and natural aggregates, creating a mutually beneficial outcome for the construction sector and the environment [6].

Nowadays, various green concretes with improved properties have been developed by blending WCT powder with OPC or replacing fine and coarse natural aggregates with crushed ceramic tiles. The incorporation of WCTs into concrete was found to appreciably enhance the durability and strength of the resulting products. It has been realized that environmentally amiable and sustainable construction components can be a suitable alternative to traditional concrete. This recycling of ceramics into concrete has two significant benefits, such as solving the waste dumping issues of ceramic industries and the development of sustainable binders and aggregates [7,8]. Numerous studies [5,8–10] have shown that waste ceramics possess strong resistance to biological degradation. Owing to their high silica–alumina content and crystalline structure, ceramics have emerged as a viable alternative to cement, capable of improving both the strength and durability of concrete [11–13]. Despite this potential, the use of ceramics as a replacement for ordinary Portland cement (OPC) in construction remains limited [7,10]. Expanding the application of waste ceramics in concrete production is therefore essential for developing high-performance, sustainable concrete at a reduced cost. As major consumers of ceramic waste, construction industries worldwide have the potential to play a crucial role in addressing various environmental challenges [14]. The cement industry can integrate ceramic materials without requiring substantial modifications to the existing production or application methods. Additionally, the reuse of ceramic waste can help lower landfill expenses, preserve natural resources, and reduce energy consumption, thereby promoting environmental sustainability. The extensive research over the years has highlighted the significant benefits construction industries can gain by recycling industrial ceramic waste into cement-free concrete, contributing to long-term sustainable development [15–18].

According to the existing research, ceramic waste shows considerable potential as a substitute for sand and natural aggregates in concrete and mortar applications [5,19]. In recent years, extensive studies have been conducted on the use of various types of waste ceramics in concrete, including wall and floor tiles [20], sanitary ceramics [21], electrical insulators [22], bone China ceramics [23], and ceramic bricks [24]. López et al. [25] demonstrated that replacing 10–50% of conventional sand with white ceramic in concrete results in tensile and flexural strengths comparable to those of concrete made with river sand, while significantly increasing the compressive strength (CS). Similarly, Pacheco-Torgal et al. [7] found that using ceramic sand instead of natural sand does not compromise strength and improves impermeability, resistance to chloride ion penetration, and performance under accelerated aging conditions. Senthamarai et al. [8] evaluated the CS and chloride ion resistance of mortars containing ceramic waste aggregates, concluding that fine ceramic aggregates enhance CS and substantially increase resistance to chloride ion penetration. They identified 20% ceramic aggregate as the optimal proportion for maximizing both mechanical performance and durability. According to [26], replacing 15% to 30% of natural sand with ceramic fine aggregates in high-performance concrete improves both compressive strength and chloride ion resistance, likely due to an internal curing effect. In a study conducted by Binici et al. [26], the authors reported that incorporating 40% ceramic fine aggregate can increase the compressive strength, improve abrasion resistance, and decrease chloride ion penetration depth.

Given the significant potential of waste ceramics as viable alternatives to OPC and natural aggregates in producing high-performance mortars and concretes, this study aims to provide a comprehensive review of innovative technologies, advancements in sustainable concrete performance, and development trends in the recycling of waste ceramics in the concrete and construction industries. The flow chart of recycling WCTs as a binder and filler replacement in green concrete is presented in Figure 1. Firstly, utilizing WCTs as a partial replacement for OPC and its effect on workability, strength, and durability performance has been widely evaluated. Secondly, the effect of WCTs as a replacement for fine and coarse natural aggregates on fresh and engineering properties is reviewed. Finally, the environmental benefits of recycling waste ceramics in the concrete industry and their contribution to combating climate change and achieving sustainability goals are also discussed. It is suggested that the broad adoption of environmentally friendly modified concrete with waste ceramic tiles could play a key role in reducing carbon emissions.

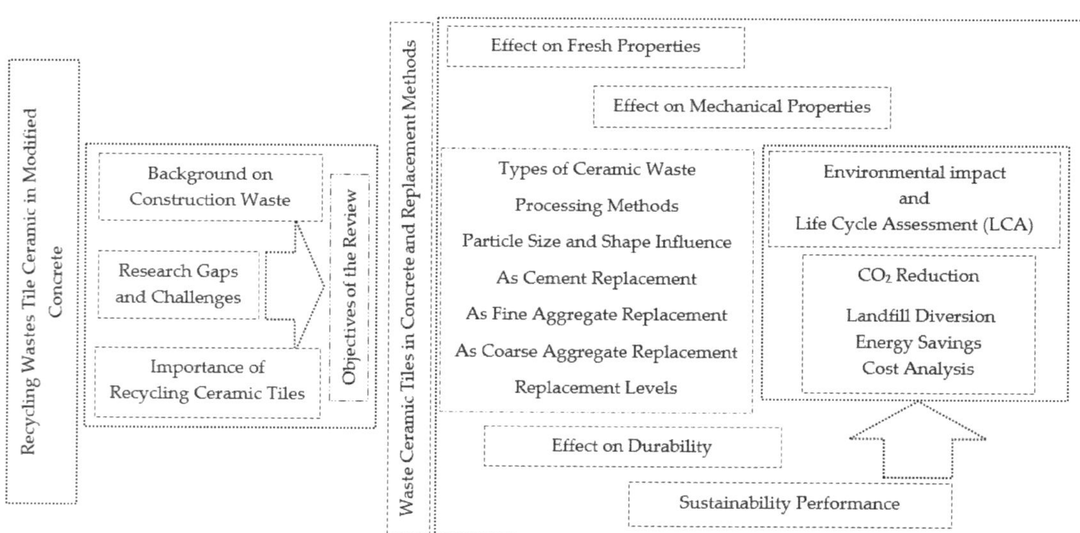

Figure 1. Flow chart of recycling waste tile ceramics in modified concrete.

2. Significance of Recycling Industrial Waste in Concrete

The number of building projects is constantly increasing to meet the needs of various industries, but this also generates waste that must be managed. A sizable proportion of artificial waste is generated during building construction and cleanup operations, which also produce a substantial amount of construction waste. Professionals are highly concerned about the improper disposal of this trash, as it has serious negative consequences for society and the environment [27]. One necessary type of waste is construction and demolition waste, such as floor and wall ceramic tiles. Construction waste accounts for roughly half of the waste that is landfilled around the world [28]. Municipal waste management has led to increased recycling rates of 70% in the U.S. and 90% in Europe [29]. Nonetheless, municipal trash management has not been successful in many countries where recycling rates are less than 10%. Ceramic waste is one of the most common types of waste generated during construction and demolition, and it can cause significant environmental pollution when disposed of in landfills [30]. Moreover, ceramic products are key components of construction, with clay, carbides, silica, metal oxides, and other earthy elements being the key materials used to produce ceramics [31]. Examples of commonly manufactured ceramics include wall tiles, floor tiles, sanitary ware, and household ceramics [32]. The global ceramic tile manufacturing size increased to 18.2 billion m^2 in 2021, from approximately 12.6 billion m^2 in 2019 [33]. Up to 30% of the ceramics produced worldwide are expected to be wasted. Most such waste is disposed of in disorganized piles and landfills, even though it could be immediately repurposed (for example, to fill excavation trenches) [34]. Over 45% of this waste is expected to be ceramics [35]. As ceramic materials are brittle by nature, manipulating and fastening them on building sites results in wasteful fragments. Every year, production errors and demolition materials contribute significantly to the levels of WCTs produced. WCTs originate mainly from two sources: the ceramic industry, due to production processes or defects, and construction activities, either during the building or demolition phase. Most WCTs are produced throughout the stages of manufacturing, construction, and demolition within the building life cycle. Moreover, dumping WCTs in landfills can contaminate the air, soil, and groundwater. Many researchers have highlighted the adverse impacts that WCTs have on the environment [36]. As WCTs are non-decomposable, long-term problems have emerged, including environmental pollution, illegal dumping, and health problems [37]. Therefore, to prevent environmental issues that could result from the inappropriate disposal of WCTs, the optimal management of WCTs must be investigated. Reusing this waste in concrete reduces the environmental effects that cement and WCTs generate, which is an interesting and long-term green option [38]. The strategy of redirecting WCTs from the dumpsite and using them to produce concrete is in line with the circular economy, which helps to preserve valuable land for better uses than waste disposal and reduces the burden of waste management on the construction industry. Furthermore, utilizing WCTs in sustainable building materials may result in lower construction costs, reduced energy consumption, and lower pollution, while requiring fewer raw materials.

3. Waste Ceramic Tiles for Cement Replacement

3.1. Physical Properties and Chemical Composition

As shown in Figure 2, there are two WCT colors, namely white [39] and red [40]. The red color is believed to be caused by high levels of iron oxide (Fe_2O_3) [7]. The waste ceramic tile powder (WCTP) particles appear to be rounder and more uniform after fine grinding takes place. WCTPs have a specific gravity in the range of 2.30 to 2.80. Due to variations in the grinding method used (such as the quantity of material in a ball mill and the duration of grinding), the stated physical attributes of the WCTPs vary between studies. The results from scanning electron microscopy reveal that the WCTPs are made up of uneven, angular particles that resemble cement. Figure 3a illustrates the surface morphology of WCTPs [41]. As presented in Figure 3b, the grain size gradation of WCTPs eventually becomes comparable to that of cement [39]. The physical properties of WCTPs, including the medium particle size, specific surface area, and specific gravity, were found to be 3.3–35 μm [2,41–43], 12.2–12.8 m^2/g [41,43], and 2.35–2.65 [31,41,43–45].

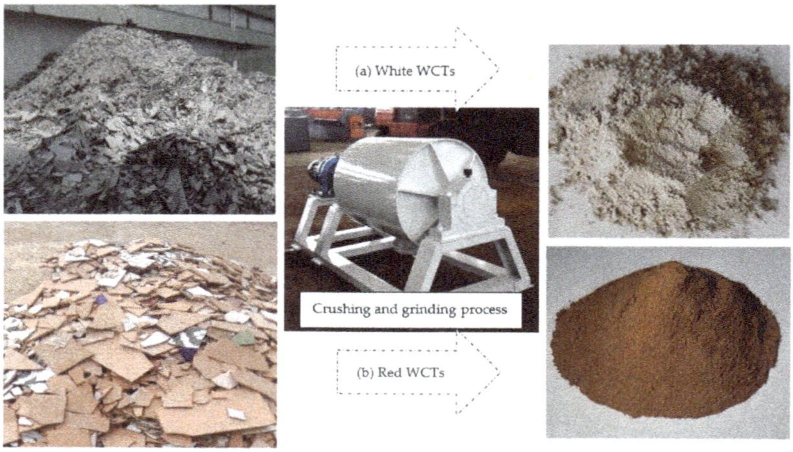

Figure 2. WCTs' colors: (**a**) white resource [39]; (**b**) red resource [40].

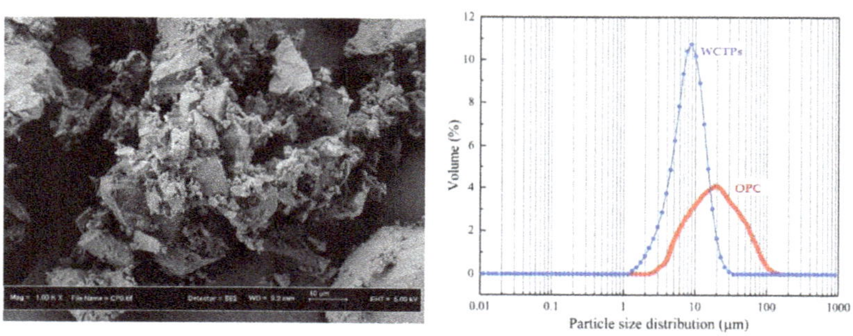

(a) (b)

Figure 3. (**a**) WCTPs' surface morphology [41]; (**b**) particle size distribution of WCTPs compared to normal cement [39].

Based on the material composition, waste tile ceramics are generally categorized into two types: white and red. White tile ceramics are manufactured using purified clays or kaolin, which contain low levels of iron oxide (typically less than 1%). In contrast, red tile ceramics are usually produced from natural clays with a high iron oxide content (normally greater than 5%). The chemical composition of tile ceramics is primarily influenced by the type and source of raw materials used for their production. In Table 1, the chemical composition of WCTPs is presented through a compilation of data from different studies. It is evident that the content of alumina and silica is high in WCTPs, while small amounts of magnesium and sodium oxides was also detected in the evaluated WCTPs. In most studies summarized in Table 1, white ceramics were found to have a higher Al_2O_3-SiO_2 content compared to red ceramics. However, numerous studies have indicated that the calcium oxide (CaO) content in the chemical composition of tile ceramics—particularly white ceramics—is generally very low, typically ranging from 0.02% to 6.67%. Additionally, the reported oxide composition of waste ceramic tile powders (WCTPs) varies across different studies, primarily due to differences in the raw materials used during the manufacturing process, such as their mineralogical and chemical characteristics [46]. WCTPs' ignition loss is less than 4%. To some extent, the difference in loss on ignition is related to the concentrations of SiO_2 and Al_2O_3. A high clay mineral content results in a greater loss on ignition, accompanied by higher concentrations of SiO_2 and Al_2O_3 [47]. Previous studies [48–50] have found that WCTPs contain more than 70% silica, alumina, and iron oxide, meeting the ASTM 618 criteria [51] and making them suitable for use as pozzolanic materials. Silica and alumina are readily available, which promotes the pozzolanic reaction and strengthens concrete. The superior pozzolanic properties of WCTPs stem from their oxide nature. The pozzolanic characteristics of WCTPs have prompted numerous investigators to investigate their potential as sustainable cementitious materials (SCMs) for use in concrete production in order to promote sustainability.

Table 1. The WCTPs' chemical composition utilizing X-ray fluorescence (XRF) analysis.

Refs	SiO_2	Al_2O_3	Fe_2O_3	CaO	MgO	Na_2O	LOI	Other	$SiO_2 + Al_2O_3$	$SiO_2 + Al_2O_3 + Fe_2O_3$
[51]	61.72	22.31	1.24	6.67	0.65	0.96	3.96	2.49	84.03	85.27
[52]	59.90	18.80	7.64	6.47	0.72	1.41	1.16	3.90	78.7	86.34
[41]	72.60	12.20	0.80	0.02	0.99	0.73	1.05	11.61	88.8	89.60
[30]	78.30	15.90	0.10	0.90	-	1.45	1.78	1.57	94.20	94.30
[50]	68.85	18.53	4.81	1.57	0.72	2.01	0.48	3.03	87.38	92.19
[53]	74.10	17.80	3.57	1.11	-	0.01	0.49	2.92	91.90	95.47
[14]	67.83	19.68	0.86	7.42	0.34	0.22	0.18	3.47	87.51	88.37
[54]	68.60	17.10	0.80	1.70	2.5	0.03	1.78	7.49	85.70	86.50
[55]	67.30	19.80	2.50	2.30	-	0.01	0.03	8.06	87.10	89.60
[56]	67.51	16.92	0.75	1.33	-	4.8	2.54	6.15	84.43	85.18
[57]	60.50	28.09	1.02	4.20	0.89	-	1.20	4.10	88.59	94.79
[58]	57.40	17.98	6.20	5.72	3.16	2.37	-	7.17	75.38	81.58
[59]	66.57	21.60	1.41	2.41	2.0	2.9	-	3.11	88.17	89.58
[60]	64.32	19.23	0.94	1.71	10.32	-	-	3.48	83.55	84.49
[61]	66.39	18.14	3.79	3.60	-	-	-	8.08	84.53	88.32
[62]	61.20	18.60	5.0	5.80	1.80	-	2.40	5.20	79.80	84.50
[63]	66.0	19.0	6.0	1.80	0.90	2.1	1.75	2.45	85.0	91.0

3.2. Leaching of Waste Tile Ceramics

Recycling waste materials, such as ceramic tiles, in the production of cement-based products offers both environmental and engineering advantages. However, it is essential to evaluate the leaching behavior of potentially harmful elements within these new materials, as they may pose risks to human health and the environment. The leaching resistance of cement-based materials at the end of their service life has become a significant focus in the recent research. It is generally agreed that the leachability of substances depends on their solubility, which is influenced by several factors, including pH, the formation of inorganic complexes, the presence of dissolved organic matter, and redox conditions. In a study conducted by Medina et al. [64], the leaching performance of concrete incorporating recycled ceramic aggregate was assessed in conditions involving direct contact with water intended for human consumption. The results indicated a slight increase in alkali concentrations (Na and K) and a decrease in the levels of other elements (B, Si, Cl, and Mg) in the water.

Importantly, all detected concentrations remained below the regulatory limits established for drinking-water quality.

3.3. Effect of Waste Tile Ceramics on Concrete Fresh Properties

The fresh properties of modified paste, mortar, and concrete—such as flowability and initial and final setting times—are significantly affected by the dosage of WCTPs used as a partial replacement for OPC. These effects are also influenced by the chemical composition, physical characteristics, and mineralogical properties of the WCTPs, leading to variations in outcomes, even at similar replacement levels. Nevertheless, a consistent trend has been observed: incorporating WCTPs up to 30% generally improves workability by reducing the viscosity of the OPC binder. In contrast, higher replacement levels result in a notable decrease in workability, primarily due to the increased water absorption capacity of the ceramic-based binders. Additionally, the initial and final setting times of binders modified with WCTPs tend to increase with a higher ceramic content. This behavior is attributed to the reduction in CaO content, which slows early hydration and leads to longer setting times compared to control samples containing only OPC. Discrepancies in the findings among researchers can also be attributed to differences in mix design parameters, such as the type of cement used, the presence of supplementary materials, water-to-cement ratios, and variations in fine or coarse aggregates or fiber content and types. As shown in Table 2, the application of WCTPs affects the workability of the concrete mixture. Generally speaking, it increases the need for water, and this requirement becomes more prominent as the WCTP content increases. The non-uniform shape and important water absorption properties of CTP are responsible for this [48,65]. Furthermore, unevenly shaped WCTP particles have been shown to increase interlocking, which increases friction between the particles and increases the energy needed for the mixture to flow (a phenomenon known as yield stress) [66]. Moreover, the workability of cement-based mixtures decreases when finer-sized WCTPs are used instead of OPC, which has a larger surface area and requires more mixing water. Hilal et al. [65] also found that the workability of mortar was reduced when cement was replaced with WCTPs. Their research findings indicated that workability decreased when the level and fineness of WCTPs were increased above those of the control samples, and this was primarily a result of the WCTPs' fineness [67]. The finer WCTP grains fill the voids and reduce the porosity of cement-based composites, ultimately densifying the matrix.

Consequently, the use of WCTPs causes the matrix to stiffen, reducing the flowability of the cement-based composite. Conversely, other research has demonstrated that the workability of cement-based composites is enhanced when WCTPs are used in place of cement. The workability of the concrete mixture is dependent on the differences in each researcher's processing procedures, particularly the size and type of WCTPs employed. The amount of WCTPs ground at a time, the equipment utilized, and the length of the grinding process all play a role in determining the particle size of WCTPs, which ultimately affects the workability of the concrete. Generally, replacing a high volume of cement with ceramic powder increases water demand, which can lead to reduced workability. To address this issue, numerous researchers [6,68] strongly recommend incorporating superplasticizers into the cement matrix to enhance workability. Superplasticizers improve the flowability and consistency of concrete, facilitating easier placement and consolidation.

The time it takes for a cement paste to set is influenced by several factors, including the water–binder ratio, fineness, mineral composition, and chemical composition [69,70]. In Table 2, it is evident that most researchers agree that applying WCTPs prolongs the setting time. It was revealed by Pitarch. et al. [62] that the cement paste's initial setting time was delayed by up to 20 min when WCTPs replaced the cement, and this was primarily due to the relatively lower cement content in the WCTP samples than 100% OPC samples, which reduces the activation phase and prolongs the setting process. Nonetheless, adding finer-sized WCTPs produces different outcomes due to the impact that fineness has on chemical reactivity. Other researchers, such as El-Kattan et al. [69], have found that setting times were reduced for mixtures containing ultrafine WCTPs in place of cement. The large surface area causes open pores to become filled, while the high pozzolanic activity of silicate-rich WCTPs reduces the setting time. Furthermore, a larger grain surface area facilitates nucleation and the development of hydration products, thereby increasing the system's reaction rate. Thus, it is reasonable to conclude that the setting time of concrete paste is impacted by the application of WCTPs with different levels of fineness.

Table 2. Effects of WCTPs on the flowability and setting times of proposed concretes.

Refs	Median Particle Size, μm	Replacement Level, %	Flowability, mm	Setting Times, min	Results
[48]	17.3	10, 20, 30	55–70	Increased	It has been found that the workability of the proposed concrete is significantly influenced by the ceramic content and the setting time trend, which increases with the increasing content of WTCPs in the matrix up to 30%.
[62]	14	15, 25, 35, 50		Initial: 147, 154, 160, 163 Final: 228, 225, 222, 217	It was found that both the initial and final settings were slightly influenced by the content of WCTPs as OPC replacement. For the initial setting time, replacing the OPC with 15–50% of WCTPs results in an increase in the setting time from 147 min to 163 min. Unlike the final setting time, the trend was to decrease with increasing the content of WCTPs. Increasing the replacement level to 50% resulted in a decrease in time from 239 min to 217 min.
[71]	<53	15, 25, 35,50	-	Increased	The results show that the inclusion of WTCPs as OPC replacement leads to an increase in the initial setting of proposed mixtures.
[69]	≤70	1, 3, 5	-	Initial: 190, 199, 210 Final: 260, 268, 280	In general, and compared to control mixtures, the inclusion of WCTPs in the matrix led to an increase in both initial and final setting times, from 190 and 260 min to 210 and 280 min, respectively.

Table 2. *Cont.*

Refs	Median Particle Size, μm	Replacement Level, %	Flowability, mm	Setting Times, min	Results
[63]	10	10, 20, 30	-	Initial: 111, 117, 117 Final: 160, 174, 175	The inclusion of WCTPs as OPC replacement slightly increases both the initial and final setting times. Compared to 87 min, replacing the OPC by 10, 20, and 30% of WCTPs leads to an extension of the initial setting time to 111, 117, and 117 min, respectively. A similar trend of results is observed for the final setting time, and the inclusion of the WCTPs leads to an increase in time from 155 min to 160, 174, and 175 min.
[57]		5, 10, 15, 20	85, 90, 92, 105	-	It has been found that replacing OPC with WCTPs enhances the workability performance, and the slump value trend increases from 80 mm to 105 mm with an increase in the level of replacement from 0% to 20%. For the replacement level in the range of 5–15, the degree enhancement is classified as medium; however, the 20% replacement level is classified as high enhancement.
[72]		5, 10, 15, 20, 25	109, 105, 96, 90, 81	-	the authors evaluated the effect of WCTPs as an OPC replacement in 1% sisal fiber-reinforced concrete. In comparison to the control mix (100% OPC), the increasing level of replacement to 5, 10, 15, 20, and 25% resulted in a drop in slump values from 120 mm to 109, 105, 96, 90, and 81 mm, respectively. The loss of workability is attributed to the lower specific gravity of WCTPs compared to OPC.

Table 2. *Cont.*

Refs	Median Particle Size, μm	Replacement Level, %	Flowability, mm	Setting Times, min	Results
[58]	≤75 μm	20, 40, 60	240, 220, 200		The results indicate that the prepared mortar workability is significantly influenced by WCTP content as cement replacement. Increasing the replacement level to 60% results in a decrease from 260 mm to 200 mm or below, depending on the particle size used in WCTPs. Specimens prepared with fine WCTPs (≤No. 45) showed a worse workability performance. The decrease in workability is primarily attributed to the characteristics and fineness of the ceramic materials used in the concrete. Typically, finer WCTP particles enhance matrix densification by filling voids and reducing the overall porosity. As a result, the incorporation of WCTPs leads to a stiffer matrix, thereby diminishing workability. The high absorption capacity of WCTPs further influences this reduction. Moreover, the use of very fine ceramic particles increases the surface area, which in turn raises the absorption rate and contributes to decreased workability.
	≤45 μm	20, 40, 60	220, 200, 180		
[73]		5, 10, 15, 20, 25, 30	120, 90, 80, 80, 70, 70		The concrete workability trend decreases with increasing the replacement level from 5% to 30%, which is attributed to the high porosity of ceramics compared to OPC, increasing the water demand in the mixture and reducing the flowability performance.

Table 2. *Cont.*

Refs	Median Particle Size, μm	Replacement Level, %	Flowability, mm	Setting Times, min	Results
[74]	98% ≤ 90 μm	5, 10, 15, 20	42, 45, 48, 50		The results indicate that the inclusion of 5, 10, 15, and 20% of WTCPs as OPC replacement leads to improved workability and achieved slump values of 42, 45, 48, and 50 mm compared to a 30 mm slump of the control mixture, respectively.
[59]	89.1% ≤ 45 μm	5, 15, 20, 25, 30	0.40: 94, 92, 89, 85, 82 0.50: 107, 108, 103, 99, 95 0.60: 118, 117, 112, 109, 106		For all the prepared concrete mixtures, the workability trend increases with the water-to-cement ratio from 0.40 to 0.50 and 0.60. However, the inclusion of WCTPs in the concrete matrix as an OPC replacement negatively affects the workability performance and leads to a decrease in slump values. For the water-to-cement ratio of 0.40, the increasing WCTPs from 0% to 30% resulted in a reduction in slump from 95 mm to 82 mm, respectively. A similar trend of results was observed for both 0.50 and 0.60 water-to-cement ratios, and the slump readings dropped from 110 and 120 mm to 95 and 106 mm, respectively, with an increasing replacement level from 0% to 30%.
[75]		10, 20, 30	50, 40, 30	-	The slump values trend to decrease from 60 mm to 50, 40, and 30 mm with increasing the replacement level of OPC by 10, 20, and 30% of WTCPs, respectively.
[76]		10, 20, 30	90, 99, 100	-	The results show a slight improvement in mortar workability with increasing levels of replacement to 10, 20, and 30% of WTCPs.

Table 2. *Cont.*

Refs	Median Particle Size, μm	Replacement Level, %	Flowability, mm	Setting Times, min	Results
[60]	14.32	10, 20, 30, 40, 50	180, 105, 30, 22, 17	-	The substituted OPC by 10, 20, 30, 40, and 50% of WTCPs resulted in a drop in slump values from 195 mm to 180, 105, 30, 22, and 17 mm, respectively. The decrease in workability is primarily due to the higher specific surface area of the ceramic material compared to cement. As a result, additional water is needed to coat the particles when this waste is incorporated into the mixture.
[61]	-	5, 10, 15, 20, 25, 30	58, 54, 50, 47, 43, 41	Initial: 30/70 Final: 540/475	The workability of concrete mixtures tends to decrease with increasing WCTP content as an OPC replacement. Compared to 60 mm, the slump values decreased to 58, 54, 50, 47, 43, and 41 mm with the inclusion of 5, 10, 15, 20, 25, and 30% of WTCPs in the matrix, respectively.

3.4. Concrete Dry Density

The CS of concrete is significantly impacted by the dry density, thus rendering it one of the most important properties. Applying WCTPs to concrete facilitated a decrease in the hardened density of the concrete, as WCTPs have an inferior particle density compared to cement. In line with this, a study conducted by Hilal et al. [65] revealed that applying WCTPs with a lower specific gravity value than cement results in the production of concrete that has a lower dry density. Other researchers have also identified similar patterns.

3.5. Compressive Strength Development

As presented in Table 3, numerous studies have investigated the impact of varying replacement levels of ordinary Portland cement (OPC) with waste ceramic tiles on the mechanical properties of modified concrete, including compressive, flexural, and tensile strengths. It is well established that strength development is influenced not only by the replacement level but also by the chemical composition, physical characteristics, and mineralogical properties of the ceramic material, as well as other components in the mix, such as OPC, aggregates, and chemical admixtures. The water demand of the concrete mix is significantly influenced by the specific surface area and chemical composition of the ceramic, necessitating adjustments to the water-to-cement ratio for each mix based on the type and characteristics of the waste ceramic used. Minor variations in the reported strength results can be attributed to these factors. Generally, incorporating 5% to 15% of waste ceramic tends to improve the compressive strength by increasing the availability of dissolved silica, which contributes to the formation of denser gel structures. However, increasing the ceramic content beyond this range may result in a slight reduction in strength, with the degree of strength loss becoming more pronounced as the replacement level increases. This decline is primarily due to the presence of a high amount of unreacted silica, which weakens the bond strength and leads to an unsatisfactory early strength performance. The researchers have discovered that when the level of substitution increases, the 28-day compressive strength of the cement-based composite decreases due to the presence of WCTPs. Nonetheless, a slight decline in compressive strength was evident when cement was substituted with up to 10% WCTP [30,55]. The analysis revealed that, over 28 days, the compressive strength of the test samples dropped by up to 15% compared to the control samples. In general, the relative fineness of WCTPs in contrast to cement may have caused an apparent decrease in the compressive strength [51]. Lower fineness levels relative to cement usually result in reduced pozzolanic and filling properties. Interestingly, as the substitution ratio increased above 20%, a significant decrease in the 28-day compressive strength was observed. The predominant diluting effect is primarily responsible for this reduction in compressive strength [77]. Nonetheless, at later phases of the curing process, particularly at 90 and 365 days, the mix's compressive strength decreased somewhat despite the addition of up to 30% WCTP. After 365 days of curing, Pitarch et al. [62] identified a 35% WCTP reduction in mortar strength relative to the reference mortar. The observed behavior can be explained by the progressive reaction that the WCTP surface experiences with calcium hydroxide (CH) at later phases [78]. This reaction results in the formation of a substantial amount of calcium silicate hydrate (C-S-H) on the surface, as illustrated in Figure 4. The cement and filler particles form a solid bond as a result of this process, which helps the subsequent stages of compressive strength growth.

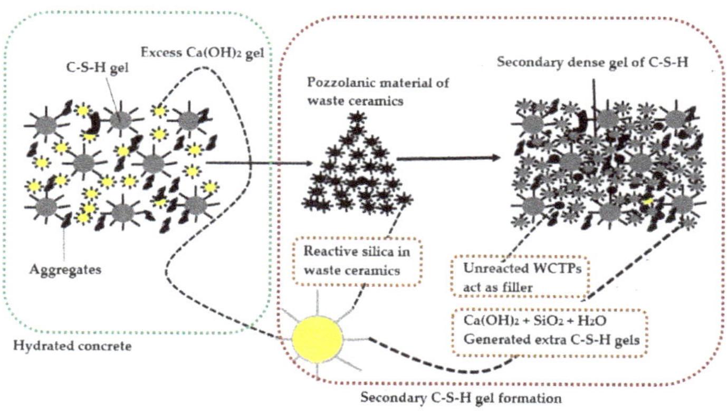

Figure 4. Benefits and mechanism of the inclusion of WCTPs in the OPC matrix to generate extra C-S-H gel [78].

The effects of white and red WCTPs as a cement substitute on mortar were investigated by Lasseuguette et al. [79]. The outcome demonstrated that, except for the 15% mortar, which was comparable to the control for up to 56 days and longer than 90 days, all blended mortars had somewhat lower compressive strengths than the OPC mortar (23.2 MPa versus 21.8 MPa). For up to 56 days, blended cement has been found to have higher concentrations of C-S-H. Meanwhile, the blended samples contained up to 12.5% C-S-H, compared to only 4% C-S-H in the normal OPC. Additionally, reactivity levels were found to be higher in white ceramics than in red ceramics. In the samples containing 15% white ceramics, 5% more C-S-H was absorbed than in the 15% red ceramic sample. This may be because the silica concentration is higher in white ceramics than in red ceramics, indicating potentially higher reactivity. Additionally, it was asserted by Pitarch. et al. [62] that there were no significant changes in the compressive strengths of mortars when up to 25% WCTPs partially replaced OPC.

In a study conducted by El-Kattan et al. [69], the researchers investigated the impact of incorporating ceramic powder at varying replacement levels (1%, 3%, and 5%) as a substitute for white cement across different curing ages. As illustrated in Figure 5, the inclusion of 1% ceramic powder in the white cement matrix notably improved the development of compressive strength at both the early and later stages. This improvement was attributed to the enhanced hydration process and the additional formation of C-S-H gel. The ultrafine waste ceramic powder is considered an amorphous silicate source, where the surplus hydration products contribute to pore refinement through a filling effect. Moreover, the smooth surface texture of ceramic powder enhances the cohesiveness and workability of the mix during the hydration process. However, as the replacement level of ceramic powder increases, the compressive strength tends to decline due to the dilution of the cementitious content and high demand for water, which results in a reduced hydration process and a higher porosity within the cement matrix.

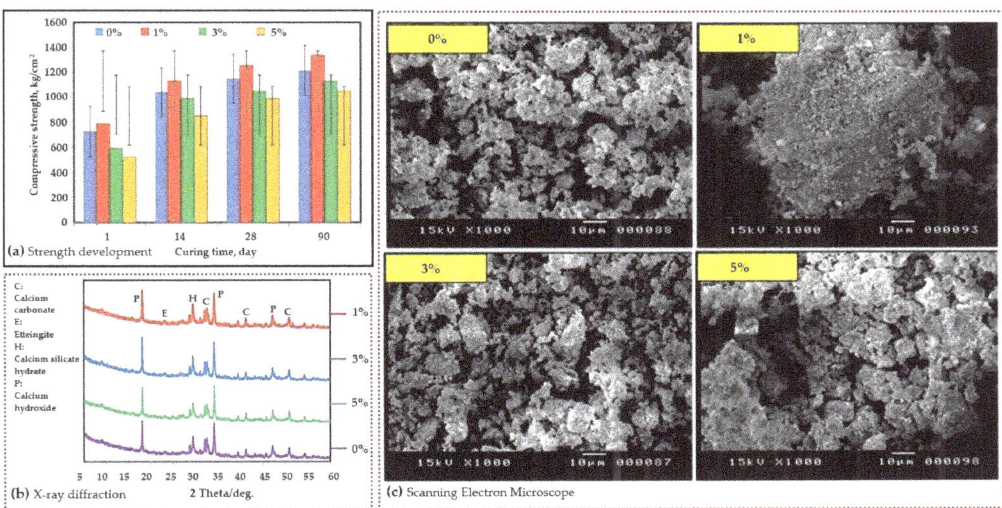

Figure 5. Effects of 1, 3, and 5% waste ceramic powder as white cement replacement on the microstructure of the proposed concrete (**a**) strength development; (**b**) X-ray diffraction; (**c**) Scanning Electron Microscope [69].

Moreover, Heidari et al. [50] fabricated concrete by employing WCTPs to replace up to 15% cement and found that there was a significant reduction in CS at an early stage. This was due to the WCTPs serving as a filler without having a pozzolanic impact at an early stage. Nonetheless, compressive strength data reveal that, after 28 days, the mixture containing WCTPs had strength comparable to that of the control samples. Similar trends in the performance of water-cured mortar were also noted by Mohammadhosseini et al. [80] after partial cement replacement with WCTPs was included. Furthermore, recent studies have revealed that heat therapies could counteract the negative effects of WCTPs on strength. For example, a study by Taher et al. [51] revealed that, when subjected to heat treatment, the compressive strength of WCTPs increased. The findings demonstrate that adding thermally treated WCTPs up to 50% produced concrete with equal strength to the control sample.

However, other research has revealed that the inclusion of WCTPs, even at a low substitution level, has a beneficial effect, as it boosts reactivity and the material's capacity to function as a micro-filler. According to earlier research, the optimal cement-to-WCTPs ratio is between 10% and 20%, which produces a notable improvement in compressive strength after 28 days, which can range from 5 to 22%. In a separate study, the replacement of cement with nano-WCTPs resulted in a notable 40% improvement in compressive strength after 28 days compared to the control group. According to Xu et al. [30], 25% WCTPs is the ideal amount since it results in specimens with the best strength value. The application of WCTPs as a cement replacement below 45% enhances the compressive strength of UHPC. This may be the result of WCTPs' pozzolanic activity and their reaction with CH, which promotes cement hydration and the formation of the C-S-H gel [2]. Densifying UHPC involves packing the pores with C-S-H gel, whereas producing more C-S-H gel enhances compressive strength and lowers porosity (Figure 6). Nevertheless, UHPC strength can be reduced by employing up to 55% WCTPs. A pozzolanic reaction can occur when there is an excessive drop in the cement content because it hinders the hydration process and produces less C-S-H gel and calcium hydroxide [81]. According to Mansoori et al. [44], the use of WCTPs in place of cement affects the compressive strength of self-compacting concrete. The findings indicate that applying 10% and 20% WCTPs results in a 7% and 19% reduction in compressive strength, respectively. As the 28-day compressive strength of the specimens was over 20 MPa, they can be considered suitable for use in structural concrete. The effects that WCTPs have on the CS of different cement-based composites can be seen in Table 3.

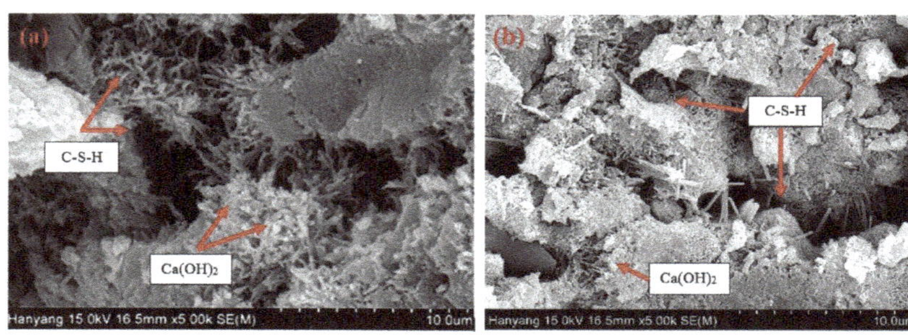

Figure 6. Surface morphology of (**a**) OPC mortar and (**b**) 40% WCTPs -OPC mortar [2].

3.6. Flexural Strength Development

Prior research has shown that when the replacement level increases, the addition of WCTPs causes the flexural strength of the cement-based composite to diminish during the 28-day curing stage. Previous studies [82,83] have shown that adding up to 10% WCTPs causes a slight decrease in flexural strength after 28 days, with the largest recorded declines occurring between 5.8% and 17%. Nonetheless, after a lengthy healing period of 56 days or more, the adverse effects of WCTPs on strength gradually diminished. The formation of more calcium silicate hydrate (C-S-H) gels may result from a subsequent hydration reaction that takes place between WCTPs and calcium hydroxide ($Ca(OH)_2$). These gels produce a denser matrix structure by efficiently filling the pores and interfacial transition zone (ITZ). Many studies have found that adding 5–35% of WCTPs in place of cement significantly increases the flexural strength of cement-based composites after 28 days [72]. The pozzolanic reactivity of WCTPs is responsible for the strength enhancement that was observed. This enhanced the densification of the microstructure, rendering the sample more pliable.

Furthermore, WCTPs are considered to be a filler material that can partially fill in the gaps and/or apertures between cement, improving the granules' physical density. Nonetheless, previous studies have shown that, when the substitution ratio surpasses

35%, there is a reduction in compressive strength. This is primarily due to the well-known dilution effect. Xu et al. [30] revealed that ultrahigh-performance concrete (UHPC) is produced when up to 35% of WCTPs is used as a replacement, and this improves the flexural strength relative to the control samples after 7 and 28 days of curing. This study also highlighted that the same mix had less strength than the control sample after three days of curing. A similar pattern was highlighted by Li et al. [82], who revealed that the strength of concrete fabricated with 10% WCTPs was lower than the control samples after 7 days of curing. It was also found to have enhanced the flexural strength following 28 days of curing.

WCTP-based concrete has low early strength due to the incompleteness of the pozzolanic reaction that is required to produce the C-S-H gel (which enhances the strength increment). Moreover, the flexural strength of concrete can be increased by mixing a suitable content of WCTPs at 6%, 10%, and 35% as a partial substitute for cement. Moreover, it was revealed by Tawfik. et al. [54] during a study of the effects that nano-WCTPs has on the flexural strength of concrete when applied as a cement substitute that using nano-sized particles in smaller percentages could effectively enhance the strength of concrete. The use of a finer material promotes a quicker chemical reaction, resulting in the formation of high amounts of C-S-H gel over a shorter time period than when larger-sized pozzolanic ash is employed [84]. The difference in the results is likely due to the diverse oxide composition and uniqueness of the approach used to process pozzolanic ash, which affects the material's particle size. Although applying large quantities of WCTPs would create a more sustainable construction material, the strength of the concrete would be compromised. In Table 3, the effects that WCTPs have on the flexural strength of different cement-based composites are presented.

3.7. Tensile Strength Development

Furthermore, not only does the splitting tensile strength of concrete provide important information about the incremental cracking patterns that occur during periods of tensile stress, but it also allows us to indirectly evaluate the load under which cracking occurs [85]. Recent studies have revealed that WCTPs can reduce the splitting tensile strength of cement-based composites over the course of 28 days [51,85]. There is also a directly proportional relationship between the magnitude of this loss and the extent of substitution. Nonetheless, only a small (13%) reduction in splitting tensile strength was achieved when the cement was substituted with 10% WCTPs. As was also the case with compressive strength, the observed reduction in tensile strength is attributed to the comparable or relatively lower fineness of WCTPs compared to cement. Interestingly, there was also a substantial decrease in tensile splitting strength after 28 days when a substitution ratio in excess of 30% was used [55]. The primary reason for this reduction in tensile strength is the predominant dilution effect. However, the addition of 30% WCTPs caused a relatively lower reduction in the splitting tensile strength toward the end of the curing process, especially after 56 and 90 days of curing. Furthermore, Mohammadhosseini et al. [80] found that adding WCTPs to the mortar can enhance the splitting tensile strength when allowed to cure for longer periods of time. When compared to plain concrete, the blended cement–WCTP mortar was found to have a somewhat reduced early strength (approximately 4%). However, after 90 days of curing, there was a 15% increase in strength. The quantity of binding gel produced is also lower when WCTPs are used, which, in turn, reduces the cement content and facilitates the hydration process. The WCTPs' pozzolanic reaction changes the microstructure of the mortar, rendering it stronger and denser than the control sample. Conversely, prior research has revealed that replacing cement with WCTPs increases the splitting tensile strength by approximately 5–25%, with such increases varying between

4% and 18% [86]. There are two primary reasons for the apparent increase in the splitting tensile strength: the pozzolanic process that occurs within the cement matrix and the filling effect generated by the minute particles in the concrete matrix. A study by Tawfik. et al. [54] found that using higher-fineness, nano-sized WCTPs increases the splitting tensile strength of concrete. When 6% nano-WCTPs were used in place of cement, the most significant increase in the splitting tensile strength was observed. It is thus essential to evaluate the enhancements that can be made to the methods of producing WCTPs with high pozzolanic reactivity, as this can facilitate the utilization of high-volume waste to produce concrete with improved strength. In Table 3, the effects of WCTPs on the splitting tensile strength of various cement-based composites are presented.

Table 3. Effect of WCTP content as OPC replacement on the strength properties of concrete specimens, such as compressive, flexural, and tensile strengths.

Refs	Replacement Level	CS, MPa	FS, MPa	STS, MPa	Findings
[57]	5, 10, 15, 20	33.26, 32.55, 31.99, 30.13	-	3.26–2.97	Findings: Compared to 28.92 MPa, the inclusion of 5–20% of WCTPs as OPC replacement leads to an improvement in the CS to 33.26, 32.55, 31.99, and 30.13 MPa for the specimens evaluated at 28 days of age with increment percentages of 15, 12.6, 10.6, and 4.2%, respectively. A similar trend of results was observed for the STS test, and the cylinder specimens prepared with WCTPs achieved an enhancement from 3.8% to 13.5%. It is a summary of the percentage of 5% WCTPs, the optimum replacement level, and the other four levels.
[72]	5, 10, 15, 20, 25	40.3, 41.7, 37.1, 34.1	5.31–6.73	4.31–3.14	Specimens prepared with a CS of 40 MPa, in which 10% of OPC was replaced by WCTPs, demonstrated a 3.5% increase in strength. However, increasing the replacement level to 25% led to a 14.8% reduction in strength. A similar trend of results was reported for FS and STS, and the optimum value was achieved with 10% of WCTP content. The differing effects of WCTPs on the CS of concrete at various replacement levels may be attributed to several factors. A key factor is the variation in the chemical composition and specific gravity between OPC and WCTPs. Typically, WCTPs contain a higher silica content and have a lower density compared to OPC. As a result, replacing OPC with WCTPs can lead to a less dense concrete mix, potentially reducing its CS. Additionally, the particle size distribution of WCTPs may influence the overall strength. The presence of larger particles can create voids or weak zones within the concrete matrix, thereby weakening its structure. Furthermore, the pozzolanic activity of WCTPs may be less effective than that of OPC, which could result in slower early-strength development. Other factors, such as the water–cement ratio, curing conditions, and the quality of raw materials, also play important roles in determining concrete strength. Therefore, optimizing these parameters is essential to achieve the desired performance when incorporating ceramic waste powder as partial cement replacement.

Table 3. *Cont.*

Refs	Replacement Level	CS, MPa	FS, MPa	STS, MPa	Findings
[58]	≤No. 75: 20, 40, 60 ≤No. 45: 20, 40, 60	24.9, 22.2, 18.7 26.6, 23.4, 20.4	- -	- -	Compared to the CS of the control specimens (23.1 MPa), replacing 20% of ordinary OPC with waste WCTPs led to a slight improvement in strength. However, increasing the replacement level to 60% resulted in a significant reduction in CS values. The results also indicate that specimens prepared with finer WCTPs (≤No. 45) exhibit a better performance than those containing coarser particles. The decrease in strength at replacement levels of 40% and above is attributed to the high ceramic content in the mix, which increases the silica concentration. This may interact with the calcium hydroxide produced during cement hydration, potentially diminishing CS. Notably, the higher strength observed in group-B specimens is likely due to the finer particle size of the WCTPs, which may contribute to a more homogeneous matrix structure.
[73]	5, 10, 15, 20, 25, 30	62.5, 59, 54.5, 50.8, 47.4, 43.9			Compared to the control specimens' CS (62.4 MPa), replacing OPC with 5% achieved almost similar strength values (62.5 MPa). However, increasing the level of replacement to 10%, 15%, 20%, 25%, and 30% leads to an increase in the loss of strength to 5.5%, 12.7%, 18.6%, 24.1%, and 29.6%, respectively.
[74]	5, 10, 15, 20	41.4, 45, 47.3, 44.2	5.91–6.25	3.12–3.39	In comparison, the CS of the control specimen (37.1 MPa) achieved after 28 days of curing age, replacing the OPC with 5, 10, 15, and 20%, leads to an enhancement of the strength and increases its values to 41.4, 45, 47.3, and 44.2 MPa, respectively.
[59]	5, 15, 20, 25, 30	38.2, 39.3, 40.1, 35.7, 30.1	5.02–5.94	4.92–5.29	The results show that the inclusion of WCTPs as OPC positively affects the strength performance at levels of 5, 15, and 20%, and the specimens achieved the highest enhancement of 6.1% with 20% of WCTPs. However, raising the replacement level to 25% and 30% resulted in a drop in strength, causing a loss of 5.6% and 20.3%, respectively. A similar trend of results was observed for FS and STS, with specimens containing 20% WTCPs as OPC replacement achieving the highest strength performance among the other ratios. A negative effect was observed with increasing the water-to-cement ratio, and the strength trend decreased with increasing the water content. Specimens prepared with a 0.50 water-to-cement ratio showed a loss in strength between 1.7% to 7.8% at 28 days. A significant loss of strength was observed for specimens prepared with a 0.60 water-to-cement ratio, ranging from a 15.9% to 17.8% loss. For a high water content (0.50 and 0.60), specimens prepared with a high content of WCTPs (30%) showed a reduced loss of strength of 1.7% and 15.9%, respectively.
[75]	10, 20, 30	21, 19, 20	-	-	Compared to concrete prepared with OPC only as a control specimen, replacing OPC with 10%, 20%, and 30% of WTCPs leads to a decrease in CS from 22 MPa to 21, 19, and 20 MPa, respectively.

Table 3. *Cont.*

Refs	Replacement Level	CS, MPa	FS, MPa	STS, MPa	Findings
[60]	10, 20, 30, 40, 50	38.4, 40.2, 39.6, 37.7, 30.1			Compared to 38.1 MPa, the inclusion of 10, 20, and 30% of WCTPs as OPC replacement leads to an enhancement of strength by 7.4, 16.2, and 13.2%, respectively. However, the greatest loss of strength (15.2%) was observed with specimens prepared with 50% of WTCPs as OPC replacement.
[61]	5, 10, 15, 20, 25, 30	36.1, 35, 29.9, 28.7, 22.6, 23.5		2.37–3.25	In comparison to the CS of control specimens (33.2 MPa), the inclusion of 5 and 10% of WCTPs as OPC replacement slightly increased the CS to 36.1 and 35 MPa. However, continuing to increase the replacement level to 15, 20, 20, 25, and 30 resulted in a drop in strength to 29.9, 28.7, 22.6, and 23.5 MPa, respectively.
[48]	10, 20, 30	22.9, 23.1, 23.7, 25.4	-	1.67–3.64	Both CS and STS were found to be enhanced by replacing the OPC with 10, 20, and 30% of WCTPs. In comparison to the 22.9 MPa CS achieved with control specimens, the inclusion of 10, 20, and 30% of WCTPs slightly improved the strength to 23.1, 23.7, and 25.4 MPa, respectively.
[62]	15, 25, 35, 50	48, 42, 38, 29	-	-	At 28 days of age, it was observed that replacing OPC with 15%, 25%, 35%, and 50% of WCTPs resulted in a drop in CS from 50 MPa to 48, 42, 38, and 29 MPa, respectively.
[69]	1, 3, 5	Increased	-	-	Specimens prepared with 1% of WCTPs as OPC replacement achieved the best strength performance by 9.8%. However, the increased replacement level to 3 and 5% caused a loss of strength by 1.9 and 14.7%, respectively.
[77]	10, 20, 30, 40, 50	21.3, 23.5, 24.6, 17.9, 17.5	-	-	The results indicate that replacing OPC with 30% of WTCPs leads to an increase in strength from 20.8 MPa to 24.6 MPa. However, the strength trend drops with increasing the level of replacement to 40% and 50%.
[54]	2, 4, 6, 8, 10	28.8, 29.3, 29.7, 29.2, 28.7	3.4–3.6	2.3–2.5	In comparison to the strength of control specimens (27.6 MPa), the replacement of OPC by 2, 4, 6, 8, and 10% of WCTPs leads to an enhancement of strength to 28.8, 29.3, 29.7, 29.2, and 28.7 MPa, respectively. A similar trend of results was observed for FS and STS, and all the specimens prepared with WCTPs displayed a better performance than the control specimens.
[87]	15, 25, 35, 50	40, 45, 40, 34	-	-	The inclusion of 25% of WCTPs as OPC replacement leads to an increase in the CS from 38 MPa to 45 MPa. However, raising the replacement level to 50% results in a lower gel formulation and a drop in strength to 34 MPa.
[31]	5, 10, 15, 20, 25	50.4, 53.9, 47.2, 45.6, 42.8	6.14, 6.15, 5.98, 5.29, 5.15		Specimens prepared with 10% WCTPs as OPC replacement achieved the highest CS (53.9 MPa) and FS (6.15 MPa) among the adopted levels, compared to control specimens (48.5 MPa and 5.68 MPa).

Table 3. *Cont.*

Refs	Replacement Level	CS, MPa	FS, MPa	STS, MPa	Findings
[55]	10, 20, 30, 40, 50	20.9, 22.7, 24.5, 18.1, 17.9	-	-	In comparison to OPC specimens, the inclusion of 10, 20, and 30% of WCTPs as OPC replacement leads to an increase in CS from 20.4 MPa to 20.9, 22.7, and 24.5 MPa, respectively. However, increasing the replacement level to 40 and 50% leads to a drop in strength below 18.1 MPa.
[51]	Unburned 10, 20 Burned 10, 20	32.9, 26.8 36.4, 34.0	5.3, 4.8 6.2, 5.8	4.5, 3.6 4.9, 4.6	At 28 days, all the specimens prepared with unburned and burned ceramics displayed lower CS, FS, and STS compared to control specimens (38.6, 6.4, and 5.2 MPa).

3.8. Sulphate Attack Resistance

If sulphate is present in soil or saltwater, it can damage the OPC concrete that comes into contact with it. Moreover, when calcium hydroxide reacts with a sulfate solution in concrete, calcium aluminate sulfate hydroxide is created. When this sulphate solution reacts with calcium sulphate, calcium sulphate ($CaSO_4$) is produced. Moreover, expansion results from the formation of ettringite [88]. As stated by Samadi et al. [2], mortar samples immersed in 5% Na_2SO_4 and containing 40% WCTPs experience a decrease in the loss of compressive strength. After being immersed for 18 months, the residual strength of samples containing 0% and 40% WCTPs dropped by 41.1% and 16.8%, respectively. Furthermore, it was reported by Mohammadhosseini et al. [80] that the OPC mortar lost 0.7% of its original mass after three months of immersion in the solution, although the WCTP mortar gained 1.7%. Similarly, Li et al. [82] revealed that mortars consisting of 5% and 10% WCTPs were less vulnerable to spalling and cracking on the surface. As illustrated in Figure 7, the authors [89] found that replacing OPC with 10 and 20% of WCTPs could significantly improve the durability of proposed concrete by reducing the expansion percentage. In fact, no spalling or cracks occurred when the WCTP content was 20%. The incorporation of WCTPs, which breaks down weak calcium hydroxide to create more C-S-H gel, makes the concrete's interior structure denser and more resistant to sulphate attack. By densifying the pore structure, C-S-H gel formation reduces the porosity of cement-based composites and enhances their resistance to sulfates. It is essential to examine the effect of varying fineness on the ability of concrete to resist a sulphate attack. Thus, future research should focus on examining the sulphate resistance of high-performance concretes that incorporate WCTPs as a partial substitute for cement.

3.9. Acid Attack Resistance

Acid attack resistance is fundamental for external concrete applications. Mohit et al. [90] examined the impacts that WCTPs (as a cement substitute) have on concrete subjected to a hydrochloric acid attack. After being left to cure for 56 days, the mass loss of the control sample was found to be 8.64%, while the mass loss of concrete with 5, 10, 15, 20, and 25% WCTPs was 7.11%, 7.74%, 7.92%, 8.37%, and 9%, respectively. Meanwhile, the compressive strength of the sample mixtures containing 5, 10, 15, 20, and 25% WCTPs decreased by 50.1%, 54.8%, 53.2%, 56.7%, and 56.8%, respectively. This was reduced by 56.2% in the control sample. As demonstrated in Figure 8, the concrete samples containing 5% WCTPs experienced the lowest loss in strength over time, while the mixtures containing 25% WCTPs displayed the worst strength loss following exposure to an acid attack. As less C-S-H gel forms due to the reduced cement content (which also has an impact on the pozzolanic reaction), the concretes fabricated using WCTPs over the recommended quantity showed poor resistance to acid attacks. Similar results were reported by Rodhia et al. [91], who found that using WCTPs as a cementing agent increased the specimens' resistance to sulfuric acid compared to the control specimens. By slowing the transfer of the acid solution and densifying the concrete's structure through a pozzolanic reaction, concrete containing WCTPs has the ability to withstand an acid attack. The acid resistance of binary, ternary, or quaternary blended cement concrete using WCTPs as a partial cement substitute is still unknown. Thus, to create a high-volume WCTP blended cement mix that demonstrates excellent resistance to acid attacks, further research is required.

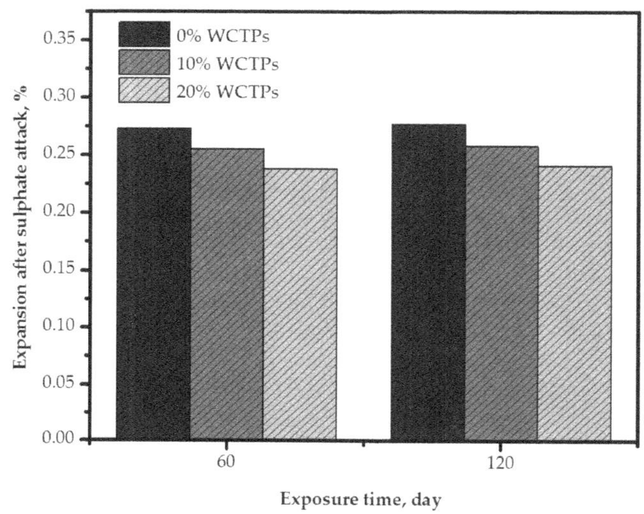

Figure 7. Effect of varying contents of WCTPs as OPC replacements on concrete specimens' exposure to sulphate attack [89].

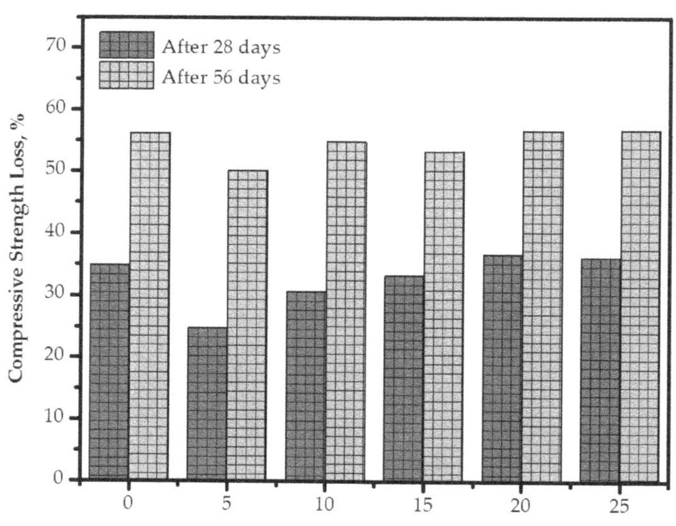

Figure 8. Effect of WCTP content on the compressive strength loss percentage of concrete exposed to hydrochloric acid with a pH of 2 [90].

Rodhia et al. [91] investigated the impact of replacing OPC with WCTPs at varying levels (10%, 30%, 50%, 70%, and 90%) on the durability performance of concrete, specifically its resistance to acid attacks. The study employed two types of 2% acid solutions—sulfuric acid (H_2SO_4) and hydrochloric acid (HCl). Concrete specimens were immersed in these solutions for 56 days, after which the reduction in compressive strength was measured. As illustrated in Figure 9, the strength performance was assessed both before and after exposure to acid. Notably, after 56 days, the concrete mixtures exposed to sulfuric acid exhibited a greater weight loss and a greater reduction in compressive strength at all replacement levels compared to those exposed to hydrochloric acid. The more severe deterioration observed in the sulfuric acid solution can be attributed to several contributing factors. The compressive strength ratio under HCl exposure was consistently higher than that under H_2SO_4 exposure, largely due to the higher concentration of hydrogen ions (H^+) in sulfuric acid. Acid attacks on foam concrete typically result in the dissolution of cement hydrates and calcium hydroxide, accompanied by the formation of calcium salts [92]. Hydrochloric acid primarily reacts with calcium hydroxide and has a relatively minimal effect on calcium silicate hydrate (C-S-H) gels. The resulting reaction products can further interact with the calcium aluminate (C_3A) present in the cement. In foam concrete containing HCl, both soluble and insoluble salts are formed, with calcium chloride ($CaCl_2$) being water-soluble. When the acid attack is superficial, as in the tested mixes, much of the $CaCl_2$ dissolves in water [93]. Although the pozzolanic activity of ceramic powder contributes to long-term strength development, it provides limited resistance against aggressive sulfuric acid environments compared to hydrochloric acid.

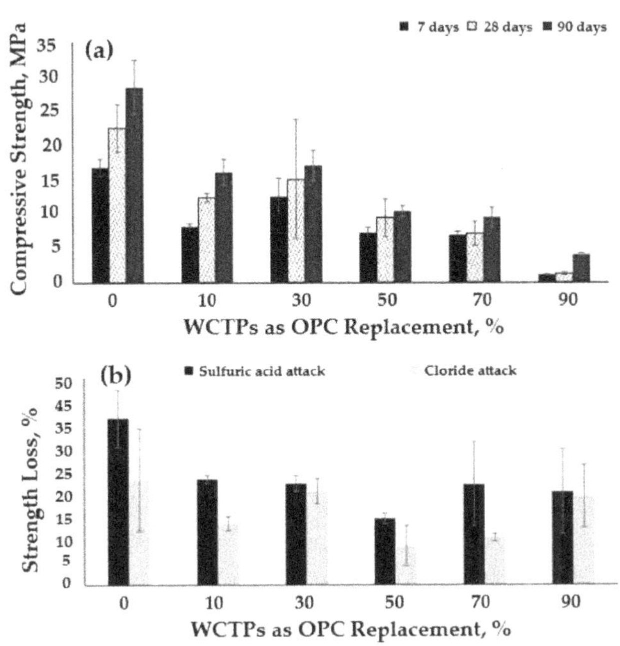

Figure 9. Effect of different acid solutions on the durability performance of concrete prepared with varying levels of WCTPs as OPC replacement: (**a**) compressive strength development without an acid solution; (**b**) loss of compressive strength under acid attack [91].

3.10. Porosity, Capillary Sorptivity, and Drying Shrinkage

Capillary sorptivity refers to a porous substance's capacity to absorb liquid water via its open and connected capillary pores. In a study conducted by Xu et al. [39], the effect of WCTPs as OPC replacements on the porosity performance of low-carbon ultrahigh-performance concrete was reported. For all the specimens, the porosity of concrete tended to decrease with increasing the curing age from 7 to 28 days; a significant enhancement of specimens' structure prepared with 55% of WCTPs was observed after 28 days of curing, which led to a drop in the porosity readings from 29.8% to 7.9%. As shown in Figure 10, the authors found that both early and late porosity reading trends to increase with increasing the content of ceramic in the concrete matrix. Li et al. [82] note that using up to 40% WCTPs to replace cement results in a decline in the sorptivity of mortar. Additionally, when 40% WCTPs are applied, there can be a reduction in the water absorption capacity of up to 60% when the percentage of WCTPs is 40%. In line with this, Chen et al. [14] replaced 10% of recycled aggregate concrete with ceramics and observed a significant reduction in secondary water absorption. This may be because WCTPs have a smaller grain size than cement, generating the micro-filler effect [14]. The pozzolanic reaction caused by adding WCTPs produces a more homogeneous mortar matrix. The formation of vast quantities of C-S-H gel after adding a suitable percentage of WCTPs can help make the concrete's interior structure more compact [94].

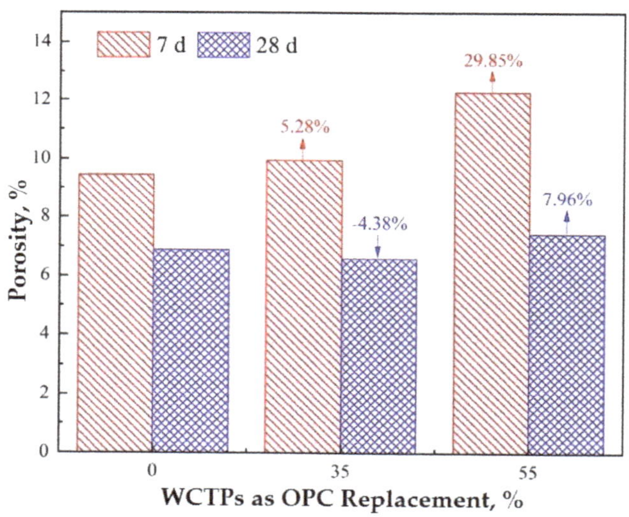

Figure 10. Effect of WCTPs as OPC partial replacement on the porosity of low-carbon ultrahigh-performance concrete [39].

Furthermore, a study conducted by El-Dieb and Kanaan [49] investigated the impacts that WCTPs had on drying shrinkage when applied as a cement substitute. The results showed that when WCTP replacement was increased, the tendency for drying shrinkage decreased for all mixes. A significant reduction in drying shrinkage was observed when the replacement was higher than 20% for all grades. When WCTPs were used in quantities over 20% (relative to the control mixture), there was a reduction in drying shrinkage in the range of 29–60%. Mixtures of 50 MPa experienced a 28–53% reduction in drying shrinkage when more than 20% WCTPs were used.

The drying shrinkage of the 75 MPa mixes decreased by 25–27%. It was revealed by Kanaan and EL-Dieb [95] that the drying shrinkage values were reduced by 5–60% compared to the control mix when WCTPs were applied in quantities of up to 40%. Meanwhile, the drying shrinkage decreased by 12–27% when high-performance concrete mixes were used. Ebrahimi et al. [42] reported that the drying shrinkage of cement mortars decreases as the content of WCTPs increases, reaching its lowest value at a replacement level of up to 40%. This reduction in drying shrinkage is attributed to the extremely fine particles of WCTPs and their pozzolanic properties. By reducing the free water content in the capillary pores, WCTPs significantly decreased the drying shrinkage of the mortar [96]. A delayed reaction in which pozzolanic ash absorbs large amounts of water on its surface and releases it into the environment during cement hydration, reducing the capillary pressure between the particles, is the primary cause of the reduction in drying shrinkage, according to Steiner et al. [97]. The other major factor is cement dilution. The reduced drying shrinkage of fiber-reinforced self-compacting concrete containing WCTPs and micro-silica as cement substitutes was also noted by Mansoori et al. [44]. It would also be beneficial to conduct further research to examine the drying shrinkage of WCTP-blended cement with fiber-reinforced concrete, high-strength concrete, and ultrahigh-performance concrete.

3.11. Resistance to Heating

The effects that WCTPs have on the fire performance of concrete have also been investigated by Mohit and Sharifi [31] who found that using WCTPs as a cement substitute at 5%, 10%, and 15% improved the compressive and flexural strength up to temperatures of 400 °C. Meanwhile, the highest compressive and flexural strengths were observed at 600 and 800 °C, respectively, with blends of 20% and 25%. This is evident in Figure 11, which displays both the residual compressive strength and surface appearance of concrete specimens exposed to 800 °C. Moreover, cracking on the surface of concrete containing WCTPs was much lower at temperatures between 600 and 800 °C, particularly for samples consisting of 25% and 20% WCTPs. Additionally, it was noted by Hilal et al. that the peak value for the compressive strength of 36 MPa was obtained for specimens containing 20% WCTPs at temperatures of up to 800 °C. Similarly, the research by Al Arab et al. [55] revealed that there was a mass loss for WCTP-based pastes of approximately 15%, compared to 25% for the control sample. Applying WCTPs thus enhances the fire resistance of concrete by forming C-S-H through the pozzolanic reaction. Generally, adding an appropriate quantity of pozzolanic ash as a partial replacement for cement enhances the fire resistance of concrete. Nonetheless, further research is required to determine the effectiveness of WCTPs with varying fineness and content in terms of fire resistance in modern cement-based concretes.

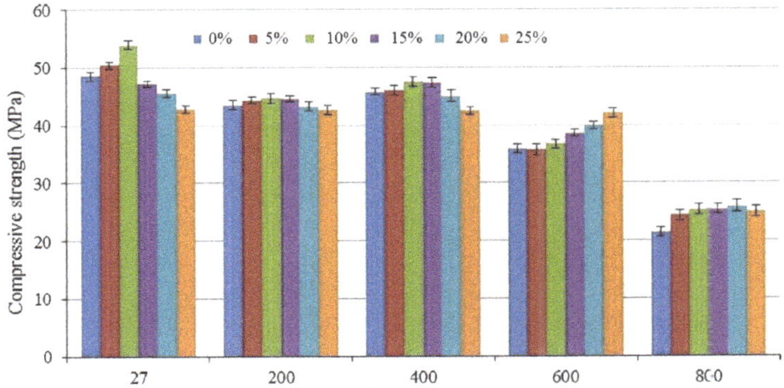

(a)

Figure 11. *Cont.*

(b)

Figure 11. Resistance to elevated temperatures of specimens prepared with varying contents of WCTPs as OPC replacements: (**a**) residual compressive strength and (**b**) surface appearance [31].

4. Waste Tile Ceramics as Aggregate Replacements

4.1. Physical Properties of Waste Ceramic Tile Aggregates

The durability and strength of concrete are significantly enhanced by the physical, mechanical, and chemical properties of waste ceramic tile aggregates (WCTAs). The WCTAs' physical properties, such as bulk density, specific gravity, crushing value, impact value, and water absorption, were found to be 1218–1468 kg/m^3 [4,20,98–101], 2.15–2.55 [4,98,100,102,103], 9.5–35.4% [4,48,71,98,100,102], 12.5–26.9% [20,48,103–105], and 0.18–2.70% [20,48,100,102,105,106]. Both the bulk density and specific gravity of WCTAs are lower than those of natural coarse aggregates (NCAs). In comparison to NCAs, WCTAs exhibited lower density and higher porosity, which resulted in greater water absorption. The bulk densities of coarse WCTAs, as reported in the literature, fell between the 1280 and 1920 kg/m^3 range that the American Concrete Institute (ACI) specifies for standard aggregates. The water absorption of WCTAs ranges between 0.55% and 14.4%, and the main reason for this substantial absorption is the material's porous structure. Compared to most NCAs, WCTAs have a higher water absorption rate that is less than 2% [107]. When compared to other waste materials, they showed a reduced absorption capability. The primary causes of the variations in WCTAs' physical characteristics are the tile composition and firing temperatures [108]. As reported in the reference [109], WCTAs have a distinguishable angular shape, and when compared to NCAs, their surface texture is determined to be rougher. The crushing procedure is what causes visible fissures. Both wall and floor tile aggregates have clearly identifiable pores, as reported by Li et al. [110], with wall tile aggregates having substantially bigger pores than natural coarse aggregates.

4.2. Workability Performance

In Table 4, the impacts that WCTAs have on the workability of cement-based composite material are illustrated. Overall, the workability of cement-based composites containing WCTAs decreases when higher levels of WCTAs are used to replace natural aggregates (NAs). The reduction in workability may be due to the increased roughness and irregular shape of the WCTAs' surface in comparison to NAs [102,111,112]. Moreover, the decrease in the workability of cement-based composites is caused by the interlocking of the WCTAs, which causes increased frictional resistance [109]. The porous nature of WCTAs also renders the mixture more capable of absorbing water. Alves et al. [113] point out that increasing the water–cement ratio can effectively address this reduction in workability. Furthermore, it is possible to mitigate the negative impacts on the workability of the cement-based composite by pre-soaking WCTAs in suitable water, which saturates the WCTAs.

Table 4. Effect of fine and coarse WCTAs on workability of proposed concrete.

Refs	Type of Replacement	Replacement Level	Slump	Discussion
[113]	Fine or coarse	0, 20, 45, 100	121, 120, 118, 116	Numerous studies have indicated that incorporating waste ceramic tiles as fine and/or coarse aggregates in both normal and self-compacting concrete tends to slightly reduce workability and decrease the slump values of the tested mixtures. The reduction in workability generally increases with higher replacement levels. A significant decline in workability is particularly evident when the replacement level reaches up to 50%. This reduction is primarily attributed to two factors. First, the high water absorption capacity of ceramic materials increases the water demand of the mixture, thereby lowering its workability. Second, the irregular shape of the ceramic aggregates, resulting from the crushing process, increases internal friction within the mix, further diminishing the workability of the concrete.
[114]	Fine or coarse	0, 10, 20, 30	78, 75, 70, 67	
[115]	Fine or coarse	0, 15, 30, 45, 60	70, 58, 50, 43, 38	
[19]	Fine or coarse	0, 10, 20, 30, 40, 50, 100	190, 178, 161, 150, 142, 139, 119	
[116]	Fine or coarse	0, 20, 50, 100	18.2, 14.8, 19.4, 9.1 18.2, 18.4, 19.6, 23.6	
[37]	Fine or coarse	0, 15, 20, 30	50, 50, 40, 30	
[117]	Fine or coarse	0, 20, 40, 60, 80, 100	120, 118, 108, 100, 100, 90	
[45]	Fine or coarse	0, 10, 20, 30, 50, 100	60, 70, 80, 98, 90, 122	
[118]	Fine or coarse	0, 5, 10, 15, 20	52, 47, 43, 35, 30	
[4]	Coarse aggregates	0, 10, 20, 30, 40, 50	0.40: 10 to 0 0.50: 45, 40, 35, 30 0.60: 120, 115, 105, 100, 90, 85	
[32]	Fine/Coarse aggregates	0, 25, 50, 75, 100	Fine: 118, 120, 117, 115, 112, 100 Coarse: 118, 115, 110, 90, 40	
[119]	Fine aggregates	0, 20, 40, 60, 80, 100	725, 710, 700, 682, 671, 664	
[120]	Fine or/and coarse aggregates	0, 10, 20, 30, 40, 50	F: 36, 25, 26, 28, 30, 31 C:36, 22, 20, 17, 16, 15 FC: 36, 20, 22, 19, 18, 16	
[101]	Coarse aggregates	0, 25, 50, 75, 100	Workability tends to decrease with the increase in ceramic coarse aggregates, and the T500 flow time increases from 4 to 4.2, 4.5, 5, and 5S.	
[121]	Coarse aggregates	0, 20, 40, 60	73, 67, 64, 60	
[112]	Fine aggregates	0, 20, 40, 60, 80, 100	559, 518, 497.5, 472, 426, 398	

4.3. Fresh and Dry Densities

The various studies in this field generally agree about how adding higher amounts of WCTAs affects the density of cement-based composites. The findings have generally revealed that fresh and dry densities decrease linearly with increased WCTA replacement [4,65,121]. As the ratio of NA to WCTA substitution increases, the density decreases. This seems to make sense, given that WCTAs have a lower density than NAs. Furthermore, a higher percentage of voids is caused by the irregular and rough shape of WCTAs [105,108]. A review of the literature revealed that when comparing WCTA mixes to a control blend, the fresh density decreased by 14% to 20% [65]. Likewise, there is a reduction in dry density from 2260 kg/m^3 (for 0% replacement) when NCAs are fully replaced by coarse WCTAs [108]. Furthermore, it has been found in other studies that the dry density of cement-based composites decreases when WCTAs are used as a replacement [100,105].

4.4. Compressive Strength Development

Compressive strength was the primary mechanical behavior examined in the research focusing on WCTA concrete. The CS performance of a cement-based composite after replacing coarse aggregate with WCTAs is shown in Table 5. When coarse WCTAs are added, the compressive strength of concrete usually increases [4,105,111,114]. The decrease in concrete's 28-day compressive strength with WCTAs replacing NCAs is shown in Table 5. According to numerous studies, replacing NCAs in the range of 5–30% with coarse WCTAs significantly increases the compressive strength from 5% to 25% [122,123]. The irregular shape and coarse texture of WCTAs facilitate more effective interlocking between the aggregates and the cured cement paste, and this ultimately causes the evident increase in compressive strength over 28 days [32]. Furthermore, it should be noted that WCTAs, being a permeable material, provide a moist atmosphere for the hydration process of the cement paste to occur. This results in a decline in autogenous shrinkage, which, in turn, increases the compressive strength of the cement-based composite [30,124].

The 28-day compressive strength of the cement-based composite also increases when fine WCTAs are used optimally. Table 5 demonstrates how the amount of substituted aggregate impacts the concrete's 28-day compressive strength, which includes fine WCTAs. In comparison to traditional concrete, the application of fine WCTAs at up to 60% in place of natural fine aggregates (NFAs) typically yields superior CS, with a 5 to 30% increase after 28 days [125,126]. The irregular shape and rough surface of the particles may contribute to the higher compressive strength of cement-based composites containing WCTAs [127]. The water content of the fine WCTAs is another factor contributing to the rise in compressive strength. This enables the cement to hydrate sufficiently through internal curing, and thus increases the hydration of the cement and enhances the quality of cement-based composites [128,129]. The pozzolanic characteristics of fine WCTAs contribute to the increase in compressive strength. Additionally, due to the mixture's consistent particle size distribution and high concentration of extremely small WCTA particles, which enhanced particle packing, the mixture containing fine WCTAs was compacted more efficiently [129,130].

4.5. Splitting Tensile Strength

Table 5 presents the change in the 28-day splitting tensile strength of concrete when WCTAs are applied (as a function of the proportion of replaced aggregates). Increased 28-day splitting tensile strength is often achieved by replacing NCAs with coarse WCTAs in the range of 5–40%, with increases ranging from 3% to 21%. It is possible that this increase in the 28-day splitting tensile strength is associated with the same factors that impact compressive strength increases, namely the irregular morphology and coarse texture of WCTAs. As a permeable material, WCTAs create a moist environment. Moreover, the incorporation of coarse WCTAs can refine the pore system, causing the volume of capillary pores to increase and the volume of macropores to decrease [21,32].

By contrast, the splitting tensile strength seems to be comparable or slightly lower when a replacement ratio of over 50% is applied. The change in the 28-day splitting tensile strength of concrete containing fine WCTAs can be seen in Table 5. Many relevant studies have found that, when WCTAs are added as a fine aggregate, there is a notable improvement in the splitting tensile strength [131]. The findings of various studies have revealed that compressive strength increases significantly (by 5% to 30%) when 60% WCTAs are used in place of NFAs [119,131,132]. The strength increases over the curing period in proportion to the fine WCTA content. Moreover, substituting fine WCTAs for 50% of the NFAs was found to enhance the splitting tensile strength by as much as 30% [32,119]. The range of 20–60% was the ideal replacement level of NFAs with WCTAs, beyond which the compressive strength steadily decreased [130]. It is also important to note that the fineness modulus is greater in the fine WCTAs than the NFAs, which causes the densification of the mixture and a decrease in macropores, rendering strength development more efficient.

4.6. Flexural Strength

Table 5 presents the change in the 28-day flexural strength (FS) of concrete containing coarse WCTAs. It has been revealed in many studies that applying 20% WCTAs in place of NCA has comparable or even enhanced FS than traditional concrete [133–135]. Furthermore, when the curing age increases, ultrahigh-performance concrete (UHPC) with WCTAs exhibits a rising trend in the ratio of FS to CS. This suggests that UHPC with WCTAs increases more quickly in its FS than in its CS. As a result, the material's toughness likewise gradually increases over time. The ratio of FS to CS reaches a maximum value of 0.143 when the WCTA content is 80%, which represents a 20% increase over the control group [112]. Moreover, as the WCTA content rises, there is a progressive increase in the flexural strength-to-compressive strength ratio, which indicates that WCTAs have a more positive impact on FS than CS.

Likewise, after 28 days of curing, concrete with fine WCTA particles frequently shows a greater FS than concrete with natural fine aggregates. The variation in the 28-day FS of concrete with fine WCTAs in relation to the degree of NCA substitution is displayed in Table 5. When compared to regular concrete, replacing NFAs with fine WCTAs in a percentage between 10% and 60% usually results in a 14.2% to 45% improvement in FS values [112,126].

Table 5. Effect of WCTAs as NFAs and NCA replacement on the mechanical properties of proposed concretes.

Refs	Type of Aggregates	Replacement Level	CS	FS	STS
[4]	Coarse aggregates	0, 10, 20, 30, 40, 50	31.2, 32.3, 31.6, 31, 29.7, 28.5	-	-
[105]	Coarse aggregates	0, 5, 10, 15, 20, 25	Cs7: 35, 37, 39, 41, 38, 36 Cs:28 42, 45, 47, 51, 48, 46	6.52, 6.63, 6.68, 6.73, 6.67, 6.64	3.25, 3.32, 3.5, 3.65, 3.48, 3.37
[118]	Coarse aggregates	0, 5, 10, 15, 20	Cs7: 20.4, 16.2, 17.5, 15.5, 14.4 CS28: 32.6, 28.3, 29.5, 27.1, 26.1	Fs7: 3.9, 4.2, 4.7, 4.3, 3.4 Fs28: 7.5, 6.2, 7, 5.1, 4.3	-
[32]	Fine/Coarse aggregates	0, 25, 50, 75, 100	FA: 24, 24.8, 25, 25.4, 28.6 CA: 24, 24.5, 24.8, 26, 32	-	-
[119]	fine aggregates	0, 20, 40, 60, 80, 100	42, 47, 51, 55, 49, 41	7.1, 7.8, 8.1, 8.6, 7.5, 6.5	5.8, 7.6, 7.9, 8.3, 6.7, 6.4
[37]	Coarse aggregates	0, 14, 20, 30	37.8, 38.5, 40.2, 38.3	-	-
[98]	Coarse aggregates	0, 10, 20, 30, 50, 100	31.1, 28.8, 24.9, 24.4, 22.2, 21.6	6.1, 6.0, 5.6, 5.1, 4.7, 4.6	3.7, 33.4, 3.25, 3.18, 3.14, 2.95
[114]	Coarse aggregates	0, 10, 20, 30	21.1, 21.4, 22.8, 24.9	-	-
[116]	Fine or Coarse aggregates	0, 20, 50, 100	36.2, 34.1, 34.9, 34.6 36.2, 35.5, 35.9, 38.7	-	2.64, 2.64, 2.48, 2.65 2.64, 2.54, 2.33, 2.47

Table 5. *Cont.*

Refs	Type of Aggregates	Replacement Level	CS	FS	STS
[120]	Fine or/and Coarse aggregates	0, 10, 20, 30, 40, 50	20.4, 19.8, 17.5, 16.1, 10.8, 8.4 20.4, 20.2, 20.8, 21.6, 20.6, 16.3 20.4, 20.5, 24.1, 19.3, 15.2, 9.9	2.98, 3.1, 3.14, 2.86, 2.64, 2.46 2.98, 3.14, 3.18, 3.22, 3.04, 2.71 2.98, 3.24, 3.36, 3.17, 2.78, 2.39	2.3, 2.14, 2.08, 1.96, 1.84, 1.55 2.3, 2.35, 2.41, 2.23, 1.98, 1.92 2.3, 2.47, 2.56, 2.43, 2.36, 2.06
[101]	Coarse aggregates	0, 25, 50, 75, 100	57.4, 61.4, 57.0, 54.0, 50.0	6.72, 7.06, 6.44, 5.99, 5.46	-
[136]	Coarse aggregates	0, 10, 20, 30	CS-28 CA-64.2, 60.4, 58.2, 54.6 CB-64.2, 71.1, 68.9, 63.9 CC-64.2, 65.7, 61.2, 58.3	-	CA-3.9, 4.1, 4, 3.9 CB-3.9, 4.7, 4.5, 4.2 CC-3.9, 4.7, 4.4, 4.3
[121]	Coarse aggregates	0, 20, 40, 60	75.9, 77.3, 80.5, 78.1	4.86, 5.1, 5.41, 5.24	6.12, 6.15, 6.31, 6.28
[112]	Fine aggregates	0, 20, 40, 60, 80, 100	126, 120, 131, 134, 124, 142	14.9, 15.3, 15.1, 17.8, 18, 18.5	-

4.7. Modulus of Elasticity

The elastic behavior of cement-based composites is significantly impacted by the addition of WCTAs in place of natural fine and/or coarse aggregates. The relationship between the percentage of aggregate substituted and the respective change in the varying ages' modulus of the elasticity of concrete with WCTAs were reported by several studies [111,121,137,138]. In a study conducted by Zareei et al. [121], the authors reported that increasing the replacement of coarse aggregates by 40% of WCTAs could enhance the modulus of concrete by 2.9%. However, increasing the replacement level to 60% caused a 1.2% drop in the modulus compared to the control specimens. Additionally, the authors observed similar behavior for the ceramic when combined with the fiber-reinforced concrete, as shown in Figure 12. From the results reported in Figure 13, the authors found [101] that the inclusion of 25, 50, and 75% of coarse WCTAs in the matrix of concrete containing calcium carbonate could contribute to enhancing the modulus of elasticity by 18, 5, and 2%, respectively. The elastic modulus of concrete consisting of WCTAs generally rises as the WCTA content increases. Moreover, the increase in the elasticity modulus of WCTA-containing concrete mixtures becomes significant when the entirety of the natural coarse aggregates is replaced by WCTAs [137,138]. This increase in the modulus of elasticity can reach 26.9% in comparison to the control concrete, with a total replacement of 100% [111]. The substantial increase in the elasticity modulus can be attributed to the increased angular morphology and textured surface of WCTAs compared to natural coarse aggregates. There is a proportional relationship between this increase and the aggregate replacement ratio. Thus, there is a clear correlation between the elasticity modulus and the number of angular aggregates [111]. Nonetheless, other studies have revealed that the application of coarse WCTAs to replace NCAs can reduce the elastic modulus of cement-based composites [139–142]. The apparent misalignments in the results of relevant studies may be due to the different physical properties of the WCTAs applied.

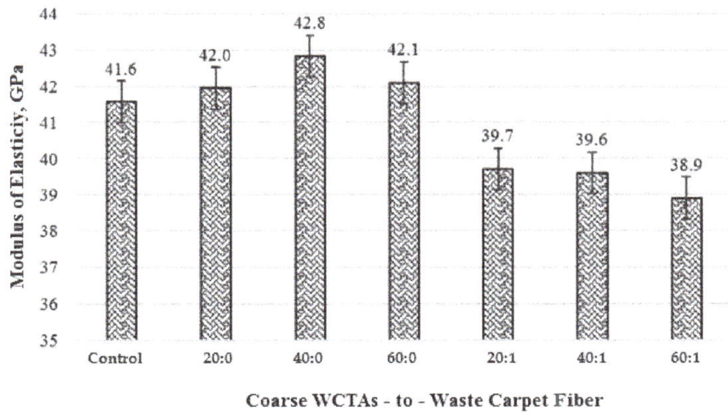

Figure 12. The effect of CTA as a substitute for NCA on the modulus of elasticity of cement-based composites [121].

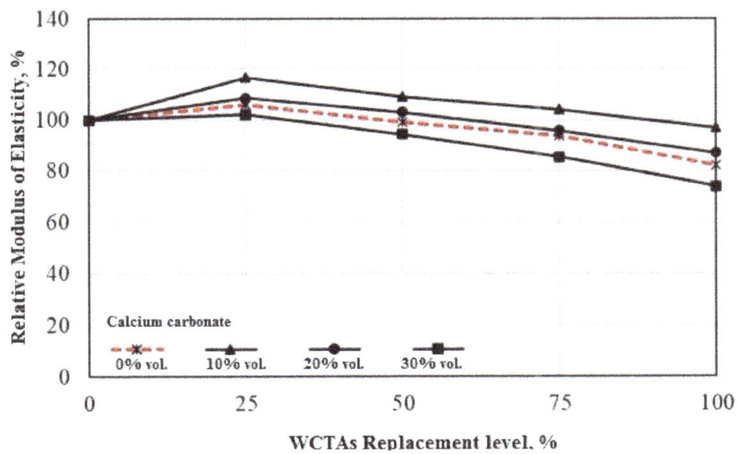

Figure 13. The effect of coarse WCTAs as a substitute for natural aggregates on the relative modulus of elasticity percentage of the proposed concrete [101].

A linear regression analysis was conducted to examine the relationship between compressive strength and the modulus of elasticity [121]. As illustrated in Figure 14, a direct correlation exists between these two properties, leading to the development of the following equation based on the analysis:

$$fc = 5.0594Ec - 135.97$$

This equation can be used to estimate the compressive strength of WCTAs and waste carpet fiber-reinforced high-strength concrete (HSC) based on its modulus of elasticity. Moreover, the R-value of 0.94 demonstrates the reliability of the linear regression analysis and confirms that the proposed equation provides a satisfactory level of accuracy.

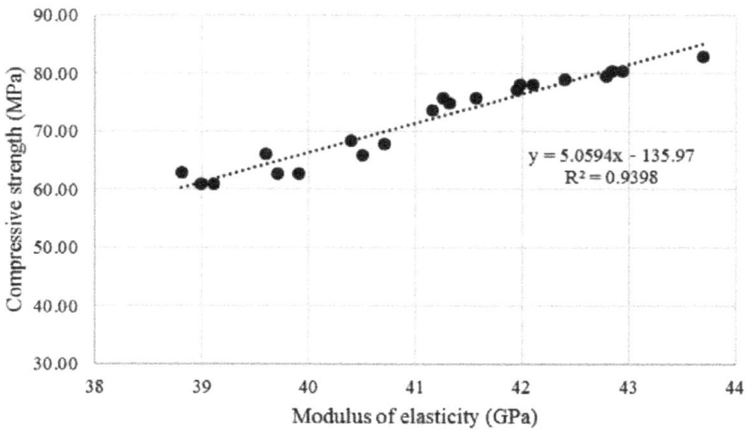

Figure 14. Relationship between the 28-day compressive strength and modulus of elasticity [121].

4.8. Water Absorption and Porosity

As the water absorption characteristics of concrete are critical to defining its quality and strength, the water absorption characteristics of WCTA-based cementitious composites are detailed in this section. It has been discovered through multiple studies that the concrete's water absorption capacity increases when the amount of WCTAs in the concrete increases [108]. Moreover, replacing NCAs in concrete with WCTAs significantly increases porosity and absorption. After 28 days, Paul et al. [45] found that, in comparison to the reference concrete, the absorption increases by roughly 130, 142, 150, 174, and 221% in concrete with WCTA contents of 10, 20, 30, 50, and 100%, respectively.

The reason why the WCTAs have a higher absorption coefficient is that they possess a better absorption capacity than the NCAs. Furthermore, the increased number of voids in mixtures containing WCTAs create an empty vessel, which renders such mixtures able to absorb higher quantities of water [65]. The research performed by Amin et al. [20] revealed that adding mineral admixtures to replace cement can reduce the water absorption of ultrahigh-performance concrete consisting of coarse WCTAs. Additionally, in comparison to normal concretes, the capillary absorption of concretes that had been prepared with coarse WCTAs exhibited noticeably higher values [143]. This increase is most likely due to the enhanced pore connectivity of the coarse-grained aggregate mixture across each grain. However, other researchers have found that water absorption gradually decreases as the coarse WCTA content rises, with a small increase observed when the degree of replacement of course WCTAs reaches a particular threshold [116,144]. Well-graded CTA is added to samples to boost the filling capacity and decrease water absorption [120]. Additionally, because WCTAs are porous, they retains a high degree of internal moisture, which promotes gel formation and pore sealing [138]. The admixture of WCTAs in place of NCAs in concrete can thus have two distinct impacts on the permeability of the cement-based composite. The impacts that WCTAs have on the water absorption of different cement composites when used in place of coarse aggregates can be seen in Figure 15. It clearly can be seen with the addition of a high amount of waste ceramic to the matrix of self-curing concrete (Figure 15a) [138] and self-compacting concrete (Figure 15b) [101] that the water absorption values tend to increase slightly.

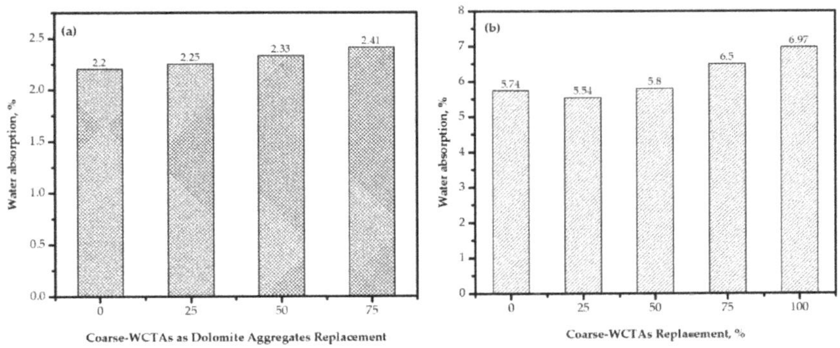

Figure 15. Influence of water absorption of different types of concrete by WCTA content: (**a**) conventional self-curing concrete [138] and (**b**) self-compacting concrete [101].

Until a certain point, the water absorption decreases as fine WCTAs are added in greater amounts to replace NFAs [145]. Meena et al. [119] found in another study that using up to 60% fine WCTAs reduced water absorption when compared to other alternatives. According to the results, less than 2% water is absorbed. Harikaran et al. [146] observed a similar outcome, demonstrating that adding 50% fine WCTAs decreased the water absorption of self-compacting concrete to 1.9% and that increasing the replacement ratio further led to a minor rise in water absorption. According to the findings by other researchers [145,147], the water absorption of different cement-based composites decreased and enhanced at greater replacement ratios when up to 20% fine WCTAs were used as NFAs. This may be caused by the presence of minuscule WCTA particles that refine the pore size [109], causing the water-to-cement ratio to decline [145]. This ratio reduction ultimately results in a decrease in the thickness and permeability of the interfacial transition zone. Meanwhile, other findings have demonstrated that the addition of fine WCTAs improves the water absorption of composites [148]. Moreover, the porous nature of WCTAs and their greater absorption capacity relative to NFAs can cause the porosity of cement-based composites to increase. It is reasonable to state that there could be two conflicting effects on water absorption by using fine WCTAs in cement-based composites instead of NFAs [126]. The relative importance of each factor thus determines how much of an influence WCTAs have on concrete's ability to absorb water. The impacts that WCTAs have on the water absorption of different cement-based composites when used in place of fine aggregates are presented in Figure 16. The results show that increasing the replacement level of fine aggregates to 30% leads to an increase water absorption from 4.58% to 5.43% [149].

4.9. Resistance to Sulphate Attack

A relationship has been observed between the resistance to expansion caused by sulphate attack and the rise in the fine fraction of WCTAs [2,150]. A portion of the fine WCTAs undergoes a pozzolanic reaction that reduces the amount of surplus ettringite that is typically created by the reaction of sulphate and calcium hydroxide [151]. Because fine WCTAs reduce the amount of secondary ettringite in concrete, it can better withstand the expansion caused by sodium sulphate. Furthermore, the inclusion of WCTA particles with an angular shape and coarse surface leads to advantageous interlocking with the cement paste, enhancing resistance to sulfuric acid attacks [151]. Importantly, the fine WCTAs' pozzolanic activity supersedes the increased porosity of concrete mixtures containing WCTAs, making them more resilient to sulfate attacks.

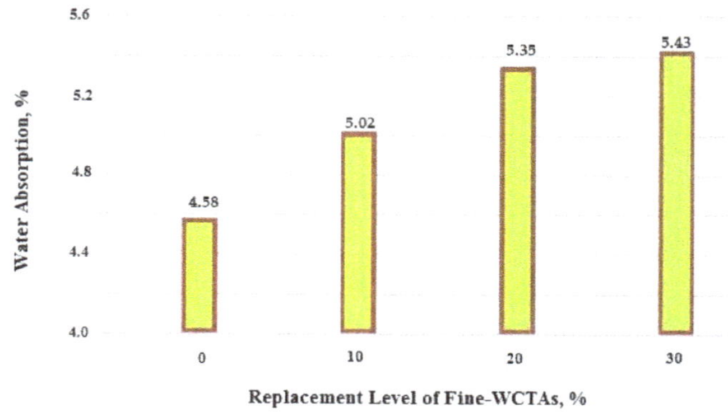

Figure 16. Water absorption and porosity of traditional concrete prepared with WCTAs as a fine aggregate replacement [149].

4.10. Drying Shrinkage

According to Elçi's [108] study, the drying shrinkage of concrete increases when WCTAs are used in place of NCAs. According to the findings, the aggregate shrinkage values for wall and floor tiles are 2.42 and 1.37 times higher than those of the control samples, respectively. WCTAs' high water absorption levels may explain this. Younis et al. [138] reported on the occurrence of early and late shrinkage in another investigation. Drying shrinkage is decreased when WCTAs are added to self-curing concrete to serve as a coarse aggregate. The "reservoir effect" is a phenomenon evident in porous WCTA materials. This effect reduces shrinkage at a constant water-to-cement ratio as the absorbed moisture is progressively released into the concrete mixture throughout the drying process [152,153]. Conversely, the cement-based composites' drying shrinkage was reduced when fine WCTAs were used in place of NFAs. Likewise, Meena et al. [109] found that replacing river sand with fine WCTAs in the production of self-compacting concrete significantly contributed to reducing drying shrinkage, as can be seen in Figure 17. The observed reduction in shrinkage suggests that the aggregates contributed to at least partially mitigating the self-desiccation process, which occurred in parallel with external drying over a specified duration. The angular particle-parking of fine WCTAs may be responsible for this decline in drying shrinkage [153].

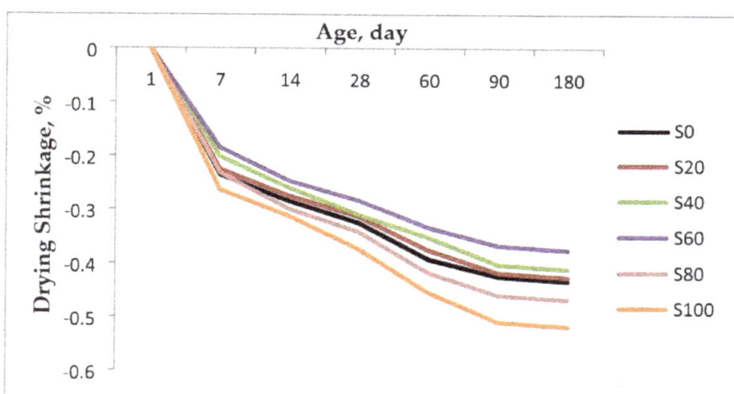

Figure 17. Effect of fine WCTAs as river sand replacements on the drying shrinkage of the evaluated concrete [109].

5. Structural Applications of Waste Ceramic Tiles

Due to their superior mechanical properties, including enhanced compressive, tensile, and flexural strength, as well as a satisfactory modulus of elasticity compared to conventional concrete, WCT-based modified concretes are well-suited for various structural applications [154]. In reference [155], the authors investigated the impact of using WCTPs, as well as fine and coarse WCTAs, as replacements for OPC and natural aggregates on the structural behavior of reinforced concrete beams. Their findings indicated that the ultimate load capacity, crack formation, and total deflection were significantly affected by the level of replacement and the amount of ceramic waste incorporated into the concrete matrix. All tested beam specimens failed in flexure, with numerous cracks appearing in the tension zone of the concrete. The initial crack was observed in the mid-span tension region between the two-point loads, where a pure bending moment is present and maximum tensile stress occurs at the bottom of the beam. As the applied load increased, both the number and depth of cracks grew. Continued loading resulted in the widening of existing cracks and the development of additional vertical cracks along the length of the constant moment region, extending upward.

The authors [155] examined the influence of WCTs on various characteristics of the tested beams, including the initiation of cracks, the number and depth of cracks, beam deflection, ultimate moment capacity, and the total number of cracks after failure, as summarized in Table 6. In the control specimen made with conventional concrete, the first crack emerged at a load of 14 kN, located at the mid-span of the effective length. This initial cracking occurred at approximately 29% of the ultimate load. As the applied load increased, the initial crack deepened, and additional cracks formed on both sides of the original one. The crack propagation gradually extended outward from the initial crack until the beam failed. In specimens where WCTAs were used to replace fine or coarse aggregates, the first crack also occurred at 14 kN, matching the control specimen. Additionally, the total number of cracks in both specimens was identical. While the general pattern of crack propagation across the effective span was similar for all specimens, the total number of cracks before failure varied. These variations in the initial crack loads and crack patterns may be attributed to the effects of the recycled ceramic aggregates on the beams' load-bearing capacity. Overall, no significant differences were observed in the number of cracks or the ultimate load-bearing capacity among the different beam types.

The control beam specimen exhibited the greatest mid-span deflection under ultimate load conditions compared to the other beam specimens. Specifically, the control beam showed a deflection of 28.46 mm, whereas the beam incorporating WCTs deflected by 17.52 mm, despite both beams sustaining nearly identical ultimate loads. This reduced deflection in the WCT-modified beam is likely attributed to the inherent physical properties of the ceramic material, which confer greater stiffness than that of the control specimen. Table 6 presents the maximum deflection and ultimate load values for all tested beams. A comparison of these deflections reveals that the WCTA beam experienced approximately 39% less deflection than the control beam, even though both demonstrated comparable flexural strength capacities.

The authors present the distribution of cracks along the span of the beam specimens [155] in Figure 18. In all cases, the initial crack appeared at the mid-span of the beam, corresponding to the pure moment region. For the control specimen, three out of sixteen total cracks (19%) were located in this region. The number of cracks within the pure moment region for the 100% C-WCTs, 100% F-WCTs, 100% WCTs, and 40% WCTPs specimens was 3, 3, 2, and 4, respectively. This corresponds to 25% of cracks for the 100% C-WCTs and 40% for the WCTPs, and 17% for the 100% F-WCTs, occurring in the pure moment region. These variations may be attributed to the properties of the ceramic material used in casting and the ultimate load-bearing capacity of each specimen. Overall, the total number of cracks among the beam specimens ranged from 12 to 16. The results indicate that as the applied load increases, both the deflection and the number of cracks also increase.

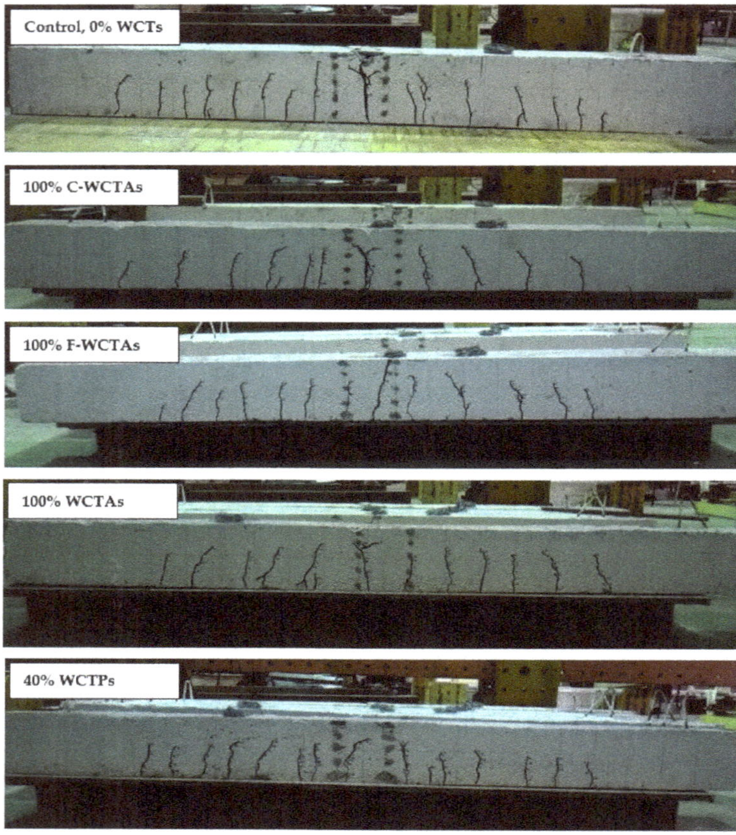

Figure 18. Crack distribution for the tested reinforced concrete beams prepared with WCTs compared to the control specimen [155].

Table 6. Structural properties of reinforced concrete prepared with WCTs as the binder and aggregates replacement [155].

Beam Code		Reinforced Concrete Beams				
		Control, 0% WCTs	**100% C-WCTs**	**100% F-WCTs**	**100% WCTAs**	**40% WCTPs**
WCTPs:F-WCTAs:C-WCTAs		0:0:0	0:0:100	0:100:0	0:100:100	40:100:100
At first crack load	Load, kN	14	10	14	14	12
	Deflection, mm	3.67	2.97	3.04	2.69	3.01
At ultimate load	Load, kN	47.6	44.9	45.8	47.1	45.7
	Deflection, mm	28.46	21.67	20.55	17.52	16.12
Total number of cracks		16	12	13	12	15
Correlation of first crack load to ultimate load, %		29	22	31	30	26

Table 6. *Cont.*

Beam Code	Reinforced Concrete Beams				
	Control, 0% WCTs	100% C-WCTs	100% F-WCTs	100% WCTAs	40% WCTPs
Ultimate moment capacity, kN.m	21.4	20.2	20.6	21.2	20.6
Theoretical moment capacity, kN.m	13	13	13	13	13
Percentage compared to the control beam, expressed as a percentage	100	94	96	99	96
Depth of neutral axis at first crack load, mm	74	71	73	61	79
Depth of neutral axis at ultimate load, mm	61	48	46	43	61

6. Convolutional-Based Deep Learning Models

The integration of computer vision algorithms and AI modeling into the assessment of waste ceramic-based sustainable concrete offers significant research and practical benefits. From a research perspective, these technologies enable a more precise, consistent, and automated analysis of concrete properties, such as crack propagation (as shown in Figure 19), surface texture, and color changes, which are key indicators of durability. By using image-based data and machine learning models, the researchers can detect early signs of deterioration, predict long-term performance, and correlate physical and microstructural changes with mechanical behavior [156]. This enhances our understanding of how ceramic waste materials influence concrete durability under various environmental and loading conditions [157,158].

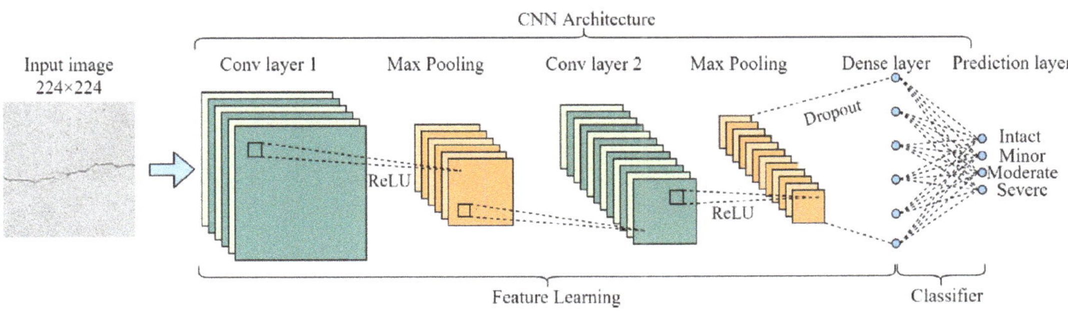

Figure 19. Convolutional neural network architecture [159].

In practical applications, AI-driven models improve the efficiency and reliability of quality control and structural health monitoring in real time [160]. These tools reduce the need for destructive testing and human error while enabling the continuous, large-scale monitoring of infrastructure made with ceramic-based concrete [161,162]. Additionally, predictive AI models can support the optimization of mix designs by simulating durability outcomes based on material inputs and exposure conditions. This leads to more sustainable construction practices, promotes the reuse of industrial waste, and supports the development of durable, eco-friendly concrete alternatives for modern infrastructure [163,164].

Kshirsagar et al. [165] used an artificial neural network and regression model for the prediction and modeling of mechanical properties of concrete modified with ceramic wastes. When evaluating the compressive and tensile strength, the authors found that the R^2 value was 0.70, whereas the ANN model achieved a higher value of 0.87. Due to its improved accuracy in predicting the strength characteristics of ceramic-based cement and structural stiffness, the ANN model demonstrates strong potential as a reliable tool for modeling different types of concrete. In a related study by Joudah et al. [6], the Random Forest Regressor—a type of ensemble learning method—was employed to predict the compressive strength of concrete mixtures incorporating high volumes of waste ceramics as replacements for cement and aggregates. The results demonstrate a strong predictive capability, with a correlation coefficient of 0.92 between the predicted and experimental compressive strength values, as illustrated in Figure 20. This finding emphasizes the significant influence of individual mix components on strength development and highlights the potential of artificial intelligence in optimizing concrete mix designs to improve performance and sustainability.

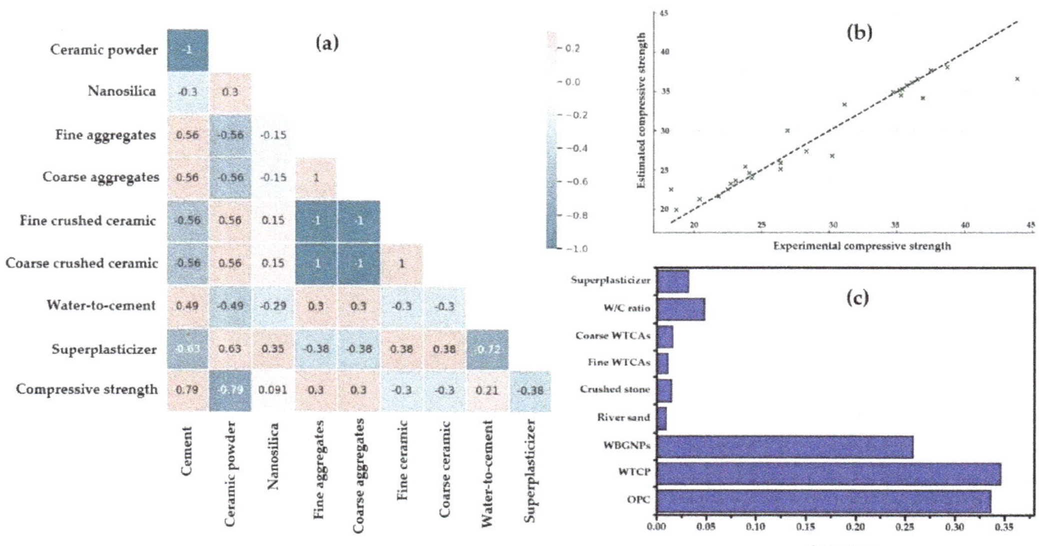

Figure 20. Artificial intelligence (AI) for estimating the compressive strength (**a**), correlation matrix plot of the dataset; (**b**) Random Forest-estimated strength; (**c**) sensitivity analysis of concrete parameters using Random Forest [6].

7. Ceramic Waste's Impacts, Risks, and Sustainable Benefits

7.1. Wastes Impact the Environment

Yearly, billions of metric tons of unsustainable industry, agriculture, and human-generated wastes are disposed in landfill, entering our environment and polluting every ecosystem around the world [166]. With the rapid development, it is expected that the generation of total waste will increase by more than 60% by 2050. In other words, we are creating more waste than ever. It is reported that 62% of global waste is collected in controlled municipal facilities, with the remaining 38% dumped, burned, or discarded. Of the total municipal waste that is collected, 19% is recycled, and the rest ends up in landfill. The increasing disposal the waste in landfill significantly leads to an increase in greenhouse gases (GHGs), which are directly released into the atmosphere. As well as the increasing waste in landfill, environmental and human health are being deeply impacted by this lack of environmental accountability and awareness.

It is well known that the construction industry produces many types of waste during the building and construction process, maintenance, and demolition stages, and this waste has a significant impact on the environment. It breaks down ecosystems, uses up natural resources, and creates clouds of pollution. Construction waste, when mismanaged, acts like a relentless flood. It keeps pouring into already brimming landfills. The overflow of waste can create many environmental hazards. It leads to toxic air and water pollution, which impacts wildlife and public health. Meanwhile, methane, a potent greenhouse gas, worsens global warming.

Improperly disposed of ceramic and glass can end up in landfill, where they take hundreds to thousands of years to break down. This not only contributes to landfill overflow but can also result in the release of toxic substances. When ceramic and glass are not recycled, the opportunity to reuse valuable materials is lost. This increases the demand for raw materials and the associated environmental impacts of their extraction and processing. One of the qualities of ceramic and glass, compared to other materials, is that they can be endlessly recycled. Effective waste management is crucial for reducing harmful environmental impacts and enhancing sustainability. By understanding what happens to ceramic and glass waste, implementing proper disposal practices, and recognizing the consequences of improper recycling, the concrete industry can contribute to a greener future while reaping the environmental and economic benefits of recycling.

Meanwhile, ceramic tile waste generated from manufacturing factories, construction sites, and building demolition projects disposed of in landfill causes land pollution and many environmental problems [49,167]. A significant percentage of this waste is recycled into current ceramic products and processes; however, it is not currently possible to recycle all the waste generated. As a result, an important amount of waste is destined for landfill or used as very low-added-value fillers. To achieve a zero waste-to-landfill strategy, the waste generated must be reduced, especially tile ceramic waste. We need to adopt sustainable design principles and practice efficient construction methods, reusing and recycling different types of waste, such as in the concrete industry, as high-volume OPC replacements or as fillers to replace natural fine and coarse aggregates.

7.2. Risk of Ceramic Waste

The major environmental impact of ceramic tiles is air pollution, which leads to several adverse impacts on the Earth, including ozone layer depletion, global warming, acidification, and eutrophication. Ceramic and porcelain tiles are composed of natural materials, like clay, feldspar, and quartz, and are generally a healthy choice. They can, however, contain toxic additives in their pigments, frits, and glazes. Historically, one major health concern has been the use of heavy metals, such as lead, in glazes.

It is the dust that ceramic tiles can emit that may cause issues. During installation or removal, dust has the potential to cause toxic effects on health. For instance, exposure to ceramic tiles results in the release of airborne dust that causes health issues, such as asthma, chronic bronchitis, silicosis, and pneumoconiosis, when inhaled. Again, this is why tilers wear masks to protect themselves from inhaling dust and wear gloves to avoid possible cuts and scratches. They also wear safety shoes or boots to avoid hurting their feet if a tile or tiles fall on them. During processes that generate dust, they use breathing protection, safety glasses, and appropriate clothing to avoid exposure to dust. However, once the ceramic is placed and left undisturbed to dry, it poses no risk.

7.3. Environmental Benefits of Recycling Waste Materials in the Concrete Industry

Recycling by-products, industrial, and agricultural waste materials, such as those in the construction industry, has many environmental and economic benefits, such as reducing landfill waste, carbon footprint, costs; conserving natural resources; and creating durable construction materials. Integrating ceramic tile waste into concrete production presents notable environmental advantages and supports sustainable development. As illustrated in Figure 21, recycling waste ceramic tiles helps to decrease the volume of refuse sent to landfills, thereby alleviating the pressure on landfill capacity and reducing environmental pollution. In the cement and concrete industry, the use of recycled materials reduces the need for extracting natural resources. Concrete produced using recycled ceramic tiles and glass bottle waste as fine and/or coarse aggregates, or as partial replacements for cement, emits fewer greenhouse gases compared to concrete made from virgin materials. Additionally, incorporating these recycled materials can lower production costs due to reduced raw material and transportation requirements. Concrete containing recycled ceramic and glass waste may also exhibit improved strength and durability compared to conventional concrete. Utilizing ceramic waste in construction materials enables the industry to lower its ecological impact and actively contribute to a more sustainable and environmentally responsible future.

Figure 21. Scheme exploring the environmental advantages of incorporating recycled waste ceramic tiles into the concrete industry as a step toward promoting sustainability and a greener world.

7.4. Life Cycle Assessment of Recycling WCTs in the Concrete Industry

A life cycle assessment (LCA) provides a systematic method for evaluating these environmental impacts, from raw material extraction through processing, use, and final disposal. The integration of waste ceramics as a partial replacement for cement, fine, or coarse aggregates in concrete has shown reductions in carbon footprint, energy consumption, and resource depletion across various LCA studies. For example, studies have shown that replacing 20–30% of cement with ground ceramic waste reduces CO_2 emissions by approximately 20–25% per cubic meter of concrete. This is attributed to the lower embodied energy of ceramic waste compared to clinker production, which emits approximately 850–900 kg of CO_2 per ton. Several case studies have demonstrated the environmental advantages of incorporating waste ceramic in concrete production through comprehensive LCA. In an Asian study conducted in Pakistan, replacing 30% of natural aggregates with ceramic waste led to a reduction in global warming potential (GWP) and significant energy savings during the production phase [114]. The authors reported that incorporating waste ceramic as a replacement for conventional aggregate contributes to the development of sustainable concrete. The higher the proportion of ceramic waste used in the mix, the greater the sustainability of the resulting concrete. A Malaysian study further confirmed these findings, showing that the use of recycled ceramic tiles as a coarse aggregate or cement replacement reduced CO_2 emissions, along with lowering water consumption and acidification potential, especially when the materials were sourced locally [2,96,168]. Similarly, the research in India revealed that substituting cement with ceramic tile powder created a reduction in CO_2 emissions and improved environmental performance across indicators, such as fossil fuel depletion and human health impacts [105,151,169]. These case studies collectively highlight the potential of waste ceramic recycling to enhance the sustainability of the concrete industry while reducing its ecological footprint.

7.5. Carbon Dioxide Emissions, Energy Saving, and Economic Benefits

Minimizing the consumption of natural resources and promoting the effective reuse of industrial waste are among the most practical strategies for achieving sustainable development and a cleaner environment. It is essential to limit the generation and disposal of waste materials that have significant ecological impacts. Incorporating solid waste into the production of construction materials can lead to the development of more sustainable and environmentally friendly products. However, these alternative materials must either be cost-competitive with conventional products or offer clear environmental benefits to construction projects.

To assess the sustainability of mortar or concrete made with recycled WCTs compared to traditional concrete, several key factors are typically considered. These include greenhouse gas emissions, production costs, and energy consumption during the manufacturing process. These indicators are primary motivators for adopting WCT-based concrete, although additional factors also play a significant role. These include technical performance, leaching behavior, water usage, the presence of harmful substances, emissions of other non-environmentally friendly gases during production, and the volume of industrial waste diverted from landfills through its use in concrete. In conducting an environmental benefits analysis, particular attention is paid to the impacts associated with the extraction and processing of raw materials and binders, as well as transportation, which are commonly evaluated within the concrete production industry.

In their study, Samadi et al. [2] investigated the impact of incorporating WCTs as both binder and aggregate replacements on the sustainability performance of proposed mortar mixtures. The findings revealed that variations in WCT content significantly affected carbon dioxide emissions, energy consumption, and production costs. As illustrated in Figure 22, sustainability and environmental performance were assessed based on total production cost, greenhouse gas emissions, and energy usage associated with WCTPs, OPC, fine WCTAs, and natural river sand. The authors observed that the grinding process for WCTPs consumed considerably less energy during manufacturing compared to the production of OPC, resulting in lower costs and reduced greenhouse gas emissions. Specifically, the energy required to produce OPC was 5.13 GJ/ton, more than four times the 1.12 GJ/ton needed for WCTPs. This disparity was attributed to the higher energy demand, cost, and emissions linked to OPC, which released 0.904 tons of greenhouse gases per ton, compared to just 0.045 tons per ton for WCTPs. Consistent with these emission trends, OPC also incurred the highest production costs among the materials studied. Therefore, reducing OPC content in the mortar mixtures was essential for minimizing energy use, lowering production costs, and enhancing overall sustainability and environmental performance.

The influence of replacing OPC and fine aggregates with WCTPs and WCTAs, respectively, on greenhouse gas emissions and energy consumption in blended cement is illustrated in Figure 22a–f. The findings indicate that increasing the proportion of WCTs in mortar mixtures leads to a reduction in both greenhouse gas emissions and energy usage. Blended cements incorporating any percentage of WCTPs consistently emitted fewer greenhouse gases compared to standard OPC-only mixtures. Specifically, producing one ton of blended cement containing 40% WCTs results in the release of only 1 m^3 of greenhouse gases—a decrease of more than 37% relative to the conventional mix. Additionally, energy consumption dropped from 3.02 GJ/m^3 to 2.13 GJ/m^3. In contrast, substituting natural river sand with WCTAs had a minimal effect on emissions and energy use, likely due to the inherently low environmental impact of river sand extraction. Overall, the low levels of greenhouse gas emissions observed in mortar samples using WCTPs as a binder highlight the potential for developing a simple and sustainable material with reduced reliance on OPC.

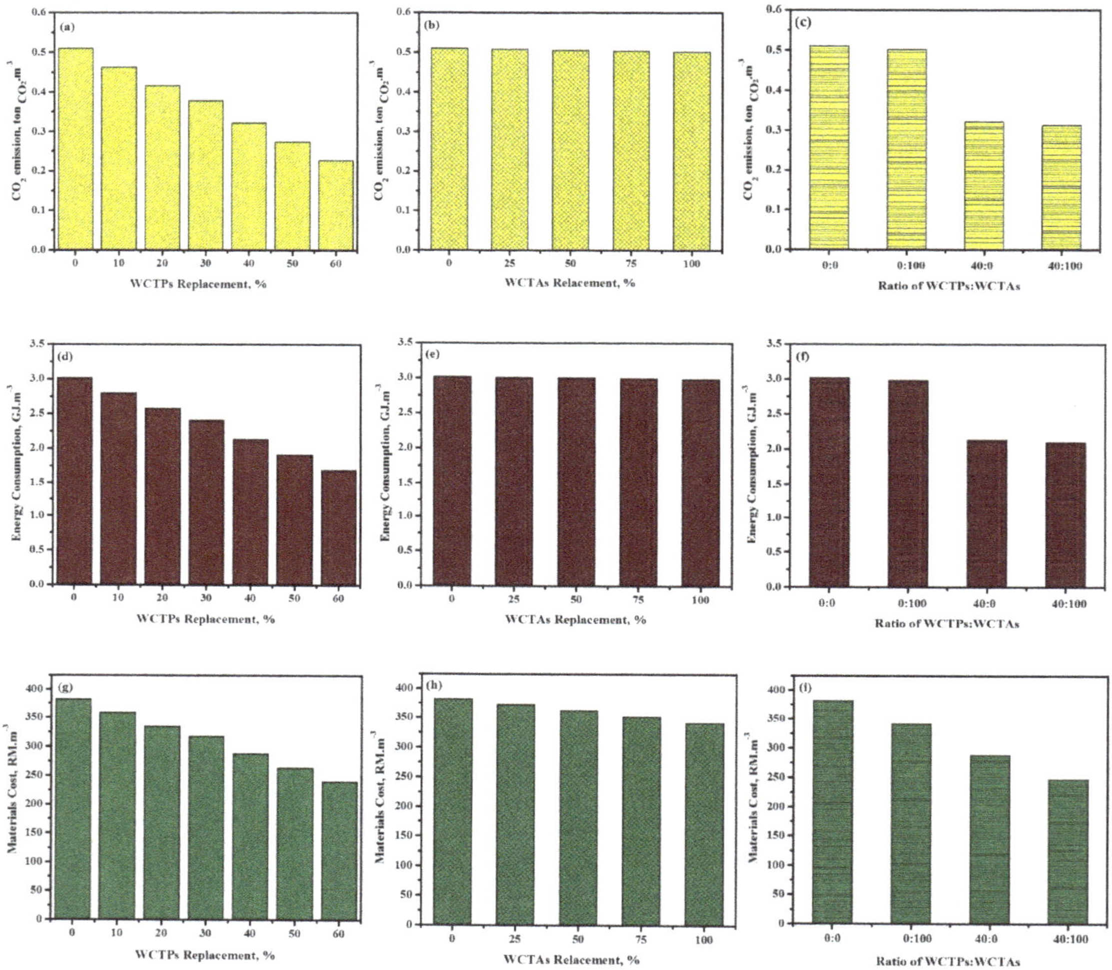

Figure 22. Environmental benefits of including waste WCTPs and WCTAs as OPC and natural aggregate replacements in sustainable construction materials; carbon dioxide emission of (**a**) WCTPs as OPC replacement (**b**) WCTAs as aggregates replacement (**c**) WCTPs and WCTAs as OPC and aggregate replacement; energy consumption of (**d**) WCTPs as OPC replacement (**e**) WCTAs as aggregates replacement (**f**) WCTPs and WCTAs as OPC and aggregate replacement; materials costs (**g**) WCTPs as OPC replacement (**h**) WCTAs as aggregates replacement (**i**) WCTPs and WCTAs as OPC and aggregate replacement [2].

The authors [2] assessed the cost-effectiveness of replacing traditional materials with WCTPs and WCTAs in the proposed mortar samples, in comparison to ordinary Portland cement (OPC) and natural river sand. The use of a high proportion of WCTPs (i.e., 40%) as a substitute for OPC resulted in significant cost savings. Incorporating WCTPs as a binder contributed meaningfully to the development of sustainable construction materials. Furthermore, the complete replacement of river sand with WCTAs showed considerable potential in promoting environmentally friendly products, thereby supporting more sustainable practices in civil engineering.

The study conducted by Ikponmwosa and Ehikhuenmen [100] investigated the impact of replacing natural coarse aggregates with varying proportions of WCTAs—specifically 0%, 25%, 50%, 75%, and 100%—on the cost-effectiveness of concrete production, as illustrated in Figure 23. WCTAs are derived from waste materials commonly generated in ceramic manufacturing processes, such as the production of tiles and pottery, and thus hold no intrinsic economic value [170]. Nonetheless, the use of WCTAs in concrete entails certain costs associated with transportation, crushing, and grinding to achieve the appropriate particle sizes. These expenses were included in the cost–benefit analysis by Ikponmwosa and Ehikhuenmen [100]. The analysis revealed that the cost per cubic meter of concrete declined as the proportion of WCTAs increased, with a full 100% replacement yielding a 13.1% reduction in cost. However, due to the adverse effects of high WCTA content on concrete properties—such as reduced workability and mechanical strength—the authors recommended limiting WCTA usage to a maximum of 75%, which still offers nearly a 10% reduction in cost per cubic meter [100]. Additionally, Gautam et al. [171] emphasized that incorporating WCTAs into construction practices not only reduces material costs, but also advances sustainability objectives. Nonetheless, the existing studies primarily focus on WCTAs as a replacement for coarse aggregates. Therefore, further research is recommended to explore the potential benefits of utilizing WCTAs as substitutes for cement or fine aggregates in concrete production.

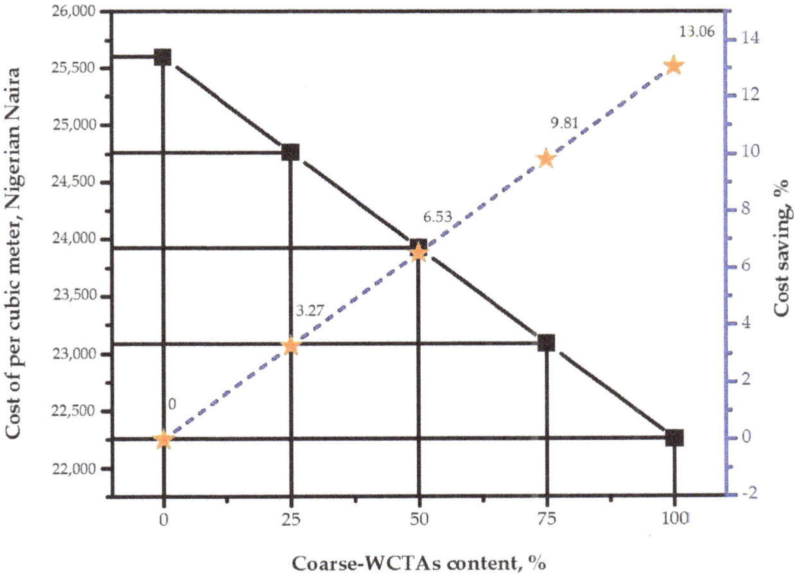

Figure 23. Cost benefits of ceramic concrete [100].

In a study by Santos et al. [60], the authors calculated the Global Warming Potential (GWP) values for various concrete mixtures, all designed to achieve a compressive strength of 54 MPa, as illustrated in Figure 24. The results reflect the material consumption required for each mix to reach this strength level. The bi value is directly related to binder consumption; therefore, mixtures with lower binder index values use less binder. Since cement contributes the highest CO_2 emissions among the concrete constituents, mixtures with a higher binder content resulted in greater GWP values. Replacing ordinary Portland cement (OPC) with varying proportions of WCTPs resulted in a reduction in the GWP of the reference concrete, primarily due to decreased cement usage. Although the consumption of other components varied among the mixtures, cement usage had a significant impact on the final GWP. Thus, incorporating WCTs offers an effective strategy for reducing greenhouse gas emissions in concrete production, while also enhancing certain concrete properties as previously discussed.

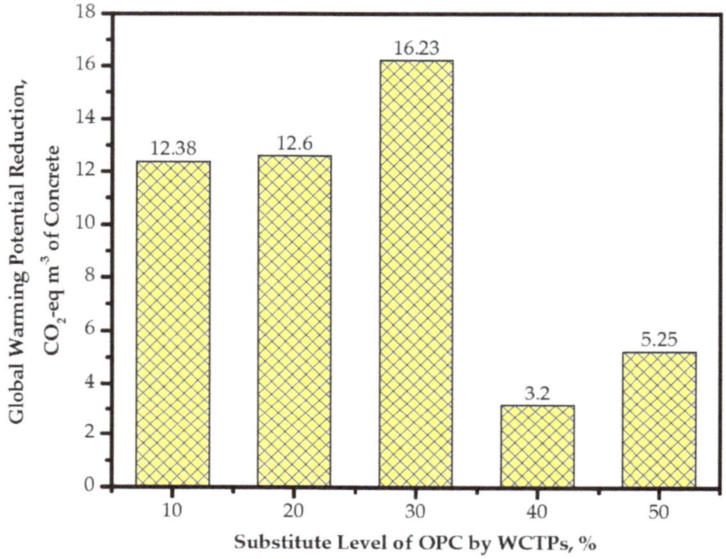

Figure 24. The global warming potential reduction of utilizing WCTPs as an OPC substitute in concrete production [60].

8. Conclusions

Based on the reviewed literature, the following conclusions can be made:

Waste ceramic tiles (WCTs) represent a significant portion of modern construction waste. Therefore, recycling them plays a crucial role in promoting sustainable construction waste management. WCTs can be effectively utilized as recycled cement, fine, and coarse natural aggregates in the concrete industry, providing a sustainable approach to minimizing waste materials.

In the production of modern concrete, WCTs have been successfully used for the production of ultrahigh-performance, self-curing, and self-compacting concretes.

The workability of proposed mortar and concrete is slightly influenced by WCT content as a binder or filler replacement. The flowability of fresh mixtures tends to decrease with increasing WCTs' replacement level, which is attributed to porosity and high water absorption, as well as the irregular shape of ceramic particles. Therefore, it is recommended to develop cement-based composites using WCT blends to achieve a consistent slump, rather than designing them based on a fixed water-to-binder ratio.

Waste ceramic tile powders (WCTPs) can be used as a partial replacement for cement, as their combined content of silicates, aluminates, and iron oxide generally exceeds the 75% threshold outlined in ASTM 618. Incorporating up to 30–35% WCTPs into cement-based composites improves their mechanical performance, including compressive strength, flexural strength, and split tensile strength. When used at a replacement level of up to 20%,

WCTPs also contribute to improved durability of the composites. Specifically, mixtures containing up to 30% WCTPs show reductions in drying shrinkage, permeability, chloride ion penetration, and carbonation, along with increased resistance to sulphate attack.

The proportion of natural fine or/and aggregates replaced with waste ceramic tile aggregates (WCTAs) can range from 0% to 100%, depending on the intended application and the required performance characteristics of the concrete.

Using waste ceramic materials as fine and/or coarse aggregates to replace natural aggregates in the production of sustainable concrete has demonstrated comparable or slightly improved strength performances compared to control specimens.

The findings indicate that recycled WCT is a highly effective substitute for traditional concrete made with cement or/and natural aggregates, and it is appropriate for both structural and non-structural applications. Using WCTs in place of cement or/and natural aggregates can greatly reduce environmental impact by minimizing the need for natural resource extraction and decreasing the amount of construction and demolition waste sent to landfills.

In structural applications involving WCTs, reinforced concrete beams incorporating WCTs as cement, fine, and/or coarse aggregates demonstrated a satisfactory performance when compared to the control beam. This indicates that ceramic waste can be efficiently repurposed to produce environmentally sustainable concrete suitable for structural applications.

Incorporating waste ceramics into the concrete industry can significantly reduce global warming potential. Replacing ordinary Portland cement with waste ceramic powder has been shown to notably decrease energy consumption and carbon dioxide emissions, primarily due to the high energy demands of cement production. Nonetheless, the challenges related to landfill use and the conservation of natural resources persist and cannot be entirely resolved. Overall, ceramic waste-blended concretes have demonstrated improved performance and energy efficiency, offering both technical and environmental benefits that may contribute to lower construction costs and enhanced sustainability.

9. Recommendations and Future Research Directions

The use of waste ceramic tiles presents several practical challenges and raises significant environmental concerns. It is now broadly recognized that increasing cement production contributes substantially to global warming. Major environmental issues related to cement manufacturing include high energy demand, emissions of pollutants, and large-scale carbon dioxide output. In response, numerous strategies have been introduced to mitigate these impacts, such as recycling industrial and agricultural waste as cement substitutes, employing carbon capture technologies, utilizing nanomaterials and nanotechnology, and promoting the development of geopolymer technology. Producing high-performance concrete that incorporates large quantities of waste ceramic powder is a crucial step toward minimizing the environmental footprint of cement and improving concrete durability. These advancements also support the creation of low-cement, eco-friendly concrete materials. The authors recommend that future studies concentrate on examining the microstructure, durability, and mechanical behavior of concrete made with high volumes of waste ceramic materials.

Future research on recycling waste tile ceramics and other types of ceramic waste in concrete and geopolymer applications should focus on optimizing their processing methods, enhancing their pozzolanic reactivity, and evaluating their long-term performance in harsh environmental conditions. Special attention should be paid to the synergistic effects of combining ceramic waste with other industrial by-products, such as fly ash, slag, or nanosilica, to develop high-performance, sustainable binders. Investigations into the durability, shrinkage behavior, and microstructural evolution of ceramic-based geopolymer composites are also crucial. A promising research topic is the development of fiber-reinforced geopolymer composites incorporating ceramic waste for structural and non-structural applications. For future review papers, a comprehensive analysis of the environmental benefits, life cycle assessment, and challenges of integrating various ceramic wastes in alkali-activated and blended cement systems is recommended.

Author Contributions: Conceptualization, G.F.H. and Z.H.J.; methodology, Z.H.J. and N.H.A.K.; software, I.F.; validation, M.H.B., I.F. and K.D.; formal analysis, Z.H.J.; investigation, G.F.H., Y.L. and X.G.; resources, Z.H.J.; data curation, Z.H.J.; writing—original draft preparation, Z.H.J.; writing—review and editing, G.F.H. and I.F.; visualization, K.D., Y.L. and X.G.; supervision, N.H.A.K.; project administration, I.F.; funding acquisition, M.H.B. All authors have read and agreed to the published version of the manuscript.

Funding: This work was supported/funded by the Guangzhou Institute of Energy Conversion, Chinese Academy of Sciences. National Natural Science Foundation of China (No. 52474445), the Science and Technology Project of Guangdong Province (No. 2023A1111120031), the Guangzhou Science and Technology Project (No. 2024B01J1407), the Beijing Natural Science Foundation-Beijing Huairou Foundation (No. L245004), and the Opening Project of State Key Laboratory of Green Building Materials (No. 2023GBM06.

Data Availability Statement: Not applicable.

Acknowledgments: This work is supported by the National Natural Science Foundation of China (No. 52474445), the Science and Technology Project of Guangdong Province (No. 2023A1111120031), the Guangzhou Science and Technology Project (No. 2024B01J1407), the Beijing Natural Science Foundation-Beijing Huairou Foundation (No. L245004), and the Opening Project of State Key Laboratory of Green Building Materials (No. 2023GBM06).

Conflicts of Interest: The authors declare no conflicts of interest.

References

1. Baylavlı, H.; Gödek, E. Use of Ceramic Tile Wastes as Raw Substitution Material in the Production of Blended Cement. *Buildings* **2024**, *14*, 2942. [CrossRef]
2. Samadi, M.; Huseien, G.F.; Mohammadhosseini, H.; Lee, H.S.; Lim, N.H.A.S.; Tahir, M.M.; Alyousef, R. Waste ceramic as low cost and eco-friendly materials in the production of sustainable mortars. *J. Clean. Prod.* **2020**, *266*, 121825.
3. Shah, K.W.; Huseien, G.F. *Recycled Ceramics in Sustainable Concrete: Properties and Performance*; CRC Press: Boca Raton, FL, USA, 2020.
4. Daniyal, M.; Ahmad, S. Application of waste ceramic tile aggregates in concrete. *Int. J. Innov. Res. Sci. Eng. Technol.* **2015**, *4*, 12808–12815.
5. Senthamarai, R.; Manoharan, P.D. Concrete with ceramic waste aggregate. *Cem. Concr. Compos.* **2005**, *27*, 910–913.
6. Joudah, Z.H.; Khalid, N.H.A.; Baghban, M.H.; Faridmehr, I.; Talip, A.R.A.; Huseien, G.F. Development sustainable concrete with high-volume wastes tile ceramic: Role of silica nanoparticles amalgamation. *Case Stud. Constr. Mater.* **2024**, *21*, e03733.
7. Pacheco-Torgal, F.; Jalali, S. Reusing ceramic wastes in concrete. *Constr. Build. Mater.* **2010**, *24*, 832–838.
8. Senthamarai, R.; Manoharan, P.D.; Gobinath, D. Concrete made from ceramic industry waste: Durability properties. *Constr. Build. Mater.* **2011**, *25*, 2413–2419.
9. Huang, B.; Dong, Q.; Burdette, E.G. Laboratory evaluation of incorporating waste ceramic materials into Portland cement and asphaltic concrete. *Constr. Build. Mater.* **2009**, *23*, 3451–3456.
10. Hussein, A.A.; Jaya, R.P.; Hassan, N.A.; Yaacob, H.; Huseien, G.F.; Ibrahim, M.H.W. Performance of nanoceramic powder on the chemical and physical properties of bitumen. *Constr. Build. Mater.* **2017**, *156*, 496–505.
11. Huseien, G.F.; Mirza, J.; Ismail, M.; Ghoshal, S.K.; Hussein, A.A. Geopolymer mortars as sustainable repair material: A comprehensive review. *Renew. and Susta. Energ. Rev.* **2017**, *80*, 54–74.
12. Huseien, G.F.; Sam, A.R.M.; Mirza, J.; Tahir, M.M.; Asaad, M.A.; Ismail, M.; Shah, K.W. Waste ceramic powder incorporated alkali activated mortars exposed to elevated Temperatures: Performance evaluation. *Constr. Build. Mater.* **2018**, *187*, 307–317.
13. Fernandes, M.; Sousa, A.; Dias, A. *Environmental Impacts and Emissions Trading-Ceramic Industry: A Case Study*; Technological Centre of Ceramics and Glass, Portuguese Association of Ceramic Industry: Coimbra, Portugal, 2004. (In Portuguese)

14. Chen, X.; Zhang, D.; Cheng, S.; Xu, X.; Zhao, C.; Wang, X.; Wu, Q.; Bai, X. Sustainable reuse of ceramic waste powder as a supplementary cementitious material in recycled aggregate concrete: Mechanical properties, durability and microstructure assessment. *J. Build. Eng.* **2022**, *52*, 104418.
15. Limbachiya, M.; Meddah, M.S.; Ouchagour, Y. Use of recycled concrete aggregate in fly-ash concrete. *Constr. Build. Mater.* **2012**, *27*, 439–449.
16. Heidari, A.; Tavakoli, D. A study of the mechanical properties of ground ceramic powder concrete incorporating nano-SiO_2 particles. *Constr. Build. Mater.* **2013**, *38*, 255–264.
17. Huseien, G.F.; Mirza, J.; Ismail, M.; Hussin, M.W.; Arrifin, M.; Hussein, A. The Effect of Sodium Hydroxide Molarity and Other Parameters on Water Absorption of Geopolymer Mortars. *Indian J. Sci. Technol.* **2016**, *9*, 1–7.
18. Huseien, G.F.; Mirza, J.; Ismail, M.; Ghoshal, S.; Ariffin, M.A.M. Effect of metakaolin replaced granulated blast furnace slag on fresh and early strength properties of geopolymer mortar. *Ain Shams Eng. J.* **2016**, *9*, 1557–1566.
19. Torkittikul, P.; Chaipanich, A. Utilization of ceramic waste as fine aggregate within Portland cement and fly ash concretes. *Cem. Concr. Compos.* **2010**, *32*, 440–449.
20. Amin, M.; Tayeh, B.A.; Agwa, I.S. Effect of using mineral admixtures and ceramic wastes as coarse aggregates on properties of ultrahigh-performance concrete. *J. Clean. Prod.* **2020**, *273*, 123073.
21. Medina, C.; Frías, M.; De Rojas, M.S. Microstructure and properties of recycled concretes using ceramic sanitary ware industry waste as coarse aggregate. *Constr. Build. Mater.* **2012**, *31*, 112–118.
22. Higashiyama, H.; Yagishita, F.; Sano, M.; Takahashi, O. Compressive strength and resistance to chloride penetration of mortars using ceramic waste as fine aggregate. *Constr. Build. Mater.* **2012**, *26*, 96–101.
23. Siddique, S.; Chaudhary, S.; Shrivastava, S.; Gupta, T. Sustainable utilisation of ceramic waste in concrete: Exposure to adverse conditions. *J. Clean. Prod.* **2019**, *210*, 246–255.
24. Juan-Valdés, A.; Rodríguez-Robles, D.; Garcia-Gonzalez, J.; de Rojas Gómez, M.I.S.; Guerra-Romero, M.I.; De Belie, N.; Morán-del Pozo, J.M. Mechanical and microstructural properties of recycled concretes mixed with ceramic recycled cement and secondary recycled aggregates. A viable option for future concrete. *Constr. Build. Mater.* **2021**, *270*, 121455.
25. Lopez, V.; Llamas, B.; Juan, A.; Moran, J.; Guerra, I. Eco-efficient concretes: Impact of the use of white ceramic powder on the mechanical properties of concrete. *Biosyst. Eng.* **2007**, *96*, 559–564.
26. Binici, H. Effect of crushed ceramic and basaltic pumice as fine aggregates on concrete mortars properties. *Constr. Build. Mater.* **2007**, *21*, 1191–1197.
27. Wu, H.; Liang, C.; Yang, D.; Ma, Z. Development of sustainable geopolymer materials made with ground geopolymer waste powder as renewable binder up to 100%. *Constr. Build. Mater.* **2023**, *400*, 132746.
28. Tanash, A.O.; Muthusamy, K.; Yahaya, F.M.; Ismail, M.A. Potential of recycled powder from clay Brick, sanitary Ware, and concrete waste as a cement substitute for Concrete: An overview. *Constr. Build. Mater.* **2023**, *401*, 132760.
29. Sáez, P.V.; Osmani, M. A diagnosis of construction and demolition waste generation and recovery practice in the European Union. *J. Clean. Prod.* **2019**, *241*, 118400.
30. Xu, F.; Lin, X.; Zhou, A. Effect of recycled ceramic aggregate on hydration heat and permeability of high performance concrete. *Cem. Concr. Compos.* **2023**, *137*, 104930.
31. Mohit, M.; Sharifi, Y. Thermal and microstructure properties of cement mortar containing ceramic waste powder as alternative cementitious materials. *Constr. Build. Mater.* **2019**, *223*, 643–656.
32. Awoyera, P.O.; Ndambuki, J.M.; Akinmusuru, J.O.; Omole, D.O. Characterization of ceramic waste aggregate concrete. *HBRC J.* **2018**, *14*, 282–287.
33. Chaudhury, R.; Sharma, U.; Thapliyal, P.; Singh, L. Low-CO_2 emission strategies to achieve net zero target in cement sector. *J. Clean. Prod.* **2023**, *417*, 137466.
34. Lu, G.; Fan, Z.; Sun, Z.; Liu, P.; Leng, Z.; Wang, D.; Oeser, M. Improving the polishing resistance of cement mortar by using recycled ceramic. *Resour. Conserv. Recycl.* **2020**, *158*, 104796.
35. Zimbili, O.; Salim, W.; Ndambuki, M. A review on the usage of ceramic wastes in concrete production. *Int. J. Civ. Environ. Struct. Constr. Archit. Eng.* **2014**, *8*, 91–95.
36. Tanash, A.O.; Muthusamy, K.; Budiea, A.M.A.; Fauri, M.A.; Jokhio, G.; Rajan, J. A review on the utilization of ceramic tile waste as cement and aggregates replacement in cement based composite and a bibliometric assessment. *Clean. Eng. Technol.* **2023**, *17*, 100699.
37. Pitarch, A.; Reig, L.; Tomás, A.; López, F. Effect of tiles, bricks and ceramic sanitary-ware recycled aggregates on structural concrete properties. *Waste Biomass Valorization* **2019**, *10*, 1779–1793.
38. Ray, S.; Haque, M.; Sakib, M.N.; Mita, A.F.; Rahman, M.M.; Tanmoy, B.B. Use of ceramic wastes as aggregates in concrete production: A review. *J. Build. Eng.* **2021**, *43*, 102567.
39. Xu, K.; Huang, W.; Zhang, L.; Fu, S.; Chen, M.; Ding, S.; Han, B. Mechanical properties of low-carbon ultrahigh-performance concrete with ceramic tile waste powder. *Constr. Build. Mater.* **2021**, *287*, 123036.

40. Cakiroglu, C.; Batool, F.; Sangi, A.J.; Fatima, B.; Nehdi, M.L. Explainable machine learning predictive model for mechanical strength of recycled ceramic tile-based concrete. *Mater. Today Commun.* **2025**, *44*, 112139.
41. Huseien, G.F.; Sam, A.R.M.; Shah, K.W.; Mirza, J. Effects of ceramic tile powder waste on properties of self-compacted alkali-activated concrete. *Constr. Build. Mater.* **2020**, *236*, 117574.
42. Ebrahimi, M.; Eslami, A.; Hajirasouliha, I.; Ramezanpour, M.; Pilakoutas, K. Effect of ceramic waste powder as a binder replacement on the properties of cement-and lime-based mortars. *Constr. Build. Mater.* **2023**, *379*, 131146.
43. Hoppe Filho, J.; Pires, C.A.d.O.; Leite, O.D.; Garcez, M.R.; Medeiros, M.H.F.d. Red ceramic waste as supplementary cementitious material: Microstructure and mechanical properties. *Constr. Build. Mater.* **2021**, *296*, 123653.
44. Mansoori, A.; Moein, M.M.; Mohseni, E. Effect of micro silica on fiber-reinforced self-compacting composites containing ceramic waste. *J. Compos. Mater.* **2021**, *55*, 95–107.
45. Paul, S.C.; Faruky, S.A.U.; Babafemi, A.J.; Miah, M.J. Eco-friendly concrete with waste ceramic tile as coarse aggregate: Mechanical strength, durability, and microstructural properties. *Asian J. Civ. Eng.* **2023**, *24*, 3363–3373.
46. Njoya, D.; Hajjaji, M.; Njopwouo, D. Effects of some processing factors on technical properties of a clay-based ceramic material. *Appl. Clay Sci.* **2012**, *65*, 106–113.
47. Boussen, S.; Sghaier, D.; Chaabani, F.; Jamoussi, B.; Bennour, A. Characteristics and industrial application of the Lower Cretaceous clay deposits (Bouhedma Formation), Southeast Tunisia: Potential use for the manufacturing of ceramic tiles and bricks. *Appl. Clay Sci.* **2016**, *123*, 210–221.
48. Awoyera, P.O.; Akinmusuru, J.O.; Ndambuki, J.M.; Lucas, S.S. Benefits of using ceramic tile waste for making sustainable concrete. *J. Solid Waste Technol. Manag.* **2017**, *43*, 233–241.
49. El-Dieb, A.S.; Kanaan, D.M. Ceramic waste powder an alternative cement replacement–Characterization and evaluation. *Sustain. Mater. Technol.* **2018**, *17*, e00063.
50. Heidari, A.; Tavakoli, S.; Tavakoli, D. Reusing waste ceramic and waste sanitary ware in concrete as pozzolans with nano-silica and metakaolin. *Int. J. Sustain. Constr. Eng. Technol.* **2019**, *10*, 55–67.
51. Taher, M.J.; Abed, E.; Hashim, M.S. Using ceramic waste tile powder as a sustainable and eco-friendly partial cement replacement in concrete production. *Mater. Today Proc.* **2023**, 1–7. [CrossRef]
52. Rashad, A.M.; Essa, G.M. Effect of ceramic waste powder on alkali-activated slag pastes cured in hot weather after exposure to elevated temperature. *Cem. Concr. Compos.* **2020**, *111*, 103617.
53. Samadi, M.; Hussin, M.W.; Lee, H.S.; Sam, A.R.M.; Ismail, M.A.; Lim, N.H.A.S.; Ariffin, N.F.; Khalid, N.H.A. Properties of mortar containing ceramic powder waste as cement replacement. *J. Teknol.* **2015**, *77*, 27–32. [CrossRef]
54. Tawfik, T.A.; Metwally, K.A.; El-Beshlawy, S.; Al Saffar, D.M.; Tayeh, B.A.; Hassan, H.S. Exploitation of the nanowaste ceramic incorporated with nano silica to improve concrete properties. *J. King Saud Univ.-Eng. Sci.* **2021**, *33*, 581–588.
55. AlArab, A.; Hamad, B.; Assaad, J.J. Strength and durability of concrete containing ceramic waste powder and blast furnace slag. *J. Mater. Civ. Eng.* **2022**, *34*, 04021392.
56. El-Dieb, A.S.; Taha, M.R.; Kanaan, D.; Aly, S.T. Ceramic waste powder: From landfill to sustainable concretes. *Proc. Inst. Civ. Eng.-Constr. Mater.* **2018**, *171*, 109–116.
57. Sondarva, P.R.; Pitroda, J.; Gujar, R.; Soni, J. An experimental investigation on the strength properties of ceramic tiles waste powder based bacterial concrete. *Mater. Today Proc.* **2022**, *62*, 7062–7067.
58. Hilal, N.; Saleh, R.D.; Yakoob, N.B.; Banyhussan, Q.S. Utilization of ceramic waste powder in cement mortar exposed to elevated temperature. *Innov. Infrastruct. Solut.* **2021**, *6*, 35.
59. El-Nadoury, W.W. Eco-friendly concrete using by-products as partial replacement of cement. *Front. Mater.* **2022**, *9*, 1043037.
60. Santos, H.M.M.; Jochem, L.F.; de Matos, P.R.; Casagrande, C.A.; Marinho, É.P.; Szeląg, M.; de Nóbrega, A.C.V. Porcelain tile polishing residue in concrete as an additive or replacement for portland cement. *Appl. Sci.* **2023**, *13*, 2824.
61. Oleng, M.; Kanali, C.; Gariy, Z.C.A.; Ronoh, E. Physical and Mechanical Experimental Investigation of Concrete incorporated with Ceramic and Porcelain Clay Tile Powders as Partial Cement Substitutes. *Int. J. Eng. Res. Technol.* **2018**, *7*, 126–130.
62. Pitarch, A.; Reig, L.; Tomás, A.; Forcada, G.; Soriano, L.; Borrachero, M.; Payá, J.; Monzó, J. Pozzolanic activity of tiles, bricks and ceramic sanitary-ware in eco-friendly Portland blended cements. *J. Clean. Prod.* **2021**, *279*, 123713.
63. Mohit, M.; Haftbaradaran, H.; Riahi, H.T. Investigating the ternary cement containing Portland cement, ceramic waste powder, and limestone. *Constr. Build. Mater.* **2023**, *369*, 130596.
64. Medina, C.; Frías, M.; de Rojas, M.I.S. Leaching in concretes containing recycled ceramic aggregate from the sanitary ware industry. *J. Clean. Prod.* **2014**, *66*, 85–91.
65. Hilal, N.N.; Mohammed, A.S.; Ali, T.K.M. Properties of eco-friendly concrete contained limestone and ceramic tiles waste exposed to high temperature. *Arab. J. Sci. Eng.* **2020**, *45*, 4387–4404.
66. de Matos, P.R.; Sakata, R.D.; Onghero, L.; Uliano, V.G.; de Brito, J.; Campos, C.E.; Gleize, P.J. Utilization of ceramic tile demolition waste as supplementary cementitious material: An early-age investigation. *J. Build. Eng.* **2021**, *38*, 102187.

67. Bheel, N.; Abbasi, R.A.; Sohu, S.; Abbasi, S.A.; Abro, A.W.; Shaikh, Z.H. Effect of tile powder used as a cementitious material on the mechanical properties of concrete. *Eng. Technol. Appl. Sci. Res.* **2019**, *9*, 4596–4599.
68. Jain, P.; Gupta, R.; Chaudhary, S. A literature review on the effect of using ceramic waste as supplementary cementitious material in cement composites on workability and compressive strength. *Mater. Today Proc.* **2022**, *65*, 871–876.
69. El-Kattan, I.M.; Abdelzaher, M.; Farghali, A. Positive impact of ultra fine-ceramic waste on the physico-mechanical features and microstructure of white cement pastes composites. *J. Mater. Res. Technol.* **2020**, *9*, 9395–9402.
70. Huseien, G.F.; Sam, A.R.M.; Shah, K.W.; Asaad, M.A.; Tahir, M.M.; Mirza, J. Properties of ceramic tile waste based alkali-activated mortars incorporating GBFS and fly ash. *Constr. Build. Mater.* **2019**, *214*, 355–368.
71. Awoyera, P.O.; Akinmusuru, J.O.; Moncea, A. Hydration mechanism and strength properties of recycled aggregate concrete made using ceramic blended cement. *Cogent Eng.* **2017**, *4*, 1282667.
72. Balaji, Y.B.; Goud, E.S.K.; Yesuratnam, G. Study of concrete behavior by partial replacement of cement with Ceramic Waste Powder in the presence of Sisal fiber. *Mater. Today Proc.* **2023**, 1–8. [CrossRef]
73. Penteado, C.S.G.; de Carvalho, E.V.; Lintz, R.C.C. Reusing ceramic tile polishing waste in paving block manufacturing. *J. Clean. Prod.* **2016**, *112*, 514–520.
74. Nalli, B.R.; Vysyaraju, P. Utilization of ceramic waste powder and rice husk ash as a partial replacement of cement in concrete. *IOP Conf. Ser. Earth Environ. Sci.* **2022**, *982*, 012003.
75. Soomro, B.; Mangi, S.A.; Bajkani, R.A.; Junejo, A.Q. Recycling of ceramic tiles and marble powder waste as partial substitution in concrete. *Neutron* **2021**, *20*, 128–137.
76. Wu, C.-H.; Wang, W.-C.; Jung, C.H. Potential pozzolanic reactivity of recycled aggregates and waste powders as cement mortar component. *J. Chin. Inst. Eng.* **2023**, *46*, 399–408.
77. Aswin, S.; Mohanalakshmi, V.; Rajesh, A.A. Effects of ceramic tile powder on properties of concrete and paver block. *Glob. Res. Dev. J. Eng.* **2018**, *3*, 84–87.
78. Raghav, M.; Park, T.; Yang, H.-M.; Lee, S.-Y.; Karthick, S.; Lee, H.-S. Review of the effects of supplementary cementitious materials and chemical additives on the physical, mechanical and durability properties of hydraulic concrete. *Materials* **2021**, *14*, 7270. [CrossRef] [PubMed]
79. Lasseuguette, E.; Burns, S.; Simmons, D.; Francis, E.; Chai, H.; Koutsos, V.; Huang, Y. Chemical, microstructural and mechanical properties of ceramic waste blended cementitious systems. *J. Clean. Prod.* **2019**, *211*, 1228–1238.
80. Mohammadhosseini, H.; Lim, N.H.A.S.; Tahir, M.M.; Alyousef, R.; Samadi, M.; Alabduljabbar, H.; Mohamed, A.M. Effects of waste ceramic as cement and fine aggregate on durability performance of sustainable mortar. *Arab. J. Sci. Eng.* **2020**, *45*, 3623–3634.
81. Ferrara, L.; Deegan, P.; Pattarini, A.; Sonebi, M.; Taylor, S. Recycling ceramic waste powder: Effects its grain-size distribution on fresh and hardened properties of cement pastes/mortars formulated from SCC mixes. *J. Sustain. Cem.-Based Mater.* **2019**, *8*, 145–160.
82. Li, L.; Liu, W.; You, Q.; Chen, M.; Zeng, Q. Waste ceramic powder as a pozzolanic supplementary filler of cement for developing sustainable building materials. *J. Clean. Prod.* **2020**, *259*, 120853.
83. Atkuri, V.K.; Rao, G.R. Strength properties of ceramic waste concrete. In *IOP Conference Series: Materials Science and Engineering*; IOP Publishing: Jakarta, Indonesia, 2021.
84. Pereira-de-Oliveira, L.A.; Castro-Gomes, J.P.; Santos, P.M. The potential pozzolanic activity of glass and red-clay ceramic waste as cement mortars components. *Constr. Build. Mater.* **2012**, *31*, 197–203.
85. Fapohunda, C.; Akinbile, B.; Shittu, A. Structure and properties of mortar and concrete with rice husk ash as partial replacement of ordinary Portland cement–A review. *Int. J. Sustain. Built Environ.* **2017**, *6*, 675–692.
86. Lim, S.K.; Lee, Y.L.; Yew, M.K.; Ng, W.W.; Lee, F.W.; Kwong, K.Z.; Lim, J.H. Mechanical properties of lightweight foamed concrete with ceramic tile wastes as partial cement replacement material. *Front. Built Environ.* **2022**, *8*, 836362.
87. Mas, M.A.; Monzó, J.; Payá, J.; Reig, L.; Borrachero, M.V. Ceramic tiles waste as replacement material in Portland cement. *Adv. Cem. Res.* **2016**, *28*, 221–232.
88. Nayana, A.; Rakesh, P. Strength and durability study on cement mortar with ceramic waste and micro-silica. *Mater. Today Proc.* **2018**, *5*, 24780–24791.
89. Bayraktar, O.Y.; Tunçtan, M.; Benli, A.; Türkel, İ.; Kızılay, G.; Kaplan, G. A study on sustainable foam concrete with waste polyester and ceramic powder: Properties and durability. *J. Build. Eng.* **2024**, *95*, 110253.
90. Mohit, M.; Ranjbar, A.; Sharifi, Y. Mechanical and microstructural properties of mortars incorporating ceramic waste powder exposed to the hydrochloric acid solution. *Constr. Build. Mater.* **2021**, *271*, 121565.
91. Rodhia, R.; Sahdeo, S.K.; Kumar, B. Mechanical, microstructural and durable characteristics of foam concrete ceramic mixes exposed to H_2SO_4 and HCl solution. *Constr. Build. Mater.* **2024**, *454*, 139080.
92. Aiken, T.A.; Gu, L.; Kwasny, J.; Huseien, G.F.; McPolin, D.; Sha, W. Acid resistance of alkali-activated binders: A review of performance, mechanisms of deterioration and testing procedures. *Constr. Build. Mater.* **2022**, *342*, 128057.

93. Chandra, S. Hydrochloric acid attack on cement mortar—An analytical study. *Cem. Concr. Res.* **1988**, *18*, 193–203.
94. Sánchez de Rojas, M.I.; Frías, M.; Sabador, E.; Asensio, E.; Rivera, J.; Medina, C. Use of ceramic industry milling and glazing waste as an active addition in cement. *J. Am. Ceram. Soc.* **2018**, *101*, 2028–2037.
95. Kanaanla, D.M.; EL-Dieb, A.S. Ceramic waste powder as an ingredient to sustainable concrete. In Proceedings of the Fourth International Conference on Sustainable Construction Materials and Technologies, Las Vegas, NV, USA, 7–11 August 2016.
96. Zhang, B.; Zhu, H.; Cheng, Y.; Huseien, G.F.; Shah, K.W. Shrinkage mechanisms and shrinkage-mitigating strategies of alkali-activated slag composites: A critical review. *Constr. Build. Mater.* **2022**, *318*, 125993.
97. Steiner, L.R.; Bernardin, A.M.; Pelisser, F. Effectiveness of ceramic tile polishing residues as supplementary cementitious materials for cement mortars. *Sustain. Mater. Technol.* **2015**, *4*, 30–35.
98. Mohammed, H.; Ahmed, S. Mechanical performance evaluation of concrete with waste coarse ceramic aggregate. In *Smart Cities—Opportunities and Challenges: Select Proceedings of ICSC 2019*; Springer: Berlin/Heidelberg, Germany, 2020; pp. 1–10.
99. Nepomuceno, M.C.; Isidoro, R.A.; Catarino, J.P. Mechanical performance evaluation of concrete made with recycled ceramic coarse aggregates from industrial brick waste. *Constr. Build. Mater.* **2018**, *165*, 284–294.
100. Ikponmwosa, E.; Ehikhuenmen, S. The effect of ceramic waste as coarse aggregate on strength properties of concrete. *Niger. J. Technol.* **2017**, *36*, 691–696.
101. Sua-iam, G.; Jamnam, S. Influence of calcium carbonate on green self-compacting concrete incorporating porcelain tile waste as coarse aggregate replacement. *Case Stud. Constr. Mater.* **2023**, *19*, e02366.
102. Awoyera, P.O.; Akinmusuru, J.O.; Ndambuki, J.M. Green concrete production with ceramic wastes and laterite. *Constr. Build. Mater.* **2016**, *117*, 29–36.
103. Singh, P.; Singla, R.K. Utilization of waste ceramic tiles as coarse aggregate in concrete. *J. Multidiscip. Eng. Sci. Technol. (JMEST)* **2015**, *2*, 3294–3300.
104. Manikandan, K.; Nanthakumar, P.; Balachandar, M.; Shankar, D.G.; Vijayakumari, G. Partial replacement of aggregate with ceramic tile in concrete. *Mater. Today Proc.* **2023**, 1–8. [CrossRef]
105. Goyal, R.K.; Agarwal, V.; Gupta, R.; Rathore, K.; Somani, P. Optimum utilization of ceramic tile waste for enhancing concrete properties. *Mater. Today Proc.* **2022**, *49*, 1769–1775.
106. Keshavarz, Z.; Mostofinejad, D. Porcelain and red ceramic wastes used as replacements for coarse aggregate in concrete. *Constr. Build. Mater.* **2019**, *195*, 218–230.
107. Neville, A.M.; Brooks, J.J. *Concrete Technology*; Longman Scientific & Technical England: London, UK, 1987; Volume 438.
108. Elçi, H. Utilisation of crushed floor and wall tile wastes as aggregate in concrete production. *J. Clean. Prod.* **2016**, *112*, 742–752.
109. Meena, R.V.; Jain, J.K.; Chouhan, H.S.; Beniwal, A.S. Mechanical and durability performance of self-compacting concrete with waste ceramic tile as a replacement for natural river sand. *Mater. Today Proc.* **2022**, *60*, 187–193.
110. Li, Q.; Liu, P.; Wang, M.; Xia, H. Effects of elevated temperature on the mechanical properties of concrete with aggregate of waste porcelain tile. *J. Build. Eng.* **2023**, *64*, 105585.
111. Anderson, D.J.; Smith, S.T.; Au, F.T. Mechanical properties of concrete utilising waste ceramic as coarse aggregate. *Constr. Build. Mater.* **2016**, *117*, 20–28.
112. Zhang, L.; Shen, H.; Xu, K.; Huang, W.; Wang, Y.; Chen, M.; Han, B. Effect of ceramic waste tile as a fine aggregate on the mechanical properties of low-carbon ultrahigh performance concrete. *Constr. Build. Mater.* **2023**, *370*, 130595.
113. Alves, A.; Vieira, T.; De Brito, J.; Correia, J. Mechanical properties of structural concrete with fine recycled ceramic aggregates. *Constr. Build. Mater.* **2014**, *64*, 103–113.
114. Rashid, K.; Razzaq, A.; Ahmad, M.; Rashid, T.; Tariq, S. Experimental and analytical selection of sustainable recycled concrete with ceramic waste aggregate. *Constr. Build. Mater.* **2017**, *154*, 829–840.
115. Sharba, A. Possibility of using waste glass powder and ceramic tile as an aggregate on the flexural behavior and strength properties. In Proceedings of the 1st International Multi-Disciplinary Conference–Sustainable Development and Smart Planning: IMDC-SDSP, Cyperspace, 28–30 June 2020; pp. 1–14.
116. Roig-Flores, M.; Reig, L.; Albero, V.; Hernández-Figueirido, D.; Melchor-Eixea, A.; Pitarch, Á.M.; Piquer, A. Utilisation of ceramic stoneware tile waste as recycled aggregate in concrete. *Buildings* **2023**, *13*, 1968. [CrossRef]
117. Giridhar, V.; Rao, H.S.; Kumar, P.S.P. Influence of ceramic waste aggregate properties on strength of ceramic waste aggregate concrete. *IJRET Int. J. Res. Eng. Technol.* **2015**, *4*, 15–24.
118. Daniel, R.J.; Sangeetha, S. Experimental study on concrete using waste ceramic as partial replacement of aggregate. *Mater. Today Proc.* **2021**, *45*, 6603–6608.
119. Meena, R.V.; Jain, J.K.; Beniwal, A.S.; Chouhan, H.S. Sustainable self-compacting concrete containing waste ceramic tile aggregates: Fresh, mechanical, durability, and microstructural properties. *J. Build. Eng.* **2022**, *57*, 104941.
120. Sivakumar, A.; Srividhya, S.; Sathiyamoorthy, V.; Seenivasan, M.; Subbarayan, M. Impact of waste ceramic tiles as partial replacement of fine and coarse aggregate in concrete. *Mater. Today Proc.* **2022**, *61*, 224–231.

121. Zareei, S.A.; Ameri, F.; Bahrami, N.; Shoaei, P.; Musaeei, H.R.; Nurian, F. Green high strength concrete containing recycled waste ceramic aggregates and waste carpet fibers: Mechanical, durability, and microstructural properties. *J. Build. Eng.* **2019**, *26*, 100914.
122. Subedi, B.; Wagle, D.K.; Basnet, K. Utilization of crushed ceramic tile wastes as partial replacement of coarse aggregate in concrete production. *Int. J. Eng. Res. Technol. (IJERT)* **2020**, *9*(7), 1572–1584.
123. Varma, M.; Pravalli, M. A study on partial replacement of coarse aggregate by ceramic tile aggregate and fine aggregate by marble dust to improve concrete strengths. *Int. Res. J. Mod. Eng. Technol. Sci.* **2022**, *12*, 1769–1779.
124. Yiosese, A.; Ayoola, A.; Ugonna, M.; Adewale, A. Partial replacement of coarse aggregate with broken ceramic tiles in concrete production. *Int. J. Sci. Eng. Res.* **2018**, *9*, 81–87.
125. Abadou, Y.; Mitiche-Kettab, R.; Ghrieb, A. Ceramic waste influence on dune sand mortar performance. *Constr. Build. Mater.* **2016**, *125*, 703–713.
126. Guendouz, M.; Boukhelkhal, D. Properties of flowable sand concrete containing ceramic wastes. *J. Adhes. Sci. Technol.* **2019**, *33*, 2661–2683.
127. Gharibi, H.; Mostofinejad, D. Thermal and mechanical properties of concrete containing porcelain ceramic tile waste as fine and coarse aggregates. *Mag. Concr. Res.* **2023**, *75*, 123–134.
128. Gonzalez-Corominas, A.; Etxeberria, M. Properties of high performance concrete made with recycled fine ceramic and coarse mixed aggregates. *Constr. Build. Mater.* **2014**, *68*, 618–626.
129. Meena, R.V.; Jain, J.K.; Chouhan, H.S.; Beniwal, A.S. Impact of waste ceramic tile on resistance to fire and abrasion of self-compacting concrete. *Mater. Today Proc.* **2022**, *60*, 167–172.
130. Evangelista, A.C.J.; Tam, V.W.; Santos, J. Recycled ceramic fine aggregate for masonry mortar production. *Proc. Inst. Civil. Eng.-Constr. Mater.* **2019**, *172*, 225–234.
131. Shareef, Z.A.; Ahmed, S.Y.; Abdulkareem, O.M. Potential Use of Wastes of Thermostone Blocks and Ceramic Tiles as Recycled Aggregates in Production of Foam Concrete. *Civ. Eng. Archit.* **2023**, *11*, 1280–1296.
132. Ray, S.; Haque, M.; Rahman, M.M.; Sakib, M.N.; Al Rakib, K. Experimental investigation and SVM-based prediction of compressive and splitting tensile strength of ceramic waste aggregate concrete. *J. King Saud Univ.-Eng. Sci.* **2021**, *36*, 112–121.
133. Agrawal, A.; Singh, A.; Imam, A. Utilization of ceramic waste as a sustainable building material. In Proceedings of the National Conference on Structural Engineering, NCRASE, Jamshedpur, India, 21–22 August 2020; pp. 1–6.
134. Bommisetty, J.; Keertan, T.S.; Ravitheja, A.; Mahendra, K. Effect of waste ceramic tiles as a partial replacement of aggregates in concrete. *Mater. Today Proc.* **2019**, *19*, 875–877.
135. Chandel, S.K.; Goyal, R. Study the effect of tiles aggregate waste as partial replace with natural aggregate in cement concrete. *Mater. Today Proc.* **2022**, *51*, 905–908.
136. Xu, F.; Lin, X.; Zhou, A.; Liu, Q.-F. Effects of recycled ceramic aggregates on internal curing of high performance concrete. *Constr. Build. Mater.* **2022**, *322*, 126484.
137. Meillyta, M.; Wahyuni, W.; Fahmi, Z. Mechanical Behaviour of Self-Compacting Concrete Using Ceramic Waste. *IOP Conf. Ser. Earth Environ. Sci.* **2023**, *1140*, 012024.
138. Younis, M.; Amin, M.; Tahwia, A.M. Durability and mechanical characteristics of sustainable self-curing concrete utilizing crushed ceramic and brick wastes. *Case Stud. Constr. Mater.* **2022**, *17*, e01251.
139. Al-Azzawi, A.A.; Al-Azzawi, A.A. Mechanical properties of green concrete. *IOP Conf. Ser. Earth Environ. Sci. Eng.* **2020**, *888*, 012022.
140. Mohan, A.; Thomas, J.; Joseph, N. Use of clay Tile Chips as Coarse Aggregate in concrete. *IOP Conf. Ser. Earth Environ. Sci. Eng.* **2018**, *396*, 012002.
141. Peter, D.; Awang, A.; Sam, A.; Ma, C.; Loo, P. Eco-efficient concrete containing recycled ceramic wastes aggregate. In *IOP Conference Series: Materials Science and Engineering*; IOP Publishing: Dnipro, Ukraine, 2020; Volume 849, p. 012035.
142. Tahwia, A.M. Eco-Friendly Concrete Made with Recycled Aggregates. *Life Sci. J.* **2017**, *14*, 121–128.
143. Arun, B.C.; Ramaswamy, P. Properties of concrete partially replaced with copper slag as fine aggregate and ceramic tile waste as coarse aggregate. *Indian. J. Sci. Technol.* **2016**, *9*, 1–5.
144. Nataraja, M.; Bhat, G.; Manoj, M.; Mallya, S. Investigation on concrete with crushed vitrified tiles as coarse aggregates. *J. Build. Pathol. Rehabil.* **2022**, *7*, 103.
145. Gautam, L.; Jain, J.K.; Alomayri, T.; Meena, N.; Kalla, P. Performance evaluation of self-compacting concrete comprising ceramic waste powder as fine aggregate. *Mater. Today Proc.* **2022**, *61*, 204–211.
146. Harikaran, M.; Boopathi, S.; Rajkannan, M.; Gokulakannan, S. Impact analysis of ceramic tile powder aggregates on self-compacting concrete. *Eng. Res. Express* **2023**, *5*, 025069.
147. Kherraf, L.; Hebhoub, H.; Abdelouahed, A.; Boughamssa, W. Comparative study on the performance of sand-based mortars from marble, floor tile and cinder block waste. *J. Build. Eng.* **2022**, *45*, 103433.
148. Ajamul, S.; Oluremi, J.; Ogunyemi, E. Impact of replacement of coarse aggregates with ceramic tile waste on the strength characteristics of concrete. *J. Res. Inf. Civ. Eng.* **2018**, *15*, 2228–2244.

149. Yahya, N.; Razak, S.; Othman, M.; Noor, S.; Muhamad, K.; Jaih, M. Mechanical and rheological properties of concrete with ceramic tile waste as partial replacement of fine aggregate. In *IOP Conference Series: Materials Science and Engineering*; IOP Publishing: Jakarta, Indonesia, 2020; Volume 743, p. 012033.
150. Ambrose, E.; Ogirigbo, O.; Ekop, I. Compressive strength and resistance to sodium sulphate attack of concrete incorporated with fine aggregate recycled ceramic tiles. *J. Appl. Sci. Environ. Manag.* **2023**, *27*, 465–472.
151. Meena, R.V.; Beniwal, A.S.; Jain, A.; Choudhary, R.; Mandolia, R. Evaluating resistance of ceramic waste tile self-compacting concrete to sulphuric acid attack. *Constr. Build. Mater.* **2023**, *393*, 132042.
152. Batikha, M.; Ali, S.T.M.; Rostami, A.; Kurtayev, M. Using recycled coarse aggregate and ceramic waste to produce sustainable economic concrete. *Int. J. Sustain. Eng.* **2021**, *14*, 785–799.
153. Rodríguez-Álvaro, R.; Seara-Paz, S.; González-Fonteboa, B.; Etxeberria, M. Study of different granular by-products as internal curing water reservoirs in concrete. *J. Build. Eng.* **2022**, *45*, 103623.
154. Adeala, A.; Olaoye, J. Structural Potential of Concrete Modified with Ceramic Tiles In The Production of Concrete. *Int. J. Women Tech. Educ. Employ.* **2024**, *5*, 149–160.
155. Samadi, M.; Baghban, M.H.; Kubba, Z.; Faridmehr, I.; Abdul Shukor Lim, N.H.; Benjeddou, O.; Ariffin, N.F.B.; Huseien, G.F. Flexural behavior of reinforced concrete beams under instantaneous loading: Effects of recycled ceramic as cement and aggregates replacement. *Buildings* **2022**, *12*, 439. [CrossRef]
156. Kabir, H.; Wu, J.; Dahal, S.; Joo, T.; Garg, N. Automated estimation of cementitious sorptivity via computer vision. *Nat. Commun.* **2024**, *15*, 9935.
157. Huseien, G.F.; Faridmehr, I.; Nehdi, M.L.; Abadel, A.A.; Aiken, T.A.; Ghoshal, S. Structure, morphology and compressive strength of Alkali-activated mortars containing waste bottle glass nanoparticles. *Constr. Build. Mater.* **2022**, *342*, 128005.
158. Faridmehr, I.; Nehdi, M.L.; Huseien, G.F.; Baghban, M.H.; Sam, A.R.M.; Algaifi, H.A. Experimental and informational modeling study of sustainable self-compacting geopolymer concrete. *Sustainability* **2021**, *13*, 7444. [CrossRef]
159. Shi, T.; Luo, H. Deep learning for automated detection and classification of crack severity level in concrete structures. *Constr. Build. Mater.* **2025**, *472*, 140793.
160. Prakash, V.; Debono, C.J.; Musarat, M.A.; Borg, R.P.; Seychell, D.; Ding, W.; Shu, J. Structural Health Monitoring of Concrete Bridges Through Artificial Intelligence: A Narrative Review. *Appl. Sci.* **2025**, *15*, 4855.
161. Faridmehr, I.; Nehdi, M.; Baghban, M.H. Novel informational bat-ANN model for predicting punching shear of RC flat slabs without shear reinforcement. *Eng. Struct.* **2022**, *256*, 114030.
162. Das, S.; Datta, A.K.; Topdar, P.; Pal, A. Innovative approaches to concrete health monitoring: Wavelet transform and artificial intelligence models. *Asian J. Civ. Eng.* **2025**, *26*, 107–120.
163. Mhaya, A.M.; Baharom, S.; Baghban, M.H.; Nehdi, M.L.; Faridmehr, I.; Huseien, G.F.; Algaifi, H.A.; Ismail, M. Systematic experimental assessment of POFA concrete incorporating waste tire rubber aggregate. *Polymers* **2022**, *14*, 2294. [CrossRef] [PubMed]
164. Marchewka, A.; Ziolkowski, P.; García Galán, S. Real-time prediction of early concrete compressive strength using AI and hydration monitoring. *Sci. Rep.* **2025**, *15*, 15463.
165. Kshirsagar, P.R.; Upreti, K.; Kushwah, V.S.; Hundekari, S.; Jain, D.; Pandey, A.K.; Parashar, J. Prediction and modeling of mechanical properties of concrete modified with ceramic waste using artificial neural network and regression model. *Signal Image Video Process.* **2024**, *18*, 183–197.
166. Zanelli, C.; Conte, S.; Molinari, C.; Soldati, R.; Dondi, M. Waste recycling in ceramic tiles: A technological outlook. *Resour. Conserv. Recycl.* **2021**, *168*, 105289.
167. Chang, Q.; Liu, L.; Farooqi, M.U.; Thomas, B.; Özkılıç, Y.O. Data-driven based estimation of waste-derived ceramic concrete from experimental results with its environmental assessment. *J. Mater. Res. Technol.* **2023**, *24*, 6348–6368.
168. Joudah, Z.H.; Khalid, H.A.; Algaifi, H.A.; Mhaya, A.M.; Xiong, T.; Alsultani, R.; Huseien, G.F. Effects of Waste Glass Bottle Nanoparticles and High Volume of Waste Ceramic Tiles on Concrete Performance When Exposed to Elevated Temperatures: Experimental and Theoretical Evaluations. *Fire* **2024**, *7*, 426. [CrossRef]
169. Suchithra, S.; Sowmiya, M.; Pavithran, T. Effect of ceramic tile waste on strength parameters of concrete-a review. *Mater. Today Proc.* **2022**, *65*, 975–982.
170. Ahmad, J.; Alattyih, W.; Jebur, Y.M.; Alqurashi, M.; Garcia-Troncoso, N. A review on ceramic waste-based concrete: A step toward sustainable concrete. *Rev. Adv. Mater. Sci.* **2023**, *62*, 20230346.
171. Gautam, L.; Jain, J.K.; Kalla, P.; Choudhary, S. A review on the utilization of ceramic waste in sustainable construction products. *Mater. Today Proc.* **2021**, *43*, 1884–1891.

buildings

MDPI

Article

Reducing the Environmental Footprint of Urban Housing in Sub-Saharan Africa: A Case Study of Cameroon

Modeste Kameni Nematchoua * and Mbani Menguissa Andre Marie

Energy and Environment Laboratory, Department of Physics, Faculty of Science, University of Yaoundé I, Yaounde P.O. Box 812, Cameroon; mbani@facsciences-uy1.uninet.cm
* Correspondence: kameni.modeste@yahoo.fr

Abstract: This study presents a comprehensive Life Cycle Assessment (LCA) of the NENGOUE residence, a multi-occupancy building located in Yaoundé, Cameroon, over an 80-year lifespan. The analysis encompasses four life cycle phases—construction, use, renovation, and deconstruction—and evaluates twelve environmental impact categories. The results reveal that the use phase contributes overwhelmingly to environmental burdens, accounting for over 96% of total impacts. To mitigate this dominance, two alternative scenarios were assessed: a sustainable transport model and the integration of a photovoltaic system. In the first scenario, environmentally friendly commuting strategies, such as increased walking, cycling, and public transport, led to a 17.10% reduction in greenhouse gas (GHG) emissions. In the second, rooftop photovoltaics offset 69.29% of the building's electricity needs, resulting in a 26.72% GHG reduction. A third combined scenario demonstrated the highest environmental gains, achieving a 42.97% reduction in GHG emissions, alongside substantial improvements across other impact categories, including acidification (−38.4%), cumulative energy demand (−28.3%), and photochemical ozone formation (−40.18%). In addition to the environmental benefits, the study highlights the importance of considering social acceptance, behavioral change, and economic feasibility for real-world implementation. The willingness of residents to adopt sustainable mobility practices, cultural preferences, safety concerns, and the initial cost barriers associated with photovoltaic technology are identified as critical factors. These findings underscore the need for integrated strategies that combine technological innovation with inclusive urban planning and stakeholder engagement. The proposed approach demonstrates that aligning environmental measures with local socio-economic realities can significantly enhance the sustainability of residential buildings, contributing meaningfully to climate change mitigation in Sub-Saharan African cities.

Keywords: environmental footprint; urban housing; Sub-Saharan Africa; Cameroon

1. Introduction

The residential building sector accounts for more than 22% of the global final energy consumption, ranking among the most energy-intensive sectors, alongside transportation and industry [1]. It is also directly responsible for over 6% of global greenhouse gas (GHG) emissions and indirectly for more than 11%, resulting in a combined GHG impact of over 17% [1]. Such a significant contribution cannot be overlooked, especially as the global environmental situation becomes increasingly alarming. According to the United Nations Environment Programme (UNEP), human activities are driving three interrelated environmental crises: pollution and waste, climate change, and biodiversity loss coupled with ecosystem degradation [2]. In an effort to address what has become one of the most pressing global concerns, several international organizations, in collaboration with national governments, have initiated actions aimed at promoting sustainable development to help preserve the "blue planet". In this context, the United Nations (UN) developed the eight Millennium Development Goals (MDGs) and the 17 Sustainable Development Goals (SDGs). It also established the Conference of the Parties (COP) at the Earth Summit in Rio de Janeiro in 1992 to reduce greenhouse gas (GHG) emissions and minimize human impact on climate change [3]. More recently, on 20 September 2023, the UN held the Climate Ambition Summit in New York, USA [4], with objectives that included limiting global warming to 1.5 °C and catalyzing "a decisive political moment to demonstrate global collective will to accelerate the pace and scale of a just transition towards a more equitable, renewable energy-based, and climate-resilient global economy" [4].

Numerous additional initiatives have emerged worldwide, with many countries adopting regulations aimed at improving energy use and protecting the environment. For example, the European Union's Ecodesign Regulation prohibits the sale of energy-related products that do not meet minimum performance standards [5], France's Energy Transition Law promotes a low-carbon economy [6], and in West Africa, energy labeling has been introduced for electric lamps and new household appliances [7], among others. Several methodologies can be utilized to evaluate the environmental performance of buildings. Among them, Life Cycle Assessment (LCA) has emerged as a widely acknowledged and internationally standardized scientific approach. It remains the only method with full scientific validation for assessing environmental impacts at the building level, offering a comprehensive and quantitative evaluation across the entire life cycle of a structure. Most peer-reviewed studies addressing environmental performance in the construction sector are grounded in this methodology [8]. The concept of LCA dates back to 1969, when the Midwest Research Institute (MRI) conducted an early assessment for the Coca-Cola Company. The objective was to provide evidence-based guidance on selecting the most environmentally appropriate packaging option. However, due to the proprietary nature of the data, the findings of this pioneering work were never released to the public [9]. In subsequent decades, particularly during the 1970s, global events such as oil crises and a surge in environmental, industrial, and natural disasters—ranging from toxic emissions and severe storms to nuclear accidents like the Chernobyl disaster—heightened global awareness of ecological issues. These crises prompted the development of energy efficiency evaluation tools, which temporarily overshadowed more comprehensive frameworks like LCA [10]. Nonetheless, progress in environmental data collection eventually led to the release of the first publicly accessible LCA database in Switzerland, known as BUWAL [11].

The international standards ISO 14040 and ISO 14044 [12,13] were revised to provide more precise guidelines for conducting Life Cycle Assessment (LCA), thereby solidifying its methodological foundations. Today, LCA is widely acknowledged as the most robust and comprehensive multi-criteria tool for evaluating environmental impacts, including at the full building scale [14]. During the 2010s, the operational phase of a typical building was found to account for approximately 60–90% of its total environmental footprint, primarily due to energy demands for heating and cooling [15]. Rossi et al. [16] performed a comparative life cycle analysis of a conventional residential building constructed using two distinct techniques alongside a metal-framed dwelling across three European nations—Belgium, Portugal, and Sweden. Despite the climatic differences among these countries, the operational stage consistently emerged as the dominant contributor to environmental impact in all cases. Citherlet [17] underscored the significance of the construction and end-of-life phases, particularly when annual energy consumption falls below 150 MJ/m^2, suggesting that the relative impact of these phases becomes more pronounced. Research by Thormark [18,19] in Switzerland investigated the environmental potential of recycling building materials, while Blengini [20] explored the demolition phase of a four-decade-old residential structure in Italy. Both studies revealed that material reuse typically yields greater environmental benefits than recycling.

Erlandsson and Levin [21] examined the ecological implications of renovating existing homes in Switzerland, concluding that renovation is generally a more sustainable approach compared to demolition and new construction. However, they noted that restrictive urban planning regulations often hinder the adoption of optimal energy efficiency strategies. One persistent challenge in neighborhood-scale climate analysis has been the limited availability of consistent and high-resolution data regarding life cycle carbon emissions and energy usage. Addressing this issue, Nematchoua et al. [22] conducted a 2022 study that compared urban, rural, and sustainable neighborhood typologies adaptable to diverse geographic regions. Their work provided strategic recommendations to curb emissions and minimize energy consumption. Notably, the combination of widespread building retrofits and rooftop photovoltaic systems was identified as a promising approach, supporting the feasibility of achieving carbon neutrality at the neighborhood level by 2050.

This study aims to minimize the environmental impacts and costs associated with a residential building throughout its entire life cycle, with the overarching goal of achieving nearly zero energy consumption. Specifically, the research focuses on the NENGOUE residence, a four-story building (ground floor plus three levels) located in the Simbock neighborhood of Yaoundé. The methodology involves conducting a dynamic thermal analysis of the building using architectural simulation software, followed by a comprehensive Life Cycle Assessment (LCA) of the case study. The environmental cost of the resulting impacts is then quantified using the Global Method Monetization (MMG) approach. Subsequently, various scenarios are developed to enhance the building's energy performance while simultaneously reducing its environmental cost. These alternative scenarios are compared to the baseline configuration, ultimately leading to the identification of the optimal scenario, which serves as an optimized model for the case study.

2. Materials and Methods

2.1. Building Description

The building analyzed in this study, referred to as "Residence NENGOUE", is situated in a suburban area of Yaoundé, at the boundary between the SIMBOCK and MBALGONG neighborhoods within the MBANKOMO District, approximately 100 m from the MEFOU River. It is a residential structure surrounded by other dwellings on both sides, with its main façade facing a narrow street. The building comprises a ground floor and three upper floors. The ground floor includes four commercial shops, an unused garage, and two residential rooms. Each of the three identical upper floors consists of one apartment containing a living room, kitchen, and two bedrooms—a master bedroom with an en-suite bathroom and a secondary bedroom with access to a shared bathroom—along with three studio apartments per floor. Constructed in 2021, the building is considered relatively new. Figure 1 presents a top-down view of the structure, geolocated using Google Maps at coordinates 3°48′27″ north latitude and 11°27′57″ east longitude.

In this study, we selected "NENGOUE residence" as a case study, because this building is located on the outskirts of Yaoundé, Cameroon's largest city, and represents a typical example of urban housing in Sub-Saharan Africa, characterized by rapid urbanization, high population density, and socio-economic diversity. The architectural typology—a reinforced concrete structure with a corrugated metal roof and limited natural ventilation—is commonly found in many urban areas across the region. Furthermore, this case study was chosen due to the availability of comprehensive data, including architectural plans, material properties, and localized climatic data, which allowed for a reliable and precise energy simulation. Most importantly, the insights gained from this study are transferable to similar urban settings in Sub-Saharan Africa, offering valuable guidance for improving energy performance and climate resilience in resource-constrained environments. The building has a total floor area of approximately 420 m^2. Energy consumption patterns are primarily driven by the use of lighting, refrigeration, fans, and, increasingly, air-conditioning systems during the hot seasons. The building is connected to the national power grid and does not currently utilize renewable energy technologies. Mobility patterns of the residents are characterized by a high dependence on individual motorized transport due to limited access to reliable public transportation, which significantly contributes to the overall environmental impacts during the use phase. This contextualization allows the findings of the study to be interpreted within a realistic urban African framework and provides insights into the applicability of sustainable interventions in similar settings.

Figure 1. Location of Residence NENGOUE (**over**) and main façade of Residence NENGOUE (**below**).

2.2. Building Modeling

To perform this modeling, we obtained a photocopy of the 2D plans of the building from the owner of the residence. Using the "Modeleur" component of Pléiades software version 6.25.3, we were able to create a digital model of the building, as shown in Figure 2 below.

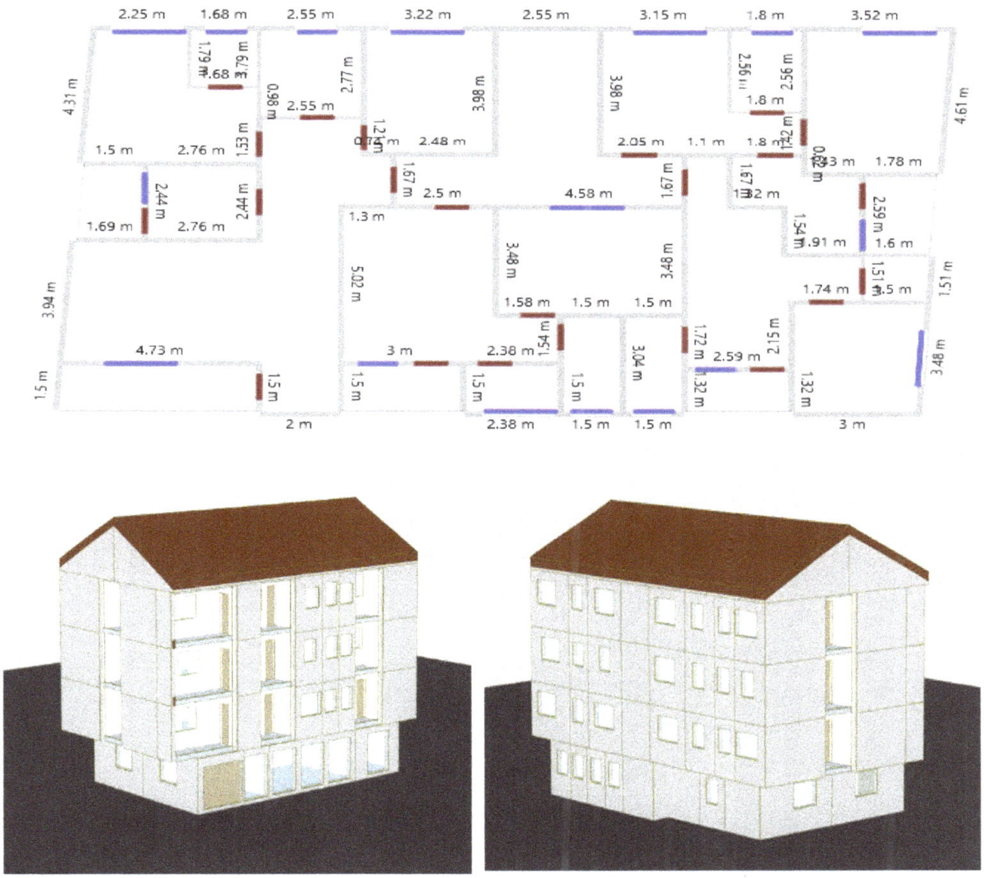

Figure 2. Layout plan of levels 1, 2, and 3 in the Pléiades Modeleur (**figure above**). North and west façades (**left side**) and south and east façades (**right side**).

2.3. Detailed Inventory of the Building Parameters

2.3.1. Construction Data

The modeling of a building in Pléiades Modeleur follows the sequential order of tabs, starting with the input of envelope elements (including wall compositions, floors, partitions, roofing, and surface states), followed by the definition of openings (doors and windows) and, finally, the thermal bridges. It is worth noting that the software provides predefined compositions for these elements, although users also have the option to modify existing compositions or create custom ones by inputting the specific characteristics of materials. Material quantities (e.g., volumes of concrete, steel, aluminum, and timber) were obtained directly from the architectural and structural drawings of the NENGOUE residence, provided by the project's engineering team. Energy use data for the operational phase were derived from dynamic thermal simulations using EnergyPlus software, version 8.4.0 calibrated with real-time meteorological data for Yaoundé and typical user behavior in residential buildings. Where available, transport patterns and distances were based on site-specific information, including the locations of suppliers and construction sites, as well as interviews with local contractors and transportation service providers. When local data were not available, we relied on regionally appropriate averages from the Ecoinvent

3.1 database, adjusted to reflect the Sub-Saharan African context. To ensure data quality, all inputs were validated through cross-referencing with published literature on similar housing projects in Cameroon and other parts of Sub-Saharan Africa.

2.3.2. Composition of the Envelope Elements

The Pléiades library allows users to select different types of predefined envelope elements, including walls; floors (ground, intermediate, and top floors); and roofs. In addition to these default elements, the software also enables users to create custom envelope compositions. To achieve this, it is crucial to define key material parameters, such as thickness (e), thermal conductivity (λ), density (ρ), and thermal resistance (R). Based on the available data, we aimed to model these elements as accurately as possible to reflect real-world conditions. The Table 1 shows some wall materials of residence evaluated.

Table 1. Composition of the walls of the NENGOUE residence.

Category	Element	Material	Thickness (cm)	λ (W/m.K)	ρ (kg/m^3)	CS (Wh/kg.K)	U (W/m^2·K)	R (m^2·K/W)
External Walls	Exterior Wall	Exterior Plaster	1.0	1.150	1700	0.278	115.00	0.01
		Hollow Bricks Type 1 (15 cm)	15.0	0.508	1250	0.278	3.39	0.30
		Interior Plaster (gypsum and sand, ρ = 1600)	1.0	0.800	1600	0.278	80.00	0.01
Internal Walls	Shower Partition	Exterior Plaster	1.0	1.150	1700	0.278	115.00	0.01
		Hollow Brick (10 cm)	10.0	0.476	690	0.250	4.76	0.21
		Mortar	1.0	1.150	2000	0.233	115.00	0.01
		Tiles	1.0	1.700	2300	0.194	170.00	0.01
Ground Floor	Ground Floor Slab	Tiles	1.0	1.700	2300	0.194	170.00	0.01
		Concrete with fine aggregates or sand (lightweight)	1.0	0.440	1300	0.278	44.00	0.02
		Standard Concrete—1 alveolus—hollow core height 12 cm	15.0	0.682	1200	0.278	4.55	0.22
Intermediate Floors	RDC+ Floor Slab	Exterior Plaster	1.5	1.150	1700	0.278	76.67	0.01
		20 cm Concrete Hollow Block	20.0	1.333	1300	0.180	6.67	0.15
		Tiles	1.0	1.700	2300	0.194	170.00	0.01
Roof	Roofing	Aluminum Alloys	0.05	160,000	2800	0.244	320,000.00	0.00
Attic	Attic Slab	Mortar	1.5	1.150	2000	0.233	76.67	0.01
		16 cm Concrete Hollow Block	16.0	1.231	1300	0.180	7.69	0.13
		Lightweight Wood	1.5	0.150	500	0.333	10.00	0.10

The "Joinery" tab of the Pléiades library allows for the input of the building's doors and windows. In our case study, the exterior and interior doors are made of wood, while the garage gate and sliding patio doors are made of aluminum. The properties, contained in Table 2, are described below.

Table 2. Properties of the Joinery of the NENGOUE residence.

Openings	Number of Glazings	Geometry		Vertical Uw Coefficient (W/(m^2·K))	Horizontal Uw Coefficient (W/(m^2·K))	Solar Factor Sw	Light Transmission Tl
		Height (m)	Width (m)				
Living Room and Bedroom Windows	1	1.15	1.8	3.74	3.74	0.54	0
Bathroom Windows	1	1.15	1	3.74	3.74	0.54	0
Interior Wooden Door	0	2.04	0.83	5	5	0.19	0
Garage Gate	0	2.04	3.2	7	7	0.17	0
Sliding Patio Door	0	2.15	2	4.51	4.51	0.67	0

Pléiades provides the ability to import weather files from major meteorological stations for dynamic thermal simulation (DTS) calculations. It natively uses weather files in .TRY format. In our case study, we use a TRY data file for the city of Yaoundé from the Meteocalc meteorological station. The LCA was conducted in accordance with ISO 14040 and ISO 14044 standards. The functional unit was defined as 1 m^2 of usable floor area of the NENGOUE residence over a 50-year lifespan, allowing for comparison with other residential buildings in similar climatic and socio-economic contexts. The system boundaries covered a cradle-to-grave approach, including raw material extraction, transportation, construction, operation (energy use), maintenance, and end-of-life stages (demolition and disposal). The analysis was performed using the Pleaides LCAv 6.25.3 tool with the Ecoinvent 3.1 database, which offers a comprehensive inventory of environmental data relevant to Sub-Saharan conditions. Key assumptions included the use of regionally typical materials (e.g., cement, steel, aluminum, and local wood); standard transport distances based on Cameroonian infrastructure; a 50-year service life; and disposal practices common in urban Cameroon (e.g., landfilling and partial recycling).

2.4. Dynamic Thermal Analysis

2.4.1. Building Zoning

Zoning is the basic structure of the thermal model. A thermal zone thus represents a volume from which scenarios such as occupancy, electrical energy dissipation, solar gains, conductive heat transfer, etc. can be applied. Depending on the needs, different zoning approaches can be used. For our study, we performed zoning based on the rooms of the building, as the thermal behaviors in different room types vary significantly. For example, the thermal behaviors in the living room, kitchen, bathroom, and bedroom differ greatly. All the thermal zones created in Pléiades Modeleur are shown in Table 3 below.

Table 3. Thermal zones for the case study.

Zone No.	Zone Name	Surface (m^2)	Volume (m^3)
1	Circulation	133.40	335.55
2	Storage	14.50	36.25
3	WC Ground Floor	5.89	14.74
4	Storage No. 2	78.39	195.97
5	Storage No. 1	11.13	27.82
6	Storage No. 3	21.02	52.54
7	Garage	17.34	43.36

Table 3. *Cont.*

Zone No.	Zone Name	Surface (m^2)	Volume (m^3)
8	Bedroom Ground Floor 1	10.22	25.56
9	Bedroom Ground Floor 2	9.07	22.69
10	WC Ground Floor 1	3.01	7.53
11	WC Ground Floor 2	3.24	8.09
12	Living Room A1	36.97	92.44
13	Kitchen S1	3.08	7.69
14	Kitchen A1	5.78	14.44
15	Bedroom PA1	13.75	34.38
16	Bedroom SA1	11.77	29.43
17	Bathroom A1	6.47	16.16
18	WC A1	2.68	6.70
19	Living Room S1	19.57	48.93
20	Bedroom S1	14.44	36.11
21	Bathroom S1	3.86	9.65
22	Kitchen S2	3.86	9.65
23	Living Room S2	15.62	39.06
24	Bedroom S2	10.19	25.48
25	Bathroom S2	2.01	5.02
26	Kitchen S3	6.26	15.64
27	Living Room S3	13.85	34.62

2.4.2. Heating Setpoint

For better thermal comfort of the building occupants, the interior temperatures of the rooms should vary depending on whether one is in a bedroom, living room, bathroom, or kitchen. In our context, the building under study is located in a tropical zone, which is sufficiently warm; therefore, there is no need to apply heating setpoints.

2.4.3. Building Occupancy

To obtain the occupancy profile, we conducted a survey among the building's residents, which allowed us to create an overview of the occupancy for each part of the building. The Pléiades Modeleur software enabled us to model these behaviors through occupancy scenarios specific to each of the thermal zones concerned. The values in the various tables represent the number of people present in the room during the considered time slot.

(a) Description of the Occupancy Scenarios

Since the building is a four-story structure, its occupancy scenarios depend not only on the type of dwelling (bedroom, studio, or apartment) but also on the behavior within each considered thermal zone.

2.5. Life Cycle Analysis

Pléiades software enabled the execution of a Life Cycle Assessment (LCA) by evaluating twelve environmental impacts. To achieve this, the software uses the building geometry from the dynamic thermal simulation (DTS) performed in Pléiades Modeleur and transfers it to the EQUER component for the LCA.

2.5.1. Calculation Parameters

In the absence of detailed information regarding the materials used in the building envelope, the default parameters provided by Pléiades software have been adopted. These include an estimated 5% surplus of materials during construction and predefined lifespans for various components: exterior and interior doors, as well as glazing, are assumed to last 30 years, while interior and exterior surface coatings are expected to have a lifespan of 10 years. The service life of the building equipment is set at 20 years. The habitable surface area (SHAB) is generally considered to be 20% less than the net floor area (SHON) [23]. Additionally, the average transportation distance between the production site and the construction site is set at 100 km. For thermal comfort analysis, the discomfort range is defined between 24 °C (lower limit) and 28 °C (upper limit) [23].

2.5.2. Project Association

For Life Cycle Assessment (LCA) calculations, Pléiades software uses the EQUER calculation engine, developed by the Center for Energy Efficiency of Systems at Mines ParisTech. This engine allows for the evaluation of embodied energy and, of course, its contribution to global warming. It can utilize the calculation results from the COMFIE dynamic thermal simulation (DTS) module, the RT2012 regulatory module, or perform an independent analysis from scratch. The calculations rely on the Ecoinvent environmental database version 3.1.

2.5.3. Energy

According to KB SAP CONSULTING [23], the energy mix is distributed as follows: 62% from hydropower, 24% from thermal power plants, and 14% from natural gas.

2.5.4. Water

The national water policy published in 2019 [24] showed that the daily water consumption in Cameroon per capita is 50 L/day.

2.5.5. Waste

In Pléiades software, we entered the following waste-related data: average daily waste generation per inhabitant: 850 g [25], recycling rates: 73% for glass and 43% for paper, incinerated waste: around 30%, with an energy recovery efficiency of 5%, and distance to landfill: as determined by Google Maps, the distance from the building to the NKOLFOULOU landfill is 23 km.

2.5.6. Transportation

The NENGOUE residence is situated in the Simbock suburb, approximately 10.4 km from Yaoundé's administrative center, located at the National Museum (according to Google Maps data). In Pléiades software, the transportation-related parameters were defined as follows: approximately 80% of the building's occupants engage in daily commuting, primarily for work or educational purposes (students and pupils). The average weekly distance for home-to-commerce trips was set at 3.1 km—corresponding to the distance between the Bocom Service Station and the Simbock Entrance—where the majority of commercial activities and public transportation access points are located. Additionally, the average daily home-to-work travel distance was estimated at 5 km, reflecting the typical commuting patterns of the residents.

2.6. Simulation

The Life Cycle Assessment (LCA) was conducted over a period of 80 years, evaluating twelve environmental impacts for a building with 40 occupants. The Pléiades Results module also allows for the generation of various types of graphs and curves, enabling easy comparison of the results.

2.7. The MMG (Global Method Monetize)

Table 4 summarizes the environmental costs that will be used throughout this study.

Table 4. Summary of the environmental cost per unit of the CEN indicators [26].

Environmental Indicator (100 Years)	Value (€)	Conversion (FCFA)
Greenhouse Effect (€/t CO_2 eq.)	50	32,797.85
Acidification (€/kg SO_2 eq.)	1.01	662.52
Cumulative Energy Demand (€/GJ)	0	0.00
Water Usage (€/m^3)	0.079	51.82
Abiotic Resource Depletion (€/kg Sb eq.)	1.56	1023.29
Eutrophication (€/kg eq.)	40	26,238.28
Photochemical Ozone Production (€/kg C_2H_4)	3.3	2164.66

Pléiades is a proprietary software developed by the French company IZUBA Énergies (Paris, France) which core expertise lies in energy and environmental optimization in the building sector [27].

3. Results

3.1. Analysis and Result of the Initial Scenario

Table 5 presents the results of the environmental impact assessment derived from the Life Cycle Assessment (LCA) of the initial scenario.

Table 5. Calculation of environmental impacts for the initial scenario.

Impact Over 80 Yeras	For 40 Inhabitants Located on 880.89 m^2	Per Occupant	Per m^2
Greenhouse Gas Emissions (t CO_2 eq.)	3132.31	78.31	3.56
Acidification (kg SO_2 eq.)	25,976.99	649.42	29.49
Cumulative Energy Demand (MWh)	93,898.64	2347.47	106.60
Water Usage (m^3)	212,899.34	5322.48	241.69
Inert Waste Generated (t)	7167.96	179.20	8.14
Abiotic Resource Depletion (kg E-15)	7.19	0.18	0.01
Eutrophication (kg PO_4 eq.)	9799.97	245.00	11.13
Photochemical Ozone Formation (kg ethylene eq.)	9786.90	244.67	11.11
Aquatic Ecotoxicity (m^3)	83,721,205.26	2,093,030.13	95,041.61
Radioactive Waste (dm^3)	139.51	3.49	0.16
Human Toxicity (kg)	52,221.19	1305.53	59.28
Odor (Mm^3 air)	15,151.23	378.78	17.20

By calculating the average of the twelve environmental impacts from Table 6, we observe that the usage phase is the stage of the LCA that contributes to almost all environmental impacts, accounting for 96.53%, compared to only 2.32% for construction, 0.34% for renovation, and 0.79% for deconstruction.

Table 6. Detailed results of the LCA of the initial scenario.

Impact	Unity	Construction	Usage	Renovation	Deconstruction	Total
Greenhouse Gas Emissions	(t CO_2 eq.)	164.22	2956.89	8.13	3.07	3132.31
	%	5.24%	94.40%	0.33%	0.10%	
Acidification	(kg SO_2 eq.)	691.80	25,163.95	86.13	35.11	25,976.99
	%	2.66%	96.87%	0.51%	0.14%	
Cumulative Energy Demand	(GJ)	2447.70	90,925.24	475.48	50.22	93,898.64
	%	2.61%	96.83%	0.19%	0.05%	
Water Use	(m^3)	1772.00	210,707.11	396.68	23.56	212,899.34
	%	0.83%	98.97%	0.61%	0.01%	
Inert Waste Produced	(t)	45.45	6475.71	43.60	603.20	7167.96
	%	0.63%	90.34%	0.00%	8.42%	
Abiotic Resource Depletion	(kg E-15)	0	7.19	0	0	7.19
	%	0.00%	100.00%	0.07%	0.00%	
Eutrophication	(kg PO_4 eq.)	66.87	9720.90	6.71	5.49	9799.97
	%	0.68%	99.19%	0.39%	0.06%	
Photochemical Ozone Production	(kg ethylene eq.)	420.67	9289.55	38.52	38.16	9786.90
	%	4.30%	94.92%	0.36%	0.39%	
Aquatic Ecotoxicity	(m^3)	2,065,609.41	81,252,451.92	302,282.69	100,861.24	83,721,205.26
	%	2.47%	97.05%	0.70%	0.12%	
Radioactive Waste	(dm^3)	3.34	135.02	0.97	0.19	139.51
	%	2.39%	96.78%	0.62%	0.14%	
Human Toxicity	(kg)	1152.48	50,700.21	326.29	42.20	52,221.19
	%	2.21%	97.09%	0.17%	0.08%	
Odor	(Mm^3 air)	586.35	14,535.18	26.25	3.46	15,151.23
	%	3.87%	95.93%	0.17%	0.02%	
	Average Percentage	2.32%	96.53%	0.34%	0.79%	

Results of the MMG. Based on this method, the results of the initial scenario for each of the seven environmental impacts are detailed in Table 7.

Table 7. Results of the MMG for the initial scenario for the seven environmental indicators of the LCA.

Impact (Initial Scenario)	Total (Initial)	Costs over 80 years in €	Percentage
Greenhouse Effect (t CO_2 eq.)	3132.31	125,292.46	25.10%
Acidification (kg SO_2 eq.)	25,976.99	20,989.41	4.20%
Cumulative Energy Demand (GJ)	93,898.64	0.00	0.00%
Water Used (m^3)	212,899.34	13,455.24	2.70%
Depletion of Abiotic Resources (kg E-15)	7.19	8.97	0.0018%
Eutrophication (kg PO_4 eq.)	9799.97	313,598.98	62.82%
Photochemical Ozone Production (kg ethylene eq.)	9786.90	25,837.43	5.18%

By examining the environmental cost related to greenhouse gas (GHG) emissions at each phase, we observe that the usage phase accounts for 94.40%, while the construction, renovation, and deconstruction phases represent 5.24%, 0.26%, and 0.099%, respectively. Furthermore, the overall impact cost of GHG emissions and eutrophication amounts to 25.10% and 62.82%, respectively.

3.2. Validation of the Results

These results are consistent with the literature, which estimates that the usage phase accounts for 60–90% of environmental burdens [28]. This aligns with our study, where we observe a value of 94.40%. Furthermore, the impact indicators for greenhouse gas (GHG) emissions and eutrophication together represent 87.92% of the building's environmental cost. This confirms the findings of Nematchoua and Reiter [29] in their study on the evaluation of environmental costs comparing traditional neighborhoods with eco-districts, where they obtained values of 55.4% for eutrophication and 31.2% for GHG emissions, resulting in a combined cost of 86.6%.

3.3. Impact of the Building Lifespan and Resident Transportation

3.3.1. Impact of the Building Lifespan

Depending on whether the Life Cycle Assessment (LCA) is conducted over a long period (80–100 years), a medium-term period (25–50 years), or a shorter period (5–10 years or even just 1 year), the results vary significantly. Certain life cycle phases contribute more, depending on the lifespan, and some impact indicators become more pronounced as the lifespan increases or decreases.

It is important to note that the only phase that remains unchanged, regardless of the considered lifespan, is the construction phase. However, as the lifespan extends, human activity related to the building increases. Naturally, the usage phase generates the vast majority of the building's environmental impacts. All indicators are particularly significant during this phase; the impacts from the construction phase are compounded by those of other phases, with the usage phase accounting for at least 90% of total impacts. Our analysis showed that, even within a lifespan of just ten years, the usage phase already accounted for more than 70% of the total environmental impacts of the building.

3.3.2. Scenario of Resident Transportation

It is important to recall that the building is primarily occupied by families and some students. Additionally, nearby grocery stores facilitate easy access to daily supplies. We assume that students attend schools and universities within the vicinity.

To implement this scenario, it is assumed that nearly 90% of residents adopt environmentally friendly transportation methods, such as bicycles, and live in close proximity to their workplaces and essential supply stores. The planning criteria include a maximum distance of 500 m from homes to stores and to the public transportation network and a 3-km distance from homes to workplaces. Additionally, 90% of occupants are expected to commute daily, primarily using buses as their mode of public transportation. Table 8 presents the results of the calculation of the environmental impacts of the LCA of the transport variant.

When we average the contributions of the twelve environmental impacts in Table 8, it is very clear that the use phase is the most contributory, with a rate of 96.26% on all impacts, while construction, renovation, and reconstruction constitute 2.69%, 0.39%, and 0.82%, respectively. Furthermore, the value of the greenhouse effect impact indicator decreased from 3132.31 t CO_2 eq. to 2596.72 t CO_2 eq., a decrease of 17.1% from its initial value. According to these results, the NENGOUE residence emits 32.46 t CO_2 eq./year on the 880.89 m^2 of its surface area or 36.85 kg CO_2 eq./m^2/year.

The shift in transportation modes leads to an overall reduction in environmental impacts, which, in turn, significantly lowers the building's environmental footprint. In this variant, the following reductions are observed compared to the initial scenario: greenhouse gas emissions decrease by 17.10%, acidification drops by 12.74%, and cumulative energy demand is reduced by 9.31%. Water usage and inert waste decrease by 1.98% and 1.94%, respectively, while abiotic resource depletion remains unchanged. Eutrophication is reduced by 2.33%, photochemical ozone formation drops by 22.60%, and aquatic ecotoxicity falls by 21.19%. Additionally, radioactive waste decreases by 3.01%, human toxicity is reduced by 14.18%, and odors decline by 5.16%.

Table 8. Detailed results of the LCA of the transport variant over 80 years.

Impact	Construction	Usage	Renovation	Deconstruction	Total
Greenhouse Effect (t CO_2 eq.)	164.22	2421.29	8.13	3.07	2596.72
	6.32%	93.24%	0.31%	0.12%	
Acidification (kg SO_2 eq.)	691.8	21,854.22	86.13	35.11	22,667.27
	3.05%	96.41%	0.38%	0.15%	
Cumulative Energy Demand (GJ)	2447.70	82,187.30	475.48	50.22	85,160.71
	2.87%	96.51%	0.56%	0.06%	
Water Used (m^3)	1772.00	206,498.95	396.68	23.56	208,691.19
	0.85%	98.95%	0.19%	0.01%	
Inert Waste Produced (t)	45.45	6475.71	43.6	603.2	7028.74
	0.65%	92.13%	0.62%	8.58%	
Abiotic Resource Depletion (kg E-15)	0	7.19	0	0	7.19
	0.00%	100.00%	0.00%	0.00%	
Eutrophication (kg PO_4 eq.)	66.87	9492.11	6.71	5.49	9571.18
	0.70%	99.17%	0.07%	0.06%	
Photochemical Ozone Production (kg ethylene eq.)	420.67	7077.72	38.52	38.16	7575.07
	5.55%	93.43%	0.51%	0.50%	
Aquatic Ecotoxicity (m^3)	2,065,609.41	63,507,915.01	302,282.69	100,861.24	65,976,668.35
	3.13%	96.26%	0.46%	0.15%	
Radioactive Waste (dm^3)	3.34	130.83	0.97	0.19	135.32
	2.47%	96.68%	0.72%	0.14%	
Human Toxicity (kg)	1152.48	43,292.85	326.29	42.2	44,813.83
	2.57%	96.61%	0.73%	0.09%	
Odor (Mm^3 air)	586.35	13,753.53	26.25	3.46	14,369.59
	4.08%	95.71%	0.18%	0.02%	
Average Percentage	2.69%	96.26%	0.39%	0.82%	

The results of the MMG for the transport scenario are detailed in Table 9.

Table 9. Annual results of the MMG for the transport variant.

Impact	Construction	Operation	Renovation	Deconstruction
(Unit)	Costs (FCFA)	Results	Costs (FCFA)	Results
Greenhouse Effect (t CO_2 eq.)	164.22	4308,888.22	2421.29	63,530,551.57
Acidification (kg SO_2 eq.)	691.80	366,664.17	21,854.22	11,583,088.13
Cumulative Energy Demand (GJ)	2447.70	0.00	82,187.30	0.00
Water Usage (m^3)	1772.00	73,460.90	206,498.95	8560,620.51
Abiotic Resource Depletion (kg E-15)	0	0.00	7.19	5883.58
Eutrophication (kg PO_4 eq.)	66.87	1403,602.54	9492.11	199,245,302.11
Photochemical Ozone Formation (kg ethylene eq.)	420.67	728,482.65	7077.72	12,256,677.93
Total Cost (80 years)	6,881,098.48	295,182,123.83	483,035.36	281,558.47

The environmental cost over 80 years has decreased by 7.58%.

3.4. Renovation and PV Scenarios

The choice of insulation materials is particularly critical in achieving the Q-ZEN standard of the 2021 PEB, which is one of the objectives set in this study. The following criteria will guide our selection process: (i) Thermal conductivity: This parameter determines the energy performance of an insulating material; the closer it approaches zero, the higher the quality of the insulation. However, thermal conductivity alone does not account for factors that may disqualify a material, such as its contribution to greenhouse gas emissions. (ii) Relevant applications: This refers to the specific areas of a building where the insulation material is most suitable for use. (iii) Contribution to greenhouse gas emissions: The lower the material's environmental impact, the more suitable it will be for selection. (iv) Thermal phase shift: This represents the time required for heat to pass through the insulation. This criterion is particularly crucial in our case, as it ensures that the internal temperature of the building remains stable without rapid fluctuations. (v) Embodied energy: The higher the embodied energy, the lower the insulation's environmental performance. This factor is closely linked to greenhouse gas emissions, as both tend to increase in parallel. By prioritizing insulation materials with a low contribution to greenhouse gas emissions, we have selected the materials listed in Table 10 for application in passive design strategies.

Table 10. Characteristics of the insulators selected in this study.

Insulation Material	Thermal Conductivity λ (W/(m·K))	Average Price per m^2 (€)	Embodied Energy (kWh/kg)	Relevant Applications	Average Thermal Phase Shift for 20 cm (h)	Contribution to Greenhouse Gas Emissions
Cellulose Wadding	0.038–0.042	15 €	1–2	• Attics • Floors • Roofs • Walls	10	Low
Wood Fiber or Wood Panels	0.038–0.049	20 €	1–3	• Attics • Floors • Roofs • Walls	15	Low and Very Low
Expanded Cork and Cork Panels	0.037–0.044	80 €	2–7	• Attics • Floors • Roofs • Walls • Partitions	13	Very Low

In this table, we observe that the insulation materials combining the most criteria to ensure indoor comfort are natural or bio-based. Among them, cellulose wadding has the lowest average thermal phase shift.

3.4.1. Impact of Renewable Energy: Use of a Photovoltaic System

In the initial scenario, the building is solely powered by the national electricity provider. In this scenario, we install a photovoltaic system to supply the building with electricity. Photovoltaic panels are installed on two-thirds of the roof, covering a total photovoltaic field area of 159.84 m^2. The selected inverter is the Sunny Boy SB 3000 TL, with a nominal power output of 3.2 kW and a maximum voltage of 550 V. The chosen photovoltaic modules are monocrystalline, specifically the E19-318E model, with a peak power of 318 W. They are arranged in 14 strings, each consisting of seven modules connected in series. The modules are oriented due south and tilted at an angle of 30°, as recommended by experts for optimal energy production in the city of Yaoundé.

The use of a photovoltaic panel system significantly reduces electricity costs and provides flexibility to enhance indoor comfort. With an annual energy consumption of 72,585 kWh, the photovoltaic system generates 50,294 kWh per year, covering approximately 69.29% of the building's energy needs. Thus, the system supplies nearly 70% of the total annual consumption.

The annual energy production of the photovoltaic system is illustrated in Figure 3 below.

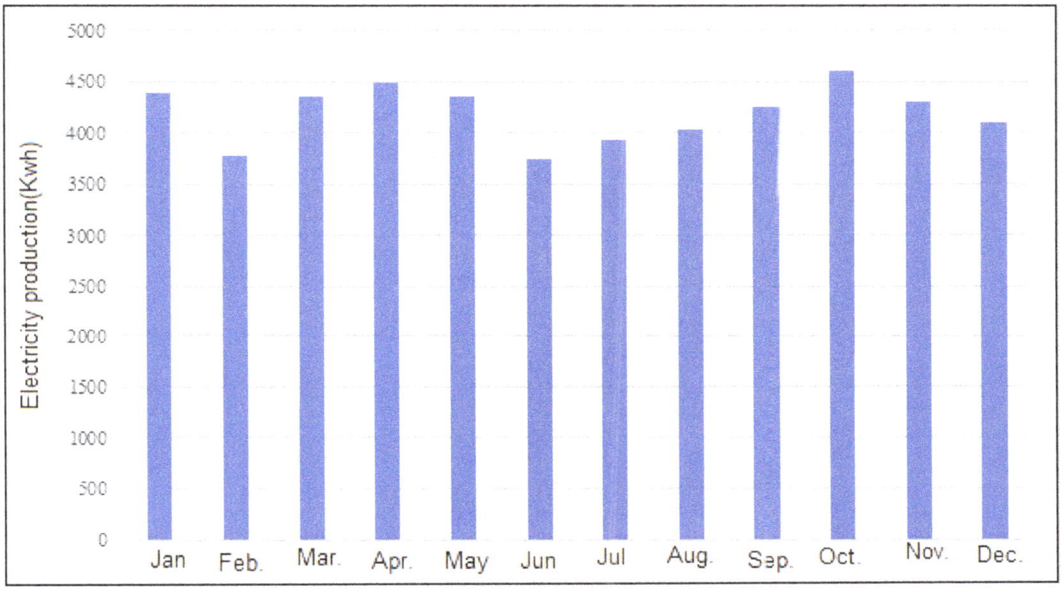

Figure 3. Annual electricity production by the PV system.

The results below illustrate a generalized decrease in almost all impacts. Let us now proceed with the environmental classification of the NENGOUE residence. Its quantity of CO_2 emissions per year amounts to 28,691 kg CO_2 eq. Consequently, the building emits 32.57 kg CO_2 eq/m^2/year, which corresponds to class D. Despite the decrease in absolute value of the various impacts, in terms of percentages, the values remain close to the initial and transport scenarios.

In terms of environmental impacts, the reductions achieved in this scenario compared to the initial scenario are as follows: 26.72% reduction in greenhouse gas emissions, 26.61% reduction in acidification, 21.88% reduction in cumulative energy demand, 19.91% reduction in water consumption, 0.80% reduction in inert waste production, 38.48% reduction in abiotic resource depletion, 3.93% reduction in eutrophication, 20.17% reduction in photochemical ozone formation, 22.09% reduction in aquatic ecotoxicity, 22.29% reduction in radioactive waste generation, 20.39% reduction in human toxicity, and 36.19% reduction in odor emissions (see Figure 4).

By applying the renewable energy impact scenario, the NENGOUE residence achieves an overall environmental cost savings of 11.87% compared to the initial scenario. The distribution of this gain across different phases is as follows: 15.88% savings during the construction phase, 11.92% savings during the operational phase, 66.05% increase during the renovation phase, and 0.69% savings during the demolition phase. Noting that the operational phase remains the most significant, the overall gain is very close to this phase. This comparison is limited to only seven impact categories, whereas the Pléiades software provided results for twelve. In terms of individual contributions, greenhouse gas emissions (20.87%) and eutrophication (68.49%) account for the highest impact reductions.

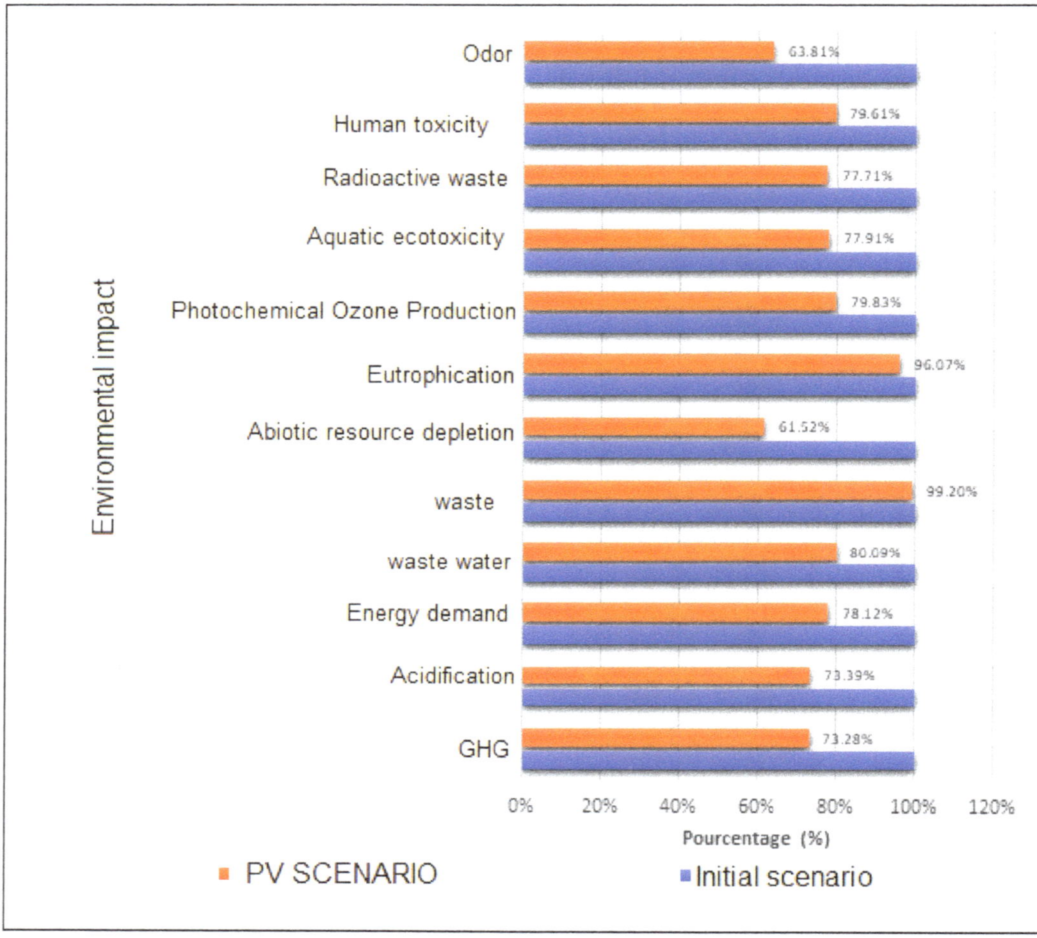

Figure 4. Comparison between the initial scenario and the renewable energy impact scenario.

3.4.2. Renovation

For the external wall coating, we will use two types of coatings (interior and exterior): mortar and lime ($500 < \rho < 750$), according to Pléiades, with a thermal conductivity of $\lambda = 0.3$ W/(m·K) and a thickness of one centimeter. The interior coatings will have the same thickness but will be made of gypsum plaster with bulk perlite, with a conductivity of $\lambda = 0.18$ W/(m·K), leading to thermal resistances of Re_ext = 0.034 (m^2·K)/W and Re_int = 0.05 (m^2·K)/W. Thermal resistance calculations and heat transfer coefficients for these walls can be quickly determined. We will retain only the bricks and replace the remaining materials. For the external walls, the total thermal resistance (including wall and coatings) is R = 0.339 (m^2·K)/W, and the thermal transmittance or thermal conductance is U = 2.95 W/(m^2·K). These values indicate very poor insulation performance. For proper thermal insulation, R should be high (greater than 4.15), and U should be close to zero (U_max = 0.24 W/(m^2·K) to meet the Q-ZEN standard). Similarly, for the internal walls, the total thermal resistance (including wall and coatings) is R = 0.249 $(m^2 \cdot K)/W$, with a thermal transmittance U = 4.76 $W/(m^2 \cdot K)$. Considering the balance between insulation quality, cost, greenhouse gas emissions, and thermal phase shift, cork panels prove to be the best choice for both interior and exterior walls. We will not use wood fiber panels in the walls due to space constraints, despite their superior thermal phase shift properties. Instead, their use will be prioritized for floors (both upper and lower), attics, and roofing.

Regarding surface finishes, we will favor those with an infrared emissivity close to 1, a solar absorptivity close to 0, and optical properties with a reflectance percentage close to 100%. Based on the Life Cycle Assessment (LCA) results obtained, the building emits 2607.27 t CO_2 eq over 80 years for this variant, corresponding to 32,590.875 kg CO_2 per year or 39.97 kg CO_2 eq/m^2/year, placing it in energy class D.

(i) Comparison between the initial scenario/scenario Impact of wall insulation and change in surface conditions

By applying this scenario, its effect is direct on several environmental impacts; in fact, the greenhouse effect (−16.76%), the inert waste produced (−17.27%), the photochemical ozone production (−27.05%), the aquatic ecotoxicity (−21.56%), and the human toxicity (−14.08%) decrease. The results are shown in Figure 5.

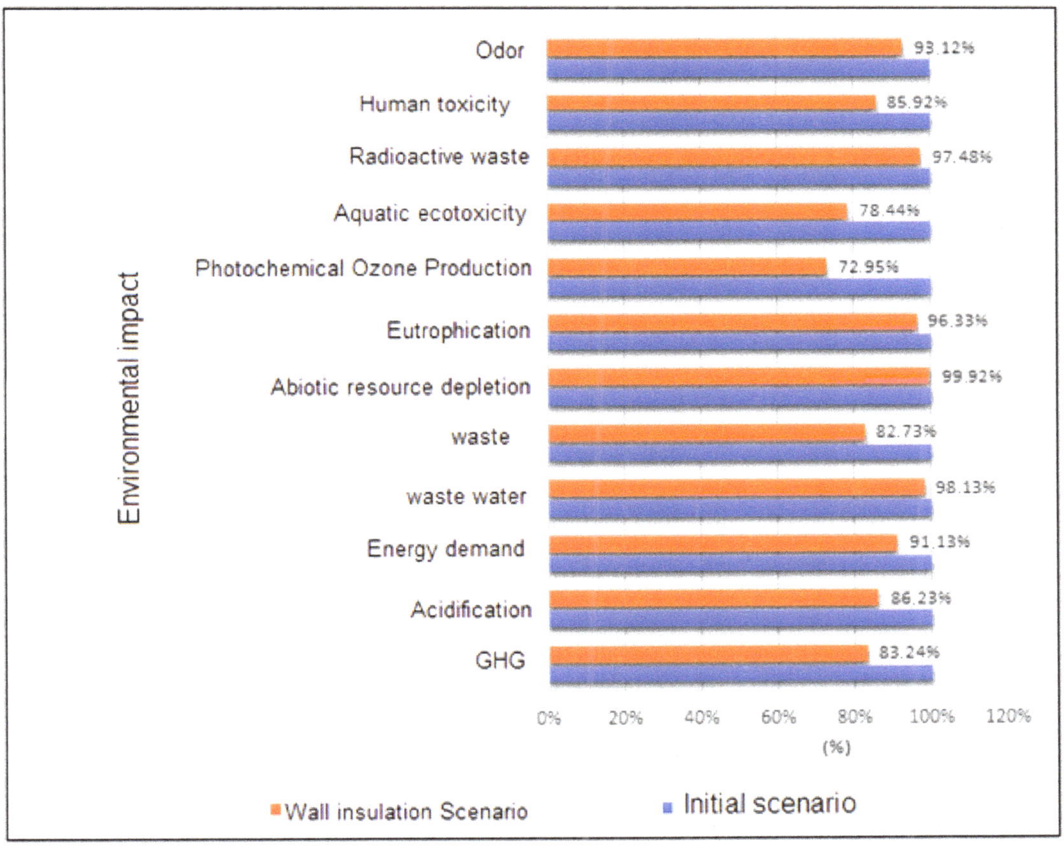

Figure 5. Comparison of the initial scenario/Impact of wall insulation and change in the surface conditions.

By making a comparison via different costs, we see that, as always, the use phase is the one that produces the greatest contribution of impacts. The environmental cost depends greatly on that of the use phase. There is a decrease in the environmental cost of 8.54%, distributed as follows: construction (−11.94%), use (−8.52%), renovation (+19.77%), and deconstruction (+0.96%).

In this research, we chose to use wood fiber panels to insulate the roofs, because we want to limit air infiltration as much as possible. We considered for the attic the slabs, and we added our wood fiber panels, which thickness we determined. The calculations show that a thickness of 28.3 cm of the wood fiber panels would be needed to insulate the attic. Overall, the roof was made of an aluminum alloy with a thickness of 0.05 mm, thermal conductivity λ = 160 W/(m·K), i.e., a thermal resistance of 0.00 (m^2·K)/W, and a thickness of wood fiber panels of 29.5 cm, which ensured good thermal insulation. Referring to the LCA results, we deduce that the building emits for this variant 3113.53 t CO_2 eq over 80 years or 38,919.125 kg CO_2 per year or 44.18 kg CO_2 eq/m^2/year, which corresponds to class D. Figure 6 shows the comparison between roof insulation and the initial scenarios.

(ii) Comparison between the initial and roof insulation scenarios

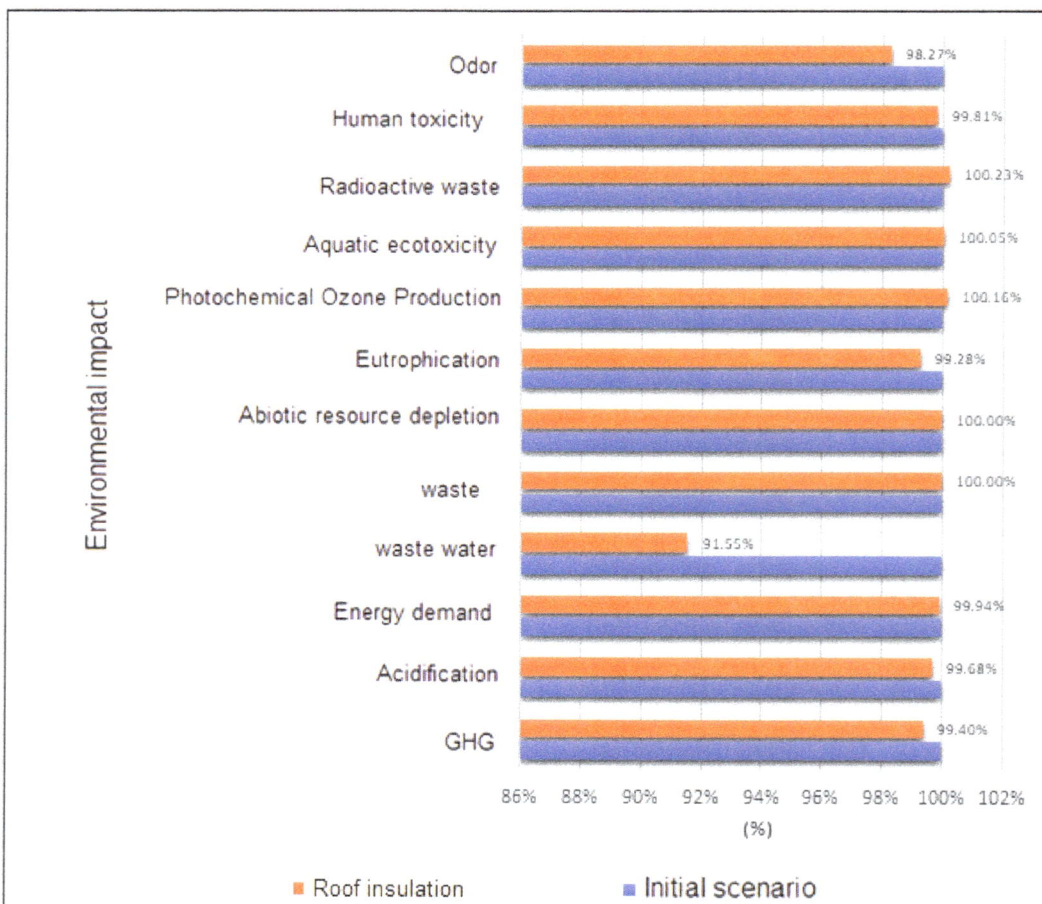

Figure 6. Comparison between the scenarios/impacts of attic and roof insulation.

Regarding these results, we deduce that the impacts of roof insulation are low compared to the other strategies. The stage use remains the most important.

3.4.3. Mixed Scenario

The optimal building model sought and meeting the Q-ZEN standard is a building that meets the following requirements: (i) Exterior and interior wall insulation with a thermal transmittance coefficient not exceeding Umax = 0.24 W/(m^2·K). (ii) Insulation of the ceiling floors, each with the same Umax = 0.24 W/(m^2·K). (iii) Attic and roof insulation, each not exceeding Umax = 0.24 W/(m^2·K). (iv) Lightweight doors and façades with a value not exceeding Umax = 2 W/(m^2·K). (v) Windows with a frame not exceeding Umax = 1.5 W/(m^2·K) and with glazing not exceeding Umax = 1.1 W/(m^2·K). (vi) Use of renewable energy. The construction of this building is the result of a combination of all the previously developed scenarios. The results are given in Figure 7 below.

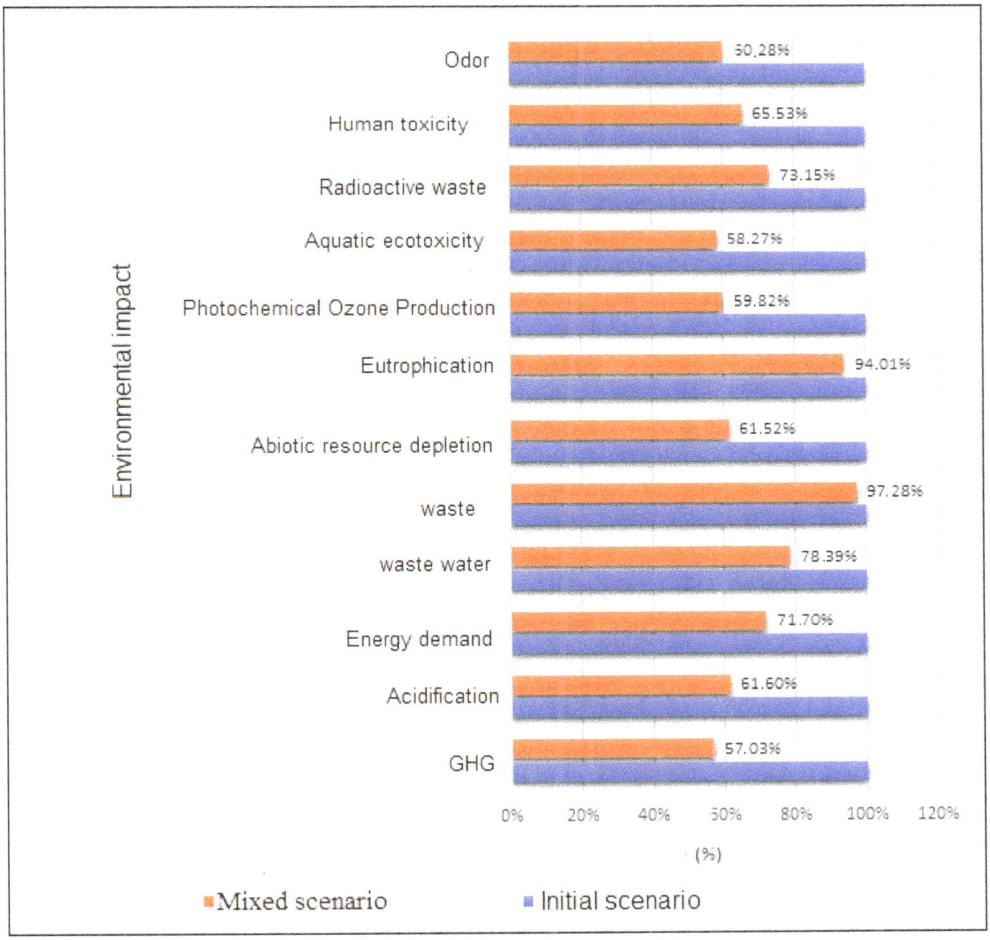

Figure 7. Comparison between the initial and optimal scenarios.

Figure 7 shows a significant decrease in the various environmental impacts of the building under the combined scenario compared to the initial scenario. The following reductions can be noted: 42.97% for the greenhouse effect, 38.4% for acidification, 28.30% for the cumulative energy demand, 21.61% for water used, 2.72% for inert waste produced, 38.48% for abiotic resource depletion, 5.99% for eutrophication, 40.18% for photochemical ozone production, 41.73% for aquatic ecotoxicity, 26.85% for radioactive waste, and 34.47% for human toxicity.

4. Discussion

The results obtained are consistent with those reported in previous research, as observed in the literature. Although the operational phase remains the most environmentally costly, its impact was reduced by nearly 20% compared to the initial scenario. As expected, the impacts associated with the construction and renovation phases increased by 8.44% and 104.35%, respectively, while the deconstruction phase showed a slight decrease of 0.35%. The overall environmental cost was also reduced by approximately 20%, which is close to the 30% reduction commonly reported in the literature [30]. Furthermore, Martin Bodmer observed a reduction in operational energy consumption ranging from 30% to 80% following renovation efforts [31]. In this study, operational energy consumption decreased from 90,925 kWh before renovation to 62,102.14 kWh after renovation, representing a reduction of 31.70%. These results are satisfactory and align well with previous studies. The implementation of this optimized building design resulted in an 18.82% reduction in the initial environmental cost, amounting to a total savings of USD 102,764.70 over 80 years or approximately USD 1285 annually. With this optimized design, the annual environmental cost decreased from USD 6825.50 to USD 5541. Overall, the optimized building scenario offers both financial savings and substantial environmental benefits.

4.1. Dominance of the Use Phase in the Life Cycle

The analysis reveals that the use phase dominates the building's environmental impacts, accounting for approximately 96.53% of the total life cycle impacts. This strong contribution is physically linked to the sustained energy consumption for heating, cooling, lighting, and appliances throughout the 80-year operational period. In addition, regular occupant transport to and from the building further amplifies energy and emissions footprints over time. These findings are consistent with earlier Life Cycle Assessments of residential and tertiary buildings, which typically report that the use phase represents between 80% and 90% of the total environmental burdens. For instance, a study evaluating passive houses highlighted that, despite improvements in the construction phase, the operational energy demand remained the predominant factor due to long-term energy use. Similarly, other researchers concluded that, unless renewable energy or passive systems are employed, the use phase will continue to drive the majority of environmental impacts. This alignment with the literature reinforces the need to prioritize operational efficiency in sustainable construction strategies and to consider passive design, high-performance envelopes, and optimized building systems.

4.2. Impact of the Sustainable Transportation Scenario

When a sustainable transportation scenario is introduced, the share of the use phase slightly decreases to 96.26%, yet noticeable reductions are observed across several indicators. Climate change-related emissions drop by 17.10%, while the acidification potential decreases by 12.74%, and photochemical ozone formation falls by 22.60%. These reductions are physically attributed to a lower dependence on individual motor vehicles and the greater use of public transport, bicycles, and walking.

Comparatively, the literature on urban mobility and low-carbon neighborhoods suggests that transportation behavior significantly influences the overall environmental impact.

Similar studies in European urban districts showed that shifting 30% of trips to public transit and cycling could cut CO_2 emissions by 15–25%, which aligns with the present results. The physical explanation lies in the high energy intensity of private car use compared to shared and non-motorized modes, particularly when the electricity grid is not fully decarbonized. This scenario confirms that promoting compact urban planning and investing in active transport infrastructure can complement building-level interventions.

4.3. Contribution of Renewable Energy

In the scenario where 69.29% of annual electricity consumption is met through on-site solar photovoltaic systems, the results show substantial environmental improvements. Greenhouse gas emissions are reduced by 26.72%, the cumulative energy demand by 21.88%, and abiotic resource depletion by 38.48%. This is largely due to the substitution of conventional grid electricity—often derived from fossil fuels—with clean, renewable power. The literature consistently emphasizes the benefits of integrating photovoltaics into building design. For instance, studies of nearly zero energy buildings across Europe have shown that PV systems can reduce operational emissions by up to 40%, depending on location and system efficiency. Another analysis revealed that self-generated renewable energy significantly reduces upstream environmental loads, including those related to resource extraction and pollutant generation. Thus, the present findings are in strong agreement with published results, underscoring the importance of on-site renewables in reducing long-term environmental impacts and enhancing the energy autonomy of buildings.

4.4. Synergy in the Combined Scenario

When both renewable energy and sustainable transportation measures are combined, their effects amplify. This scenario achieves the most favorable environmental profile, with a 42.97% reduction in greenhouse gas emissions, a 38.40% decline in acidification potential, and a 28.30% decrease in cumulative energy demand. Freshwater ecotoxicity and human toxicity indicators are also improved, with reductions of 41.73% and 34.47%, respectively. This synergistic improvement exceeds the individual contributions of the isolated measures, confirming the advantage of integrated strategies. Previous work in the field of sustainable urbanism shows similar synergies. Studies combining building-integrated renewables with behavioral changes and sustainable transport indicate total environmental improvements between 35% and 50%. This convergence validates the present methodology and suggests that system-level thinking—rather than single-solution approaches—is crucial in sustainability planning. The resulting emissions intensity of 32.57 kg CO_2-eq/m^2/year corresponds to a class D energy performance level. While this rating still leaves room for improvement, it represents a meaningful advance compared to the baseline and highlights the effectiveness of combined interventions in dense urban environments.

4.5. Comparative Summary with the Literature

The implementation of rooftop photovoltaic systems in Cameroon presents both promising economic opportunities and certain constraints. The average cost of installing a residential photovoltaic system in the country is currently estimated at USD 1000–1500 per kWp based on local market data and regional suppliers. For an average household requiring approximately 3 kWp, the initial investment would therefore range from USD 3000 to 4500. However, with an average solar irradiation of 5–6 kWh/m^2/day, such installations can generate a substantial amount of electricity, with a payback period estimated between 6 and 9 years, depending on electricity tariffs and the availability of subsidies. Few comparison of results were showed into the Table 11.

Table 11. Comparison of results with the literature.

Study/Reference	Main Approach	GHG Reduction Achieved (%)	Comparison with This Study
Ramesh et al. [32]	Thermal envelope improvement	20–30%	Comparable
Fthenakis et al. [33]	Photovoltaic integration	30–40%	Slightly lower
Creutzig et al. [34]	Urban design + sustainable mobility	20–25%	Comparable
This study (Combined scenario)	PV + Soft mobility + Compact urbanism	42.97%	Superior

Although there is a lack of strong national incentive policies, some local initiatives and international cooperation programs—such as those supported by GIZ, the European Union, or the African Development Bank—offer partial financing or low-interest loans to support energy transition. Furthermore, the continued global decline in the cost of solar equipment could significantly enhance the economic viability of these solutions in the near future.

Regarding sustainable transportation, the development of electric public transport networks or soft mobility systems (e.g., bike-sharing programs and pedestrian pathways) is currently hindered by inadequate urban infrastructure and budgetary constraints. Nevertheless, intermediate solutions such as the conversion of conventional motorcycle taxis to electric versions or the gradual introduction of hybrid buses in major cities like Douala or Yaoundé may represent a cost-effective alternative in the medium term. Although electric vehicles generally involve 30–50% higher upfront costs compared to internal combustion models, these are offset by lower operational and maintenance costs, as well as economies of scale, as fleets expand.

Ultimately, the success of such strategies will require the establishment of incentive-based regulatory frameworks, adapted fiscal policies, and increased public awareness of the long-term economic and environmental benefits of clean technologies. The adoption of sustainable practices such as rooftop photovoltaics and sustainable mobility in urban African settings, including Cameroon, cannot be fully understood without accounting for the complex social, behavioral, and cultural dynamics at play. Social perceptions and trust in technology often influence the uptake of renewable energy solutions. In many urban communities, solar technologies are still perceived as unreliable or suitable only for rural or low-income households. This perception can be reinforced by poorly maintained demonstration projects or a lack of technical support after installation. Similarly, the lack of awareness and environmental education means that many citizens are unaware of the long-term financial and ecological benefits of energy efficiency and renewable energy. On a behavioral level, daily habits and mobility routines are deeply entrenched. The preference for informal and flexible transportation options, such as motorcycle taxis ("benskin"), often outweighs the interest in structured public transport systems, especially when the latter are perceived as inefficient or unsafe. Even when alternatives exist, resistance to change—driven by convenience, status symbols associated with certain types of vehicles, or lack of perceived immediate benefit—can slow down adoption. From a cultural perspective, collective decision-making within households or communities can also delay technological uptake.

4.6. Limitations and Perspectives

While this study provides valuable insights into the environmental performance of the NENGOUE residence through Life Cycle Assessment, several limitations must be acknowledged:

- Data Representativeness: Some input data, especially for material quantities and operational energy use, were based on design assumptions and national averages due to the unavailability of measured data. This may affect the accuracy of the Life Cycle Inventory, particularly for transport behavior and occupancy profiles.
- Behavioral Variability: The adoption of sustainable transport modes relies heavily on user behavior, cultural acceptance, safety perceptions, and infrastructure availability. These factors introduce significant variability and uncertainty into the predicted environmental benefits of the transport scenario.
- Photovoltaic System Performance: The modeled efficiency of the PV system assumes optimal solar radiation and minimal maintenance degradation. In practice, performance can be affected by dust accumulation, shading, poor maintenance, and inverter losses, potentially reducing the anticipated energy generation and impact reductions.
- Generalizability: The study focuses on a single urban residential case in Douala, Cameroon. While it offers relevant insights, the findings may not be directly transferable to other building types or climatic contexts without adaptation.
- Tool and Database Limitations: The environmental impact results depend on the Ecoinvent database and the ReCiPe method. While robust, these tools reflect European datasets, which may not perfectly capture local construction practices or energy mix characteristics in Cameroon.
- Future work should address these limitations by incorporating field measurements, more detailed user surveys, and locally calibrated environmental datasets.

5. Conclusions

This study provided a comprehensive Life Cycle Assessment (LCA) of the NENGOUE residence, evaluating different mitigation scenarios to reduce its environmental impacts throughout all life stages—construction, operation, renovation, and deconstruction. The results clearly indicate that the use phase is, by far, the most environmentally impactful, contributing over 96% of the total emissions and impacts across all scenarios. This finding underscores the critical role of daily energy use and mobility behaviors in shaping a building's life cycle footprint. The first scenario, focused on low-impact transportation planning, led to a 17.10% reduction in greenhouse gas emissions and improvements in acidification (−12.74%) and photochemical ozone formation (−22.60%). These results demonstrate the importance of urban planning that promotes mixed use neighborhoods, pedestrian-friendly infrastructure, and public transport connectivity Policymakers should therefore encourage zoning regulations that reduce commuting distances while investing in bike lanes, green corridors, and affordable, efficient public transport systems to support behavioral shifts toward low-carbon mobility. The second scenario, which integrated rooftop photovoltaics, covered approximately 70% of the residence's annual electricity demand and led to a 26.72% reduction in greenhouse gas emissions and a 21.88% decrease in the cumulative energy demand. These findings support the implementation of incentive-based policies, such as feed-in tariffs, tax credits, or subsidized installation programs for residential solar PV systems. Urban planning regulations should also encourage or mandate the inclusion of renewable energy systems in new developments, particularly in sunny regions such as Cameroon.

The combined scenario (transport + photovoltaics) yielded the most substantial environmental gains, with a 42.97% decrease in greenhouse gas emissions and notable improvements across nearly all impact categories. This highlights the synergistic benefits of combining behavioral change with technological innovation.

Based on these findings, the study offers the following urban planning and policy recommendations to guide real-world implementation:

- Integrate LCA-based indicators into local and national urban planning guidelines to support low-carbon infrastructure choices.
- Promote compact urban design to shorten travel distances and enhance accessibility to essential services.
- Adopt building codes and incentives that mandate or encourage renewable energy integration and energy-efficient technologies in residential construction.
- Launch public awareness campaigns and education programs to influence energy and mobility behaviors at the household level.
- Develop financing mechanisms (e.g., green mortgages and microcredit for solar PV) to enhance affordability and access to sustainable solutions for low- and middle-income households. Achieving high environmental performance in residential buildings in Sub-Saharan African cities requires a multidimensional strategy, one that combines urban design, policy innovation, and behavioral change. Future work should expand this type of multi-scenario LCA approach to different urban contexts and building typologies to support data-driven policymaking for equitable and climate-resilient urban development.

Author Contributions: Conceptualization, M.K.N. and M.M.A.M.; methodology, M.K.N.; software, M.M.A.M.; validation, M.K.N.; formal analysis, M.K.N.; investigation, M.M.A.M.; resources, M.M.A.M.; data curation, M.K.N.; writing—original draft preparation, M.K.N.; writing—review and editing, M.K.N. and M.M.A.M.; visualization, M.K.N.; supervision, M.K.N.; project administration, M.M.A.M. All authors have read and agreed to the published version of the manuscript.

Funding: This research received no external funding.

Data Availability Statement: The data presented in this study are available on request from the corresponding author.

Conflicts of Interest: The authors declare no conflicts of interest.

References

1. GlobalABC. Rapport Sur La Situation Mondiale Des Bâtiments Et De La Construction En 2020: Vers Un Secteur Des Bâtiments Et De La Construction à émission Zéro, Efficace Et Résilient. 2020. Available online: https://globalabc.org/sites/default/files/2021-01/Buildings-GSR-2020_ES_FRENCH.pdf (accessed on 10 June 2024).
2. ONU Programme Pour L'environnement. Programme Des Nations Unies Pour. Stratégie Du Programme Des Nations Unies Pour L'environnement Pour 2022–2025 Visant à Remédier Aux Problèmes Des Changements Climatiques, De La Dégradation De La Nature Et De La Pollution. 2022. Available online: https://www.unep.org/fr/resources/pour-lhumanite-et-pour-la-planete-strategie-du-pnue-pour-2022-2025 (accessed on 14 August 2024).
3. Compte CO_2. Les COP: Une Brève Histoire De La COP1 à La COP27. 2024. Available online: https://www.compteco2.com/article/historique-cop-conference-des-parties (accessed on 23 April 2024).
4. Nations Unies. Promouvoir Le Développement Durable. 2024. Available online: https://www.un.org/fr/our-work/support-sustainable-development-and-climate-action (accessed on 23 April 2024).
5. Union Européenne. Directive 2009/125/CE Du Parlement Européen Et Du Conseil Du 21 Octobre 2009 Établissant Un Cadre Pour La Fixation D'exigences En Matière D'écoconception Applicables Aux Produits Liés à L'énergie. 2009. Available online: https://eurlex.europa.eu/LexUriServ/LexUriServ.do?uri=OJ:L:2009:285:0010:0035:fr:PDF (accessed on 5 August 2024).
6. République Française. Loi N° 2015-992 Du 17 Août 2015 Relative à La Transition Énergétique Pour La Croissance Verte. 17 Août 2015. Available online: https://www.legifrance.gouv.fr/loda/id/JORFTEXT000031044385 (accessed on 5 August 2024).
7. Union Economique et Monétaire Ouest Africaine. La Directive N° 04/2020/CM/UEMOA Portant étiquetage énergétique Des Lampes électriques Et Des Appareils électroménagers. 26 Juin 2020. Available online: https://www.pseau.org/outils/biblio/resume.php?d=11833 (accessed on 5 August 2024).

8. Peuportier, B.; Popovici, E.; Troccmé, M. Analyse du cycle de vie à l'échelle du quartier, bilan et perspectives du projet ADEQUA. *Build. Environ.* **2013**, *3*, 17.
9. Hunt, R.; Franklin, W.; Hunt, R. LCA—How it came about. *Int. J. Life Cycle Assess.* **1996**, *1*, 4–7. [CrossRef]
10. Bribian, I.Z.; Uson, A.A.; Scarpellini, S. Life cycle assessment in buildings: State-of-the-art and simplified LCA methodology as a complement for building certification. *Build. Environ.* **2009**, *44*, 2510–2520. [CrossRef]
11. Jolliet, O.; Saadé, M.; Crettaz, P. *Analyse Du Cycle De Vie: Comprendre Et Réaliser Un écobilan*; Presses Polytechniques Et Universitaires Romandes: Lausanne, Switzerland, 2005; p. 242.
12. *ISO 14040:2006*; Environmental Management—Life Cycle Assessment—Principles and Framework. ISO: Geneva, Switzerland, 2006.
13. *ISO 14044:2006*; Environmental management—Life Cycle Assessment—Requirements and Guidelines. ISO: Geneva, Switzerland, 2006.
14. Reiter, S. Life cycle assessment of buildings—A review. In Proceedings of the Arcelor-Mittal International Network in Steel Construction, Sustainable Workshop and Third Plenary Meeting, Bruxelles, Belgium, 7 July 2010; p. 1.
15. Buyle, M.; Braet, J.; Audenaert, A. Review on LCA in the Construction Industry: Case Studies. *Math. Model. Simul. Appl. Sci.* **2013**, 98–104. Available online: http://www.wseas.us/e-library/conferences/2012/Rovaniemi/INEE/INEE-16.pdf (accessed on 18 June 2025).
16. Rossi, B.; Marique, A.F.; Glaumann, M.; Reiter, S. Life-cycle assessment of residential buildings in three different European locations, basic tool. *Build. Environ.* **2012**, *51*, 395–401. [CrossRef]
17. Citherlet, S.; Defaux, T. Energy and environmental comparison of three variants of a family house during its whole life span. *Build. Environ.* **2007**, *42*, 591–598. [CrossRef]
18. Thormark, C. A low energy building in a life cycle—Its embodied energy, energy need for operation and recycling potential. *Build. Environ.* **2002**, *37*, 429–435. [CrossRef]
19. Thormark, C. The effect of material choice on the total energy need and recycling potential of a building. *Build. Environ.* **2006**, *41*, 1019–1026. [CrossRef]
20. Blengini, G.A.; Di Carlo, T. The changing role of life cycle phases, subsystems and materials in the LCA of low energy buildings. *Energy Build.* **2010**, *42*, 869–880. [CrossRef]
21. Erlandsson, M.; Levin, P. Environmental assessment of rebuilding and possible performance improvements effect on a national scale. *Build. Environ.* **2005**, *40*, 1459–1471. [CrossRef]
22. Nematchoua, M.K.; Sadeghi, M.; Reiter, S. Strategies and scenarios to reduce energy consumption and CO2 emission in the urban, rural and sustainableneighbourhoods. *Sustain. Cities Soc.* **2021**, *72*, 103053. [CrossRef]
23. Groupe Kedibuild. Situation énergétique Au Cameroun. 2021. Available online: https://drive.google.com/file/d/14ppOS44StkpGKno03s5OotWBHplBMNWY/view (accessed on 10 October 2024).
24. Ministry of Water and Energy. LA-Paulitique-Nationale-de-LEau-au-Cameroun. 2019. Available online: https://minee.cm/wp-content/uploads/2024/02/LA-Paulitique-Nationale-de-lEau-au-cameroun-30-nov-2019-SGPR.pdf (accessed on 10 August 2024).
25. Ministry of the Environment, Nature Protection and Sustainable Development. National Waste Management Strategy. 2021. Available online: https://minepded.gov.cm/2021/09/29/strategie-national-de-gestion-des-dechets/ (accessed on 10 August 2024).
26. Nematchoua, M.K.; Reiter, S. Evaluation of bioclimatic potential, energy consumption, CO_2-emission, and life cycle cost of a residential building located in Sub-Saharan Africa; a case study of eight. *Solar Energy* **2021**, *218*, 512–524. [CrossRef]
27. Kemajou, A.; Tseuyep, A.; Egbewatt, N.E. Le confort thermique en climat tropical humide vers un réaménagement des normes ergonomiques. *J. Renew. Energ.* **2012**, *15*, 427–438. [CrossRef]
28. Smith, V.T. Eutrophication; Impacts of Excess Nutrient Inputs on Freshwater, Marine and Terrestrial Ecosystem. *J. Environ. Pollut.* **1999**, *100*, 179–196. [CrossRef] [PubMed]
29. Nematchoua, M.K.; Reiter, S. Analysis, reduction and comparison of the life cycle environmental costs of an eco-neighborhood in Belgium. *Sustain. Cities Soc.* **2019**, *48*. [CrossRef]
30. Effy Connect. Comment Bien Choisir Son Isolant Thermique? 2022. Available online: https://www.effy.fr/travaux-energetique/isolation/isolant (accessed on 9 October 2024).
31. Buyle, M.; Braet, J.; Audenaert, A. Life cycle assessment in the construction sector: A review. *Renew. Sustain. Energy Rev.* **2013**, *26*, 379–388. [CrossRef]
32. Ramesh, T.; Prakash, R.; Shukla, K.K. Life cycle energy analysis of buildings: An overview. *Energy Build.* **2010**, *42*, 1592–1600. [CrossRef]
33. Fthenakis, V.M.; Kim, H.C.; Alsema, E. Emissions from photovoltaic life cycles. *Environ. Sci. Technol.* **2008**, *42*, 2168–2174. [CrossRef] [PubMed]
34. Creutzig, F.; Jochem, P.; Edelenbosch, O.Y.; Mattauch, L.; van Vuuren, D.P.; McCollum, D.; Minx, J. Transport: A roadblock to climate change mitigation? *Science* **2015**, *350*, 911–912. [CrossRef] [PubMed]

buildings

MDPI

Article

Developing a Framework for Building Condition Assessment of Schools in Osijek-Baranja County

Hana Begić Juričić * and Hrvoje Krstić

Faculty of Civil Engineering and Architecture Osijek, Josip Juraj Strossmayer University of Osijek, 31000 Osijek, Croatia; hrvoje.krstic@gfos.hr
* Correspondence: hbegic@gfos.hr

Abstract: This study introduces a novel approach to building condition assessment (BCA) by combining traditional manual grading with machine learning models—artificial neural networks (ANNs) and random forests (RFs). Individual building components (e.g., windows, roofs, and floors) were assessed based on their remaining useful life using an Excel-based system. The resulting total building grades were used to train and validate ANN and RF models. Performance was evaluated using R^2, mean squared error (*MSE*), root mean squared error (*RMSE*), coefficient of variation of *RMSE* (*CVRMSE*), and mean absolute percentage error (*MAPE*). The ANN model outperformed RF in the training set (R^2 = 0.987, *MAPE* = 0.50%) and showed high accuracy in validation (R^2 = 0.940, *MAPE* = 2.55%). The RF model also performed well (R^2 = 0.942, *MAPE* = 2.66%), confirming its viability. External validation on data from outside Osijek-Baranja County confirmed model robustness, with ANN again achieving better performance (R^2 = 0.799, *MAPE* = 7.71%) than RF (R^2 = 0.747, *MAPE* = 9.17%).

Keywords: building condition assessment; facilities management; school buildings; built environment

1. Introduction

Osijek-Baranja County school buildings, despite their significant economic, cultural, and historical importance, are showing signs of degradation. This is mostly due to a lack of accurate and comprehensive information about their condition, which is essential for effective building facilities maintenance management [1,2]. Also, the literature states that 90% of the European building stock was built before 1990 [3,4]. Furthermore, studies emphasize that a significant portion of the participants involved in the construction sector, including designers, contractors, suppliers, and owners, are expending a substantial sum of money in their search for, verification of, and recreation of facilities information that should already be easily accessible [5]. Additionally, a significant portion, namely 75–80%, of the expenses associated with a building are incurred throughout the operational and maintenance stage of its life cycle [6,7]. Moreover, it is commonly recognized that the operational and maintenance phase of the building life cycle has the most significant environmental impact [8].

Building degradation is quantitatively assessed by building condition assessments (BCAs), which collect information to determine the timing and kind of predictive, preventative, or corrective measures required to uphold the desired level of performance [8]. Considering that costs tend to increase consistently due to either inefficient maintenance methods or a lack of routine maintenance, this is a significant issue, particularly for buildings with a projected lifespan of 50 years or more [8–11]. BCA is the initial step in the asset management and maintenance process. It involves evaluating the physical components and conditions of buildings to determine appropriate maintenance and improvement methods [12,13]. There are various definitions of a building condition assessment, but it can be said that in general, a BCA is an assessment that identifies significant deficiencies for all of the systems in a facility [14]. Multiple methodologies exist for conducting a BCA since it can also be conducted at the component level, wherein each significant piece of equipment is assessed, and the value and remaining service life of the equipment are evaluated [14]. Overall, it has been noticed that a significant obstacle to the advancement of an effective BCA procedure is the inherent subjectivity and resulting lack of precision [14].

Facility managers make decisions on maintenance and renovation strategies based on limited information about the current condition of the building facilities' components [2,15]. The limited data available are inadequate for monitoring and predicting indicators of wear, tear, and aging in building facilities components [16]. This limits efforts to improve benchmarking outcomes and identify significant changes in circumstances that may impact the efficiency of these components. This problem emphasizes the need for a condition assessment framework that can provide a proactive representation of the repair and maintenance needs [17]. Given the critical role that building conditions play, this assessment process is particularly significant for educational institutions since the condition of school buildings is a crucial factor in influencing the overall school environment, impacting not only the safety and health of students and staff but also the quality of education and student performance [18]. Therefore, the condition of these buildings must be appropriately and thoroughly evaluated to ensure they provide a conducive, secure, and supportive learning atmosphere [19]. It is even reported that the school building condition is connected to school absenteeism due to the poor conditions [20].

This study focuses on evaluating the condition of school buildings, a critical process in the life cycle of comprehensive asset and facility management. The primary goal of this paper is to identify school buildings that are a high priority for renovation due to their poor conditions. Instead of only grading multiple components, this study seeks to also assign a single overall grade to each building using advanced machine learning models, which allows for a clearer assessment of their overall condition. This evaluation is crucial to ensure that the buildings can effectively support the school's core operations and provide a quality learning environment for all users.

The remainder of this paper is organized as follows: Section 2 presents a literature review of conducted research on BCAs regarding various educational buildings; Section 3 presents the methodology used for developing the BCA framework; Section 4 provides the analysis of the school sample; Section 5 presents the Excel-based element-level grading system; Section 6 presents the developed ANN and RF models; Section 7 provides the results; Section 8 presents the validation of the developed framework on an external set of schools; in Section 9, a discussion is presented; and in Section 10, conclusions are made.

2. Overview of BCA Research in Educational Buildings

Evaluating the condition of educational buildings is essential for efficient resource allocation, ensuring safety, and prioritizing maintenance. The literature has explored a variety of methodologies ranging from matrix-based approaches to machine learning models, each contributing to advancements in building condition assessment (BCA). Matrix-based frameworks have been widely employed in building evaluations. For example, the CSP1 Matrix proved useful in analyzing the conditions of smart schools in Malaysia. It demonstrated reliability in identifying deficiencies but had limitations in generating detailed inspection reports [21,22]. Kejeh et al. extended this idea with a 3×3 matrix in Nigeria. For the facility's present condition, ten key elements were used in the evaluation process.

These elements included whether the building appeared to be in good condition, whether the water system was functional, whether the waste system was operational, and other similar criteria. Each element was rated on a scale from 1 to 3, based on its current state. The average of these ten scores represented the overall present condition score of the facility. To assess the maintenance state of the building, a 3 × 3 matrix—a simple table consisting of three rows and three columns—was used. This matrix cross-tabulates the facility's present condition score with its maintenance practice score to categorize the overall maintenance state. Once both scores are determined, the corresponding position in the matrix indicates the maintenance status of the building [23]. In Croatian schools, a similar structured approach was adopted. Tijanić Štrok et al. developed a division model based on visual inspections and systematic questionnaires, revealing that mechanical systems, particularly cooling systems, were in critical condition. This study emphasized the financial constraints affecting building maintenance in Croatian schools [24]. Decision-making models such as the Analytic Hierarchy Process (AHP) and Analytical Network Process (ANP) have added quantitative rigor to BCA evaluations. Marzouk and Awad applied an AHP-fuzzy model to Egyptian schools, producing comprehensive indexes for various building components such as flooring, windows, and doors [25]. Similarly, Eweda et al. used AHP and ANP in Canada to develop condition indexes for school spaces, integrating physical and environmental factors into assessments. Their assessment framework incorporated both physical and environmental factors. The environmental factors considered included indoor air quality, thermal comfort, lighting adequacy, noise levels, and overall classroom ambiance—elements that influence the learning environment and occupant well-being. Based on the evaluation, buildings were assigned performance grades ranging from A (excellent) to F (failure) [26–28]. Automation in BCA has gained interest. Ahluwalia proposed a framework for Toronto schools that incorporated digital drawings, field data automation, and capital replacement indicators. This approach reduced inspection time and costs while enhancing data accuracy [13]. Similarly, Elhakeem and Hegazy developed a visual guidance system that supported staff with limited expertise, improving efficiency and reducing subjectivity [29]. Machine learning models have significantly advanced BCA by enabling predictive analytics and automating evaluations. Mohamed and Marzouk combined ANN with Structural Equation Modelling (SEM) to assess building components in Cairo. Their methodology identified critical maintenance factors and prioritized repairs based on their importance [2]. Hassan et al. demonstrated ANNs' robustness in predicting the deterioration of educational building elements, achieving high accuracy in case studies [30]. Sadick and Issa employed statistical and environmental models to assess schools, finding that relative humidity was a critical factor influencing building envelope condition [31]. The reviewed literature illustrates the evolution of BCA methodologies, from manual matrix-based frameworks to automated, machine learning-driven approaches. Each method offers unique advantages and limitations. Matrix-based frameworks are simple to use and provide an accessible way to assess building conditions, but they may oversimplify the evaluation process and lack precision. Decision-making tools like AHP and ANP introduce more quantitative depth, allowing for systematic assessments, but they rely on subjective judgment and can be computationally demanding. Machine learning approaches, such as artificial neural networks (ANNs) and random forests (RFs), enhance predictive precision by analyzing large datasets and identifying complex patterns, yet they require extensive data and computational resources and can be harder to interpret compared to traditional methods. Transitioning to machine learning techniques offers the potential to overcome some of the limitations of the earlier approaches, providing more accurate and dynamic assessments of building conditions. As is well known in the field of BCA, the assessment can be conducted in a variety of ways. It can be conducted at the component level, where each element is assessed to determine its worth and remaining lifespan, while it can also be conducted at the building level, where the emphasis is on evaluating the state of the whole building and assigning a value to it, as opposed to evaluating each of its constituent parts [32]. To the best of the authors' knowledge, ANNs have been used in only three papers regarding BCA [2,30,33], while RFs have not been applied to BCA yet. This is likely due to a combination of factors, including the technical limitations of earlier data collection methods, the complexity of implementing RFs in this context, and the lack of exploration into their potential applicability. Despite their extensive use in the field of energy efficiency in buildings [34–39], RFs' ability to handle complex, non-linear relationships in building data might not have been fully recognized or leveraged in BCA studies until now.

While the primary focus of this study is on the structural and material condition of building elements, it is important to note that building condition assessment is often interlinked with indoor environmental quality. Among these, thermal comfort plays a significant role in occupant satisfaction and overall building performance. Thermal comfort has emerged as a critical aspect of indoor environmental quality (IEQ), particularly in educational buildings where occupant satisfaction and performance are closely linked to environmental conditions. Several studies emphasize that thermal comfort is influenced by a combination of structural, environmental, and human factors, all of which significantly impact building energy use and occupant behavior pattern [40]. Within higher education settings, thermal comfort is consistently cited as one of the least satisfactory IEQ parameters, despite growing alignment with international green building standards [41]. The recent literature highlights the integration of facility management, smart technologies, and sustainable practices as essential for maintaining optimal thermal environments in learning spaces [42]. Strategies such as adaptive thermal control models and AI-driven management systems have demonstrated improvements in both energy efficiency and user comfort in school buildings [43]. Moreover, thermal comfort inconsistencies due to seasonal ventilation challenges, such as those observed during and after COVID-19, suggest the need for more dynamic and user-centered approaches to indoor climate control [44]. These findings support the inclusion of thermal comfort considerations in building condition assessments, as they directly affect the usability and sustainability of educational facilities [45].

This study aims to advance BCA methodologies by addressing gaps in the literature, specifically regarding the application of machine learning techniques. While traditional BCA methods like matrix-based frameworks and decision-making tools offer reliability and accessibility, machine learning has demonstrated potential to enhance predictive accuracy and reduce subjectivity. Despite the promise of these techniques, ANNs have seen limited application in BCA, and RFs have not been utilized for this purpose to date. In this study, a systematic Excel-based model will be developed to grade each building component individually, providing a detailed and transparent method for component-level evaluation which will aggregate these grades to derive an overall building condition grade. These grades will serve as targets to train the ANN and RF prediction models of the total school grade, not grading each element separately. To evaluate the effectiveness of machine learning approaches, ANN and RF model accuracy will be compared in predicting the total building grade. Accuracy is chosen as the primary metric because it provides a straightforward measure of the overall correctness of the models in predicting the total building grade.

The methodology combines the strengths of manual grading with advanced computational techniques, aiming to accomplish the following: 1. assess the performance of an ANN and RF in handling complex datasets and predicting overall grades; 2. provide insights into the comparative advantages and limitations of these machine learning models in BCA; 3. address the gap in the literature regarding the use of RFs in BCA, contributing novel findings to the field.

3. Methodology

3.1. Research Area: Osijek-Baranja County in Croatia

Osijek-Baranja County, located in eastern Croatia, presents a unique setting for evaluating the condition of school buildings within the context of Croatia's Long-Term Strategy for National Building Stock Renovation by 2050 [46]. This strategy, which aims to improve the energy performance, sustainability, and overall condition of the building stock in Croatia, is critical for addressing both the challenges of aging infrastructure and the need for energy efficiency improvements in public buildings, including schools. In this context, BCA plays a crucial role. With a significant portion of Croatian buildings being older, many need substantial renovations to meet current standards for energy efficiency and safety. The region of Osijek-Baranja County, with its mix of urban and rural areas, reflects this challenge. Urban areas such as Osijek have a higher concentration of older buildings that are likely to require more urgent renovations and ongoing maintenance. On the other hand, rural schools may face difficulties with less frequent access to resources for repairs and assessments, making it harder to address deteriorating conditions. Croatia's long-term renovation strategy highlights the necessity of a systematic, nationwide approach to building renovation. The BCA process provides a structured, empirical way to monitor the condition of school buildings and prioritize those that require immediate attention. The region, with its mix of historical schools and newer educational institutions, faces the ongoing challenge of keeping buildings in good condition to ensure a safe and conducive learning environment.

3.2. Data Collection

The Ordinance on Systematic Energy Management in the Public Sector (Official Gazette 18/2015) was adopted in 2015 which prescribes the obligation to manage energy and water consumption, the analysis of consumption, the method of reporting on energy and water consumption, and the methodology of systematic energy management in the public sector [47]. Also, the Energy Management Information System (EMIS) was developed to provide support for strategic planning of energy and sustainable management of energy resources in buildings that are owned or used by cities, counties, or the Government of the Republic of Croatia, as well as in buildings of other government budgetary and extra-budgetary users and public authorities [48]. The authors requested these data from the EMIS department for research and obtained them in December 2023. Besides the data on the energy consumption of each building, the EMIS system also contains information regarding the year of construction, constructive elements of buildings, and their specifications. Every building has information regarding door and window material, external wall material, roof construction material, roof cover material, and floor material. Since the data that are input depend on the responsibility of building managers, specific information was found to be missing for some buildings. To address this issue, the missing data were filled out using the Croatian Methodology of Energy Audit of Buildings, which provides a standardized approach to identifying the characteristic constructive elements associated with specific construction periods [49]. This methodology was chosen because it offers a reliable and systematic framework for inferring missing building details based on the year of construction, ensuring that the imputed data are consistent with commonly used materials and construction practices from that time period. By relying on a national, officially recognized methodology, the imputation process ensures a high level of accuracy and consistency in the missing data. Furthermore, it minimizes subjectivity and potential errors that might arise from assumptions made without a standardized reference, thereby enhancing the overall reliability of the data used in the analysis.

Therefore, according to the building's construction period, the missing information was filled with the characteristic materials used in that period.

Since the material is provided for each element, their lifetime was analyzed using the values of the sustainable use life of building elements. These values were obtained from the tables of sustainable use life of building elements provided in the Croatian Ordinance on property valuation methods (Official Gazette 105/15) [50]. These data were collected for 166 primary and secondary schools in Osijek-Baranja County in Croatia.

Using the provided sustainable use life for schools [50] and their year of construction, the remaining lifetime of each school was calculated by subtracting the construction year from the corresponding sustainable use life value. Similarly, the remaining lifetime of each building element was calculated using the material-specific sustainable use life values in relation to the school's construction year. In cases where a specific building component had been renovated, the year of renovation was used instead of the construction year to reflect the updated condition and extend the component's remaining life accordingly. While this approach provides a standardized and practical means of estimating remaining lifespans, it does not explicitly account for variable degradation rates or environmental influences such as humidity, exposure to harsh weather, or maintenance frequency.

The building components considered for rating the building condition were the age of the building, doors, windows, external walls, roof, roof cover, and floor. The sustainable use life of each evaluated component is presented in Table 1 [50].

Table 1. Sustainable use life of evaluated components [50].

Component	Material	Sustainable Use Life [Years]
Age of school	-	70
Windows	Wood	40
	PVC	50
	Aluminum	50
	Mixed	45
Roof construction	Concrete	100
	Wood	120
Roof coverage	Tin	35
	Salonite	40
	Roofing tile	50
	Bitumen	30
External doors	Wood	40
	PVC	50
	Aluminum	50
	Mixed	45
External walls	Reinforced concrete	70
	Reinforced concrete and brick	80
	Brick	90
Floor construction	Brick	90
	Reinforced concrete	100
	Reinforced concrete and brick	95
	Wood	80

Since there are various remaining lifetimes of each of the analyzed components, the scale used to rate each of them is presented in Table 2.

Table 2. Condition scale used for building component evaluation.

Condition Category	Remaining Lifetime	Linguistic Representation
1	<0%	failed
2	0–25%	poor
3	25–50%	fair
4	50–75%	good
5	75–100%	excellent

Using the provided condition scale, the total building condition was calculated as a geometric mean of the conditioned categories obtained for each component. It is essential to clarify that condition category 1, which includes a remaining lifetime of less than 0%, concerns the components that have exceeded their sustainable use life.

4. Analysis of the School Sample

A descriptive statistical analysis regarding the year of construction and renovation, together with the age of both, was conducted using TIBCO Statistica® 14.1.0 [51], for the school sample and is presented in Table 3.

Table 3. Descriptive statistical analysis of the construction and renovation age for the school sample.

Variable	Valid N	Mean	Minimum	Maximum	Lower Quartile	Upper Quartile	St. Dev.
Year of construction	166	1972.61	1900	2022	1959	1994	27.82
Year of last renovation	123	2012.33	1991	2023	2006	2019	7.64
Age since construction	166	51.39	2	124	30	65	27.82
Age since renovation	123	11.68	1	33	5	18	7.64

The descriptive statistical analysis of the construction and renovation dates for the sample of schools reveals significant insights into the age and maintenance of educational facilities. The dataset comprises 166 schools with a mean construction year of 1972.61, indicating that many schools were built around the early 1970s. The oldest school dates to 1900, while the most recent one was constructed in 2022, showing a broad range of building ages. The lower quartile value of 1959 suggests that 25% of the schools were built before this year, and the upper quartile value of 1994 indicates that another 25% were constructed after this date. The high standard deviation of 27.82 years reflects the wide variation in construction years.

Regarding renovations, data are available for 123 schools, with the average last renovation occurring in 2012.33. Renovations span from 1991 to 2023, highlighting schools that have not been renovated in decades and those recently renovated. The lower quartile of 2006 and the upper quartile of 2019 values suggest that many schools underwent renovations relatively recently or a considerable time ago. The mean age since construction is 51.39 years, with schools ranging from 2 to 124 years old, and the standard deviation of 27.82 years again points to a considerable range. In contrast, the mean age since the last renovation is 11.68 years, ranging from 1 to 33 years, with a standard deviation of 7.64 years, less dispersed than the construction data. This underscores the diversity in the ages of school buildings, with some having been recently built or renovated and others potentially already overdue for renovation. This disparity highlights the need for ongoing maintenance and possibly increased investment in renovating older schools to ensure all facilities meet contemporary educational and safety standards.

Also, in Table 4, a descriptive statistical analysis regarding the remaining sustainable life use for the evaluated components was conducted for the school sample.

Table 4. Descriptive statistical analysis of the remaining sustainable life of use for the school sample.

Variable—Remaining Sustainable Life of Use	Valid N	Mean	Minimum	Maximum	Lower Quartile	Upper Quartile	St. Dev.
Age of building	166	18.61	−54	68	5	40	27.82
Windows	166	11.13	−84	46	−5	34	30.17
Roof construction	166	64.64	−4	110	49	86	28.10
Roof coverage	166	3.10	−74	46	−20	27	29.60
External doors	166	8.56	−84	46	−6	31	29.26
External walls	166	36.93	−34	88	23	59	27.65
Floor construction	166	46.81	−44	98	35	70	32.18

The analysis in Table 4 provides a detailed examination of the remaining sustainable life of various components within the school sample, offering critical insights into these structures' maintenance needs and longevity. The "Age of Building" category, which includes data for 166 schools, shows an average remaining sustainable life of 18.61 years. However, the data reveal a stark disparity with a minimum value of −54 and a maximum of 68, indicating that some buildings are already significantly beyond their expected sustainable life. In contrast, others still have considerable time left. The lower quartile is five years, meaning that 25% of the buildings have five or fewer years of sustainable life remaining, while the upper quartile is 40 years, indicating that another 25% have more than 40 years left. The high standard deviation of 27.82 underscores the broad variation in the data. For specific components, the remaining sustainable life varies widely: windows average 11.13 years but range from −84 to 46 years, with a high standard deviation of 30.17, suggesting some are long overdue for replacement. Roof construction has the most substantial average remaining life at 64.64 years, with a minimum of −4 and a maximum of 110 years and a lower quartile of 49 and an upper quartile of 86 years, indicating generally good longevity.

In contrast, roof coverage is more delicate, with an average remaining life of only 3.10 years and a range from −74 to 46 years, reflecting potential issues requiring urgent attention. External doors and walls show a similar pattern of variance, with means of 8.56 and 36.93 years, respectively, and significant standard deviations (29.26 for doors and 27.65 for walls), highlighting diverse conditions across the sample. Floor construction has a relatively stable perspective, with an average of 46.81 years remaining but, again, a wide range (−44 to 98 years) and a high standard deviation of 32.18 point to substantial variability. These statistics collectively suggest that while some components of the school buildings are in good condition, others are significantly degraded and need immediate repair or replacement, underlining the importance of targeted maintenance and renovation strategies.

5. Excel-Based Element-Level Grading

After the remaining lifetime of each component was determined, it was possible to grade each component and obtain each school's total grade as an average mean of all component grades. Table 5 shows the frequency of each condition category from 1 to 5 among each component.

The data reveal diverse conditions across the components. For instance, the building age shows many buildings in the best (63 in category 5) and worst (29 in category 1) conditions, indicating a mix of well-maintained and deteriorated structures. Similarly, the condition of windows varies, with a notable number in the worst condition category (47) but also a considerable amount in the good condition (45 in category 4). Roof construction and roof coverage exhibit more concerning distributions, particularly in the lower categories, suggesting widespread issues with these components; notably, 68 instances of roof coverage fall into category 1, implying a poor condition. The frequency of categories for external doors and walls also demonstrates variability, with substantial numbers in both the lower and upper condition categories. Floor construction appears somewhat better, with a higher frequency in the mid-to-upper categories (50 in category 3 and 63 in category 4). These distributions highlight the uneven state of maintenance across different building components, emphasizing the need for targeted interventions to address specific issues and enhance the overall condition of school facilities.

Table 5. Frequency of condition categories among components.

Condition Category	Age of Building	Windows	Roof Construction	Roof Coverage	External Doors	External Walls	Floor Construction
1	29	47	5	68	51	17	18
2	29	23	13	17	30	22	7
3	27	20	41	21	20	49	50
4	18	45	67	39	42	56	63
5	63	31	40	21	23	22	28

Furthermore, using the component grades, the total grade of each school was calculated, and the results of the total grade calculation are presented in Figure 1 and Table 6.

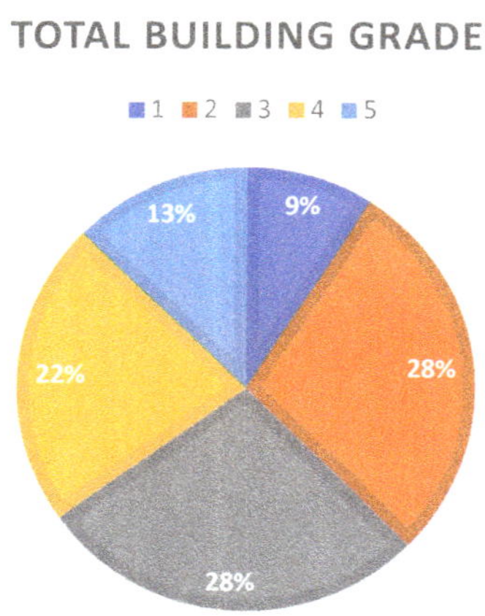

Figure 1. Share of each grade among the analyzed schools.

Table 6. Frequency of each grade among the analyzed schools.

Condition Category	Total School Grade
1	15
2	47
3	46
4	37
5	21

The distribution of total grades across the schools, categorized from 1 to 5, provides an overview of the overall condition of the school buildings as assessed by the various component grades. According to the data, 15 schools fall into the lowest grade of 1, indicating significant deficiencies across multiple building components. The largest group of schools, with a total grade of 2, comprises 47, suggesting that a substantial portion of the facilities are in poor condition but not at the worst level. The category with a total grade of 3 includes 46 schools, representing a middle where facilities are generally in fair condition but still have room for improvement. Meanwhile, 37 schools received a total grade of 4, indicating that these buildings are in relatively good condition, with most components likely well-maintained. The highest-grade category, 5, consists of 21 schools, showing a smaller group of facilities in excellent condition overall. This distribution highlights a diverse range of facility conditions, with a notable concentration of schools in the lower-to-mid-range categories, pointing to widespread needs for improvement and maintenance. These total grades are further used as a database in the next section to train the ANN and RF models to predict the total school grade.

6. Developed Models for Predicting the Total Grade

6.1. Variables

For the development of the ANN and RF models to predict the total school building grade, input variables are determined to be the remaining lifetimes of each element, without grading each specific element. The input and output variables are shown in Table 7.

Table 7. Input and output variables.

Type of Variable		Variable—Remaining Sustainable Life of Use	Label	Unit
Input	Continuous	Age of building	*Age*	year
		Windows	*Win*	
		Roof construction	*Roc*	
		Roof coverage	*Rcv*	
		External doors	*Drs*	
		External walls	*Wal*	
		Floor construction	*Flr*	
Output	Categorical	Building total grade	*Btg*	{1, 2, 3, 4, 5}

The remaining lifetimes of elements showed high correlation with the output variable as can be seen in Table 8.

Table 8. Correlation of model variables.

Variable	*Age*	*Win*	*Roc*	*Rcv*	*Drs*	*Wal*	*Flr*
Btg	0.834382	0.714653	0.820049	0.666845	0.823626	0.792957	0.739887

It is essential to observe that a positive value signifies a positive correlation, indicating that an increase in one variable results in an increase in the other [52]. Conversely, a negative value signifies a negative correlation, indicating that an increase in one variable will decrease the other. However, in this case, there are no such relationships. All of the correlations are very significant, as indicated by their high values and the red markings. These noted correlations have a significance level of $p < 0.05$.

In this study, a crucial step was taken to ensure the reliability of the developed models. The initial data set was split into two sets using a random process: the training set, which was used to create or develop a model, and the validation set, which was used solely to validate the developed models' error assessment predictions. In the training set, 117 school buildings' worth of data were chosen at random, accounting for 70.5% of the total of 166 school buildings, and 49 school buildings were in the validation set, or 29.5% of the total. Similar ratios were also used in [53,54].

6.2. ANN Model

A multilayer perceptron (MLP), a type of ANN, was employed to construct the model. MLPs consist of multiple layers of interconnected nodes, typically including an input layer, one or more hidden layers, and an output layer. Each node within a layer is connected to every node in the adjacent layer, enabling the model to learn complex relationships by adjusting the weights between the nodes through training. This architecture allows MLPs to capture non-linear patterns in data, making them particularly effective for tasks like forecasting and classification [55]. Due to the nature of the output variables, a classification ANN algorithm was used. Figure 2 shows the architecture of the optimal selected ANN model.

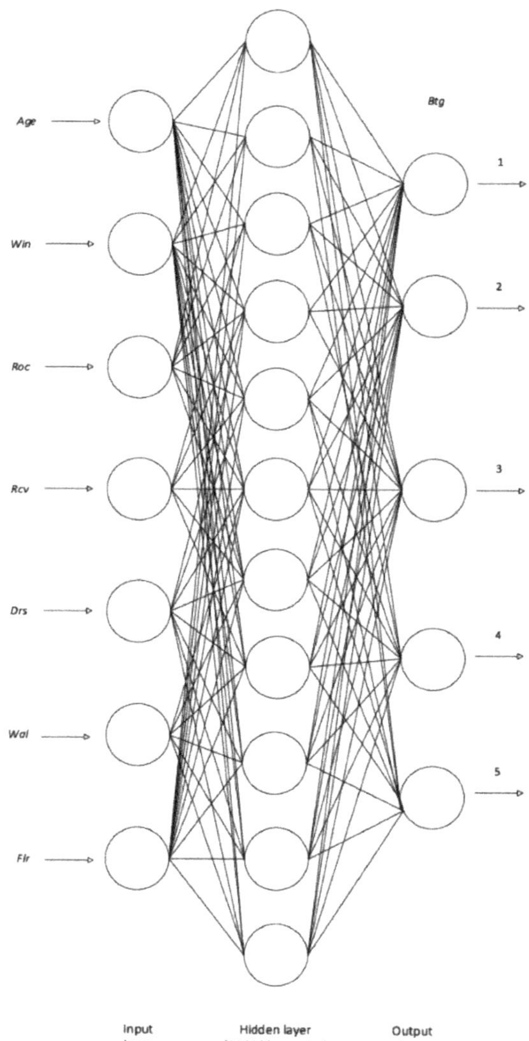

Figure 2. Optimal selected ANN model (MLP 7-11-5).

The architecture of MLP refers to the number of layers and the number of nodes in each layer. In this case, 7-11-5 indicates the following:

- If 7, there is an input layer with 7 nodes. These nodes represent the features or variables that are fed into the network;

- If 11, there is one hidden layer with 11 nodes. This layer performs the main information processing and learning within the network;
- If 5, there is an output layer with 5 nodes. This node represents the network's prediction or output value, i.e., the school grades from 1 to 5.

The number of hidden layers and neurons was determined based on empirical testing and the balance between model complexity and performance. A single hidden layer was chosen to maintain model interpretability and avoid overfitting, given the dataset size. The number of neurons (11) was selected through experimentation, comparing different configurations to identify the structure that yielded optimal predictive performance during model training and validation. While more complex architectures could be explored, the chosen configuration was found to provide a suitable trade-off between accuracy and computational efficiency for this study.

6.3. RF Model

Random forests are a larger set or ensemble of decision trees [56]. The name of the algorithm comes from the machine learning technique that the algorithm uses. The algorithm creates a large number of decision trees from the available data for learning—a forest of random trees. When constructing each tree in the forest, the algorithm randomly selects a subset of data and a subset of the input variables of the set [57,58]. RFs have many interesting features such as precise prediction power (comparable to or even outperforming ANNs and support vector algorithms), a low tendency of overtraining the model, not requiring a computer with high processing power, and having the ability to work very well with data with large dimensions [59]. One of the main advantages of the RF algorithm is resistance to training and growing a large number of random decision trees, where it does not create a risk of overlearning the model (each tree is a completely independent random experiment). The algorithm is also robust to exceptions in the data and automatically handles missing values [60].

In this case, since the output variable is the school building's total grade from 1–5, the classification RF algorithm was used. The predictor importance values for an RF model indicate how influential each variable is in predicting the output variable, in this case the school building's total grade. The number of trees was selected to be 100. It is noted in the literature that in the examined datasets, the biggest performance gain is achieved when training the first 100 trees [61]. Table 9 shows the predictor importance of the input variables. These values are typically normalized so that the most important variable has an importance of 1.

Table 9. Predictor importance.

Variable	Variable Rank	Importance
Roc	100	1.000000
Wal	93	0.929054
Drs	87	0.874778
Win	81	0.805458
Flr	78	0.779996
Age	76	0.762153
Rcv	69	0.687752

The roof construction (*Roc*) variable (rank 100, importance 1000) is the most significant predictor, making it the strongest factor in determining the building's grade. This aligns with the intuition that the condition of the roof plays a critical role in a building's overall structure and longevity. Walls (*Wal*) ranked 93 with an importance of 0.929, indicating that the condition of the walls is also a crucial factor in assessing the building's condition.

Doors (*Drs*), with a rank of 87 and importance of 0.875, and windows (*Win*), ranked 81 with an importance of 0.805, both have substantial influence, highlighting the significance of the building's exterior elements in overall grading. Floors (*Flr*), ranked 78 with an importance of 0.780, also play a notable role in the grade calculation, suggesting that the condition of the floors is a key predictor of the building's state. *Age* (rank 76, importance 0.762), though slightly lower, remains an important factor, reinforcing the idea that older buildings are more likely to experience deterioration over time. Finally, roof covering (*Rcv*), with a rank of 69 and an importance score of 0.688, while still an important predictor, has a somewhat lower impact compared to the other structural components. In conclusion, the results demonstrate that critical structural elements, particularly the roof, walls, doors, and windows, along with the building's age, are the most influential predictors in determining a building's grade. These findings suggest that predictive models, like random forests (*RFs*), that account for these variables are likely to provide reliable assessments of building condition.

7. Results

The following coefficients were used to estimate the prediction error of the developed models: mean absolute percentage error (*MAPE*), coefficient of determination (R^2), mean square error (*MSE*), root mean square error (*RMSE*), and coefficient of variation of the root mean square error (*CVRMSE*). Unlike most error metrics, *MAPE* (mean absolute percentage error) does not have a strict upper limit. The reason for this is because it involves taking the absolute value of the percentage error [62]. The closer the coefficient of determination R^2 value is to 1, the more representative the prediction model is [63]. Generally, a lower *MSE* is desirable as it signifies that the model's predictions are, on average, closer to the actual values. Conversely, a high *MSE* suggests that the model's predictions are consistently far off from the real values [64]. Typically, the upper limit for *CVRMSE* of 30% is used to measure representativeness [65]. The equations utilized to compute the statistical techniques for prediction error estimation are displayed in Table 10.

Table 10. Expressions for the calculations of statistical methods for estimating prediction error.

No.	Coefficient	Expression	Ref.
1	R^2	$R^2 = 1 - \frac{\sum_{i=1}^{n}(y_i - \hat{y}_i)^2}{\sum_{i=1}^{n}(y_i - \overline{Y})^2}$	[66]
2	*MSE*	$MSE = \frac{1}{n}\sum_{t=1}^{n}(y_i - \hat{y}_i)^2$	[67]
3	*RMSE*	$RMSE = \sqrt{\frac{1}{n}\sum_{t=1}^{n}(y_i - \hat{y})^2}$	[68]
4	*CVRMSE*	$CVRMSE = \frac{\sqrt{\frac{1}{n}\sum_{i=1}^{n}(y_i - \hat{y})^2}}{\overline{Y}} \cdot 100(\%)$	[69]
5	*MAPE*	$MAPE = \frac{1}{n}\sum_{t=1}^{n}\left\lvert\frac{(y_i - \hat{y}_i)}{y_i}\right\rvert \cdot 100(\%)$	[70]

where it is as follows:
y_i real values of the dependent variable,
$\hat{y}_i$ predicted or expected values of the dependent variable and
$\overline{Y}$ arithmetic mean of the dependent variable.

Table 11 presents the values of the aforementioned coefficients for estimating the prediction error of the developed ANN model and the RF model in the model training set with included additional metrics, i.e., precision, recall, and F1.

Table 11. Statistical analysis of the ANN and RF models' prediction errors in the training data set.

No.	Dependent Variable	Model Type	R^2	*MSE*	*RMSE*	*CVRMSE*	*MAPE*	Precision	Recall	F1
1	*Btg*	ANN	0.987	0.017	0.131	4.35%	0.50%	0.983	0.983	0.983
2		RF	0.924	0.094	0.307	10.28%	2.99%	0.920	0.906	0.907

The comparative performance of ANN and RF demonstrates the suitability of advanced machine learning methods for BCA. The ANN model exhibits better predictive accuracy, with an R^2 of 0.987 and a *MAPE* of 0.50%, indicating near-perfect alignment with the actual grades. In contrast, the random forest model, while performing adequately ($R^2 = 0.924$), has higher error metrics, such as a *MAPE* of 2.99%. These results suggest that while both models can be utilized for BCA, the ANN model provides a more reliable framework for precise grade prediction. The ANN model performs well with precision, recall, and F1 scores of 0.983, indicating high accuracy in classifying the target variable. In contrast, the RF model, with scores around 0.920 for precision and 0.906 for recall, performs slightly worse, showing more false positives and missed cases.

Table 12 presents the values of the aforementioned coefficients for estimating the prediction error of the developed ANN model and the RF model in the model validation set with included additional metrics, i.e., precision, recall, and F1.

Table 12. Statistical analysis of the ANN and RF models' prediction errors in the validation data set.

No.	Dependent Variable	Model Type	R^2	*MSE*	*RMSE*	*CVRMSE*	*MAPE*	Precision	Recall	F1
1	*Btg*	ANN	0.940	0.082	0.286	9.33%	2.55%	0.927	0.918	0.920
2		RF	0.942	0.086	0.292	9.72%	2.66%	0.920	0.915	0.915

The validation results underscore better predictive capability of the ANN model for building total grade (*Btg*) assessment. With $R^2 = 0.940$ and *MAPE* of 2.55%, the ANN achieves high accuracy and low error, making it a reliable tool for assessing building conditions. The RF model, while slightly less precise in terms of *MAPE* = 2.66%, also exhibits strong generalization performance, suggesting it is a viable alternative or complementary approach. The ANN model achieves precision, recall, and F1 scores of 0.927, 0.918, and 0.920, respectively, indicating strong performance with a slight edge over the RF model, which has scores of 0.920, 0.915, and 0.915. Both models perform similarly in terms of classification, with RF slightly lagging in recall and F1. However, both models are well suited for the task, with similar accuracy and error rates.

8. Validation of the Framework on an External Sample of Schools

As for the development, the data on schools for validation were obtained from EMIS [48]. A total of 20 schools were used for the validation of the developed models, with 10 schools from inland Croatia and 10 from the coastal region. By including schools from coastal Croatia, such as those in coastal and island areas, as well as schools from inland Croatia, the models are tested under various urban, rural, and specific regional conditions. In this way, the validation confirms how robust the models are and how ready they are for use in real-world conditions within the educational system across Croatia.

The counties of continental Croatia include the City of Zagreb, Zagreb County, Krapina-Zagorje County, Varaždin County, Koprivnica-Križevci County, Međimurje County, Bjelovar-Bilogora County, Virovitica-Podravina County, Požega-Slavonia County,

Brod-Posavina County, Osijek-Baranja County, Vukovar-Srijem County, Karlovac County, and Sisak-Moslavina County.

The counties of coastal Croatia include Primorje-Gorski Kotar County, Lika-Senj County, Zadar County, Šibenik-Knin County, Split-Dalmatia County, Istria County, and Dubrovnik-Neretva County.

Osijek-Baranja County will be excluded from the validation sample, considering that all models were developed using a sample of schools from this county. Considering that the Republic of Croatia has 21 counties, excluding Osijek-Baranja County leaves 20. The sample consists of 20 schools because one school was selected from each county to ensure diversity in the sample. Among these 20 schools, 6 are secondary schools to include larger schools with a greater number of students in the analysis.

Firstly, their grades were obtained using the Excel-based traditional grading system, and then these grades were predicted using the remaining lifetimes of elements. Table 13 presents the values of the aforementioned coefficients from Table 10 for estimating the prediction error of the developed ANN model and the RF model in the external validation set.

Table 13. Statistical analysis of the ANN and RF models' prediction errors in the external validation data set.

No.	Dependent Variable	Model Type	R^2	*MSE*	*RMSE*	*CVRMSE*	*MAPE*	Precision	Recall	F1
1	*Btg*	ANN	0.799	0.200	0.447	18.44 %	7.71 %	0.849	0.800	0.807
2		RF	0.747	0.225	0.474	19.56 %	9.17 %	0.809	0.775	0.776

The table compares the performance of the two models—the ANN and RF—in predicting the building total grade (*Btg*), and both models deliver excellent results. The ANN model slightly outperforms the RF model across all metrics. It achieves a higher R^2 value (0.799 vs. 0.747), indicating that it explains more variance in the dependent variable. It also has lower error values, with an *MSE* of 0.200 and an *RMSE* of 0.447 compared to the RF's *MSE* of 0.150 and *RMSE* of 0.388. The RF model further demonstrates greater precision in its predictions, with lower *CVRMSE* (19.56% vs. 18.44%) and *MAPE* (9.17% vs. 7.71%). However, it is worth noting that both models achieve high R^2 values and low error metrics, reflecting their strong performance and reliability in predicting the building total grade. The ANN model performs slightly better than the RF model, with precision, recall, and F1 scores of 0.849, 0.800, and 0.807, respectively. The RF model has scores of 0.809 for precision, 0.775 for recall, and 0.776 for F1, indicating slightly lower performance.

9. Discussion

The systematic management of public sector buildings is essential for sustainability and operational efficiency. Implementing the Ordinance on Systematic Energy Management in the Public Sector (Official Gazette 18/2015) in Croatia underscores the government's commitment to this goal. However, beyond regulatory frameworks, the practical application of these policies, as evidenced in the EMIS, reveals strengths and areas for improvement in public sector infrastructure management. A critical aspect of this study was using EMIS data to assess the condition of 166 primary and secondary schools in Osijek-Baranja County. These data, while comprehensive, were supplemented with historical construction characteristics from the Croatian Methodology of Energy Audit of Buildings to fill gaps, demonstrating the necessity for robust data collection mechanisms. The varying completeness of the data highlights an ongoing challenge in relying on decentralized data input, particularly in ensuring uniformity and accuracy across many buildings. The condition of the buildings was assessed using a detailed methodology that considered the sustainable use life of various building components. The first part of the study utilized an Excel-based scoring system to assign grades to individual building elements, such as windows, roofs, and floors, based on their remaining sustainable life of use. This matrix-based approach ensured simplicity and transparency, enabling practitioners to easily evaluate and grade each component systematically. While the machine learning models (ANN and RF) excel in predictive accuracy, the Excel grading framework introduces a clear, step-by-step process for initial assessments. This hybrid approach ensures that stakeholders with varying levels of technical expertise can contribute to the assessment process, bridging the gap between manual evaluation and algorithmic precision. The Excel model enabled a component-level analysis of building elements, forming the foundation for the subsequent machine learning models. The ANN and RF models utilized data from the Excel grading system to predict the overall building total grade (*Btg*). This synergy between traditional and modern techniques ensures that the assessment process retains its transparency while benefiting from advanced data-driven insights. The integration of both approaches enhances the robustness of the BCA framework, making it adaptable to a variety of use cases, including large-scale assessments and policy-driven renovations.

While the AI models demonstrated increased predictive accuracy and the ability to handle complex, non-linear relationships within the data, it is important to also recognize the strengths and limitations of traditional grading methods. Traditional systems, such as the Excel-based grading used in this study, offer simplicity and ease of implementation, making them accessible to practitioners. They provide clear and interpretable results, which can be beneficial in practical decision-making. In contrast, the AI-based models are more objective and provide higher precision in predictions, particularly as data complexity increases. They excel in identifying underlying relationships between variables that might not be immediately apparent with traditional methods. However, these models require large, high-quality datasets for effective training and can be computationally intensive, which may pose challenges for some users, particularly in resource-limited settings. Furthermore, AI models can be less interpretable, making it harder for practitioners to understand the reasoning behind the predictions. The choice between these methods should consider the specific needs, the availability of resources, and the desired level of interpretability.

The combined methodology aligns directly with Croatia's Long-Term Strategy for National Building Stock Renovation by 2050. The Excel-based grading system offers a practical tool for initial assessments of building elements, while the ANN and RF models provide detailed, predictive analyses necessary for prioritizing renovations. Together, these methods support the strategy's goals of improving public building management, ensuring that limited resources are allocated effectively. The findings also revealed significant disparities in the condition of these components, particularly in elements such as windows and roofs, which frequently exceeded their expected lifespan. This suggests a systemic issue in maintenance practices or funding allocation, where preventative maintenance may not be adequately prioritized or funded. A notable aspect of the analysis was the wide variance in the age and renovation history of the buildings. The data showed that while some buildings have been relatively recently constructed or renovated, a substantial number are significantly older, with some dating back over a century. This age disparity is crucial as it indicates a potential misalignment between the age of infrastructure and the resources allocated for their maintenance. Older buildings, often with outdated materials and construction techniques, require more frequent and intensive maintenance to meet current safety and efficiency standards. The condition ratings across the schools, with a significant number falling into the lower categories, underscore the urgent need for targeted interventions. The prevalence of poor or fair ratings in critical components like roof coverage and windows not only reflects on the immediate physical condition of the buildings but also has broader implications for energy efficiency and safety. The findings suggest that these buildings may continue to deteriorate without significant investment in renovation and maintenance, potentially compromising their usability and safety.

10. Conclusions

This study highlights the significant potential of systematic building condition management for public sector buildings, focusing on the schools in Osijek-Baranja County, Croatia. Utilizing data from EMIS, the analysis reveals considerable variations in the condition of building components, underscoring the need for targeted maintenance and renovation strategies. The findings emphasize the value of data-driven approaches in identifying and addressing deficiencies across public buildings, aiming to enhance safety, efficiency, and overall functionality.

This research presents a comprehensive framework for BCA by integrating an Excel-based component level grading system with advanced machine learning models, namely ANN and RF. The Excel grading system provides a transparent and systematic approach for evaluating individual building elements, while the machine learning models enhance the predictive accuracy of the overall building condition grade (*Btg*). The results demonstrate the better performance of the ANN model in predicting the overall building condition, with high R^2 and low *MAPE* values in both the training and validation datasets. Meanwhile, the RF model provides interpretable insights into predictor importance, complementing the ANN's precision with actionable information for stakeholders. Additionally, by performing the validation on an external data set of one school from each of the remaining 20 counties of Croatia (excluding Osijek-Baranja County on which models were developed) it can be concluded that both the ANN and RF models demonstrate excellent performance in predicting the building total grade (*Btg*), with both models achieving high R^2 values and low error metrics. The ANN model outperforms the RF slightly across all key metrics, showing better predictive power and precision. However, both models offer reliable and accurate results, making them both viable options for this type of analysis.

For school management, the ability to quickly evaluate building conditions without relying solely on visual inspections represents a significant advancement. Using EMIS data allows for rapid and comprehensive assessments, which can streamline decision-making processes, reduce reliance on subjective evaluations, and ensure that maintenance efforts are focused where they are most needed.

One significant direction for future research is the inclusion of cultural heritage buildings, which were excluded from the current analysis due to their complexity and specificity. Developing a tailored framework for managing and maintaining these buildings while balancing preservation with modern efficiency needs represents an area for future exploration. Such research could integrate conservation practices with energy management techniques to create a framework that respects historical value while ensuring functional and sustainable use. Also, the framework could be extended beyond Croatia and applied to other Eastern European countries (Slovenia, Hungary, Bosnia and Herzegovina, and Serbia) with similar construction characteristics.

Additionally, expanding the research to encompass a broader range of public buildings, such as libraries, museums, and government offices, would provide a more comprehensive understanding of infrastructure needs across different sectors. This expanded approach could lead to more effective policies and strategies for public sector building management, contributing to overall improvements in public infrastructure.

While accuracy has been the primary metric used in this study to evaluate the effectiveness of machine learning models, future research could benefit from incorporating additional performance indicators such as precision, recall, and the F1-score. These metrics are particularly valuable in situations where the data are imbalanced, as they offer a deeper understanding of how well the models perform across different categories. In cases where certain condition categories, such as "poor" or "failed" components, are more critical to identify accurately, metrics like precision and recall can help ensure that these categories are detected with higher reliability. Therefore, exploring these additional performance metrics in future studies would contribute to a more comprehensive evaluation of the machine learning models' capabilities and their practical applicability in building condition assessments.

In summary, this study provides insights into the current state of school buildings in Osijek-Baranja County and shows the applicability of the system for expanding systematic management practices across Croatia. By addressing identified challenges and pursuing new research directions, policymakers and administrators can enhance their public building maintenance and renovation strategies, ensuring long-term benefits for current and future infrastructure.

Author Contributions: Conceptualization, H.K. and H.B.J.; methodology, H.B.J. and H.K.; software, H.B.J.; validation, H.K. and H.B.J.; formal analysis, H.B.J.; writing—original draft preparation, H.B.J.; writing—review and editing, H.K. and H.B.J.; supervision, H.K. and H.B.J. All authors have read and agreed to the published version of the manuscript.

Funding: This research received no external funding.

Data Availability Statement: The data presented in this study are available on request from the corresponding author. The data are not publicly available due to privacy.

Conflicts of Interest: The authors declare no conflicts of interest.

Abbreviations

The following abbreviations are used in this manuscript:

AHP	Analytic Hierarchy Process
ANN	Artificial Neural Network
ANP	Analytic Network Process
BCA	Building Condition Assessment
CVRMSE	Coefficient of the Variation of the Root Mean Square Error
EMIS	Energy Management Information System
MAPE	Mean Absolute Percentage Error
MLP	Multi-Layer Perceptron
MSE	Mean Square Error
RF	Random Forest
RMSE	Root Mean Square
SEM	Structural Equation Modeling

References

1. Becerik-Gerber, B.; Jazizadeh, F.; Li, N.; Calis, G. Application areas and data requirements for BIM-enabled facilities management. *J. Constr. Eng. Manag.* **2012**, *138*, 431–442.
2. Mohamed, A.G.; Marzouk, M. Building condition assessment using artificial neural network and structural equations. *Expert Syst. Appl.* **2021**, *186*, 115743.
3. Galimshina, A.; Moustapha, M.; Hollberg, A.; Padey, P.; Lasvaux, S.; Sudret, B.; Habert, G. Statistical method to identify robust building renovation choices for environmental and economic performance. *Build. Environ.* **2020**, *183*, 107143.
4. Plebankiewicz, E.; Grącki, J. Analysis and prediction of universities' buildings' renovation costs using a regression model. *Appl. Sci.* **2022**, *13*, 401. [CrossRef]
5. Dejaco, M.C.; Cecconi, F.R.; Maltese, S. Key performance indicators for building condition assessment. *J. Build. Eng.* **2017**, *9*, 17–28.

6. Scaife, A.D. Improve predictive maintenance through the application of artificial intelligence: A systematic review. *Results Eng.* **2023**, *21*, 101645.
7. Marocco, M.; Garofolo, I. Integrating disruptive technologies with facilities management: A literature review and future research directions. *Autom. Constr.* **2021**, *131*, 103917.
8. Hauashdh, A.; Nagapan, S.; Jailani, J.; Gamil, Y. An integrated framework for sustainable and efficient building maintenance operations aligning with climate change, SDGs, and emerging technology. *Results Eng.* **2024**, *21*, 101822.
9. Zhao, J.; Feng, H.; Chen, Q.; de Soto, B.G. Developing a conceptual framework for the application of digital twin technologies to revamp building operation and maintenance processes. *J. Build. Eng.* **2022**, *49*, 104028.
10. Wang, J.; Wang, X.; Shen, Y.; Xiong, X.; Zheng, W.; Li, P.; Fang, X. Building operation and maintenance scheme based on sharding blockchain. *Heliyon* **2023**, *9*, e13186.
11. Hauashdh, A.; Jailani, J.; Rahman, I.A. Structural equation model for assessing factors affecting building maintenance success. *J. Build. Eng.* **2021**, *44*, 102680.
12. Linggar, S.; Aminullah, A.; Triwiyono, A. Analysis of building and its components condition assessment case study of dormitory buildings. *MATEC Web Conf.* **2019**, *258*, 03003.
13. Singh Ahluwalia, S. A Framework for Efficient Condition Assessment of the Building Infrastructure. Ph.D. Thesis, University of Waterloo, Waterloo, ON, Canada, 2008.
14. Mayo, G.; Karanja, P. Building condition assessments–methods and metrics. *J. Facil. Manag. Educ. Res.* **2018**, *2*, 1–11.
15. Besiktepe, D.; Ozbek, M.E.; Atadero, R.A. Identification of the criteria for building maintenance decisions in facility management: First step to developing a multi-criteria decision-making approach. *Buildings* **2020**, *10*, 166. [CrossRef]
16. Silva, A.; De Brito, J.; Gaspar, P.L. *Methodologies for Service Life Prediction of Buildings: With a Focus on Façade Claddings*; Springer: Berlin/Heidelberg, Germany, 2016.
17. Kwon, N.; Song, K.; Ahn, Y.; Park, M.; Jang, Y. Maintenance cost prediction for aging residential buildings based on case-based reasoning and genetic algorithm. *J. Build. Eng.* **2020**, *28*, 101006.
18. Mahli, M.; Che-Ani, A.; Yahaya, H.; Tawil, N.; Mydin, M.O. School building defect pattern. *MATEC Web Conf.* **2014**, *15*, 01007.
19. Sanoff, H. *School Building Assessment Methods*; National Clearinghouse for Educational Facilities: Washington, DC, USA, 2001.
20. Simons, E.; Hwang, S.-A.; Fitzgerald, E.F.; Kielb, C.; Lin, S. The impact of school building conditions on student absenteeism in upstate New York. *Am. J. Public Health* **2010**, *100*, 1679–1686. [PubMed]
21. Ismail, I.; Che-Ani, A.; Tawil, N.; Yahaya, H.; Abd-Razak, M. Housing Defect of Newly Completed House: An Analysis Using Condition Survey Protocol (CSP) 1 Matrix. *Int. J. Civ. Environ. Eng.* **2012**, *6*, 398–401.
22. Hamzah, N.; Mahli, M.; Che Ani, A.; Tahir, M.; Abdullah, N.; Tawil, N. The development of smart school condition assessment based on condition survey protocol (CSP) 1 matrix: A literature review. *Int. J. Civ. Environ. Eng.* **2010**, *4*, 620–625.
23. Kejeh, K.; Nwaogazie, I.L.; Samuel, S. Maintenance-State Assessment Model for Building Maintenance Forecasting. *Arch. Curr. Res. Int.* **2022**, *22*, 40–54.
24. Štrok, K.T.; Car-Pušić, D.; Marenjak, S. Elementary School Buildings Condition Assessment: Case of Primorje-Gorski Kotar County (Croatia). *Adv. Civ. Archit. Eng.* **2023**, *14*, 95–117.
25. Marzouk, M.; Awad, E. Establishing multi-level performance condition indices for public schools maintenance program using AHP and fuzzy logic. *Stud. Inform. Control* **2016**, *25*, 343–352.
26. Eweda, A.; Zayed, T.; Alkass, S. Space-based condition assessment model for buildings: Case study of educational buildings. *J. Perform. Constr. Facil.* **2015**, *29*, 04014032.
27. Eweda, A.; Zayed, T.; Alkass, S. An integrated condition assessment model for buildings. In Proceedings of the Construction Research Congress 2010: Innovation for Reshaping Construction Practice, Alberta, BC, Canada, 8 May 2010; Janaka, R., Yasser, M., Sang, H.L., Eds.; American Society of Civil Engineers: Alberta, BC, Canada, 2010; pp. 1386–1395.
28. Eweda, A.; Al-Sakkaf, A.; Zayed, T.; Alkass, S. Condition assessment model of building indoor environment: A case study on educational buildings. *Int. J. Build. Pathol. Adapt.* **2021**, *41*, 767–788.
29. Elhakeem, A.; Hegazy, T. Towards a visual guidance system for condition assessment of the building infrastructure. In Proceedings of the 1st CSCE Specialty Conference on Infrastructure Technologies, Management, and Policies, CSCE, Toronto, ON, Canada, 2–4 June 2005.
30. Hassan, A.M.; Adel, K.; Elhakeem, A.; Elmasry, M.I. Condition Prediction for Existing Educational Facilities Using Artificial Neural Networks and Regression Analysis. *Buildings* **2022**, *12*, 1520. [CrossRef]
31. Sadick, A.-M.; Issa, M.H. Assessing physical conditions of indoor space enclosing elements in schools in relation to their indoor environmental quality. *J. Build. Eng.* **2018**, *20*, 520–530.
32. Uzarski, D.R.; Grussing, M.N.; Clayton, J.B. Knowledge-based condition survey inspection concepts. *J. Infrastruct. Syst.* **2007**, *13*, 72–79.
33. Amrouni Hosseini, M.; Ravanshadnia, M.; Rahimzadegan, M.; Ramezani, S. Next-Generation Building Condition Assessment: BIM and Neural Network Integration. *J. Perform. Constr. Facil.* **2024**, *38*, 04024050.

34. Wang, Z.; Wang, Y.; Zeng, R.; Srinivasan, R.S.; Ahrentzen, S. Random Forest based hourly building energy prediction. *Energy Build.* **2018**, *171*, 11–25.
35. Ma, J.; Cheng, J.C. Identifying the influential features on the regional energy use intensity of residential buildings based on Random Forests. *Appl. Energy* **2016**, *183*, 193–201.
36. Lin, J.; Lu, S.; He, X.; Wang, F. Analyzing the impact of three-dimensional building structure on CO_2 emissions based on random forest regression. *Energy* **2021**, *236*, 121502.
37. Zekić-Sušac, M.; Has, A.; Knežević, M. Predicting energy cost of public buildings by artificial neural networks, CART, and random forest. *Neurocomputing* **2021**, *439*, 223–233.
38. Tooke, T.R.; Coops, N.C.; Webster, J. Predicting building ages from LiDAR data with random forests for building energy modeling. *Energy Build.* **2014**, *68*, 603–610.
39. Fang, Y.; Lu, X.; Li, H. A random forest-based model for the prediction of construction-stage carbon emissions at the early design stage. *J. Clean. Prod.* **2021**, *328*, 129657.
40. Arowoiya, V.A.; Onososen, A.O.; Moehler, R.C.; Fang, Y. Influence of Thermal Comfort on Energy Consumption for Building Occupants: The Current State of the Art. *Buildings* **2024**, *14*, 1310. [CrossRef]
41. Sirror, H.; Labib, W.; Abowardah, E.; Metwally, W.; Mitchell, C. Sustainability in the Workplace: Evaluating Indoor Environmental Quality of a Higher Education Building in Riyadh. *Buildings* **2024**, *14*, 2115. [CrossRef]
42. Mahmoud, A.S.; Hassanain, M.A.; Alshibani, A. Evolving Trends and Innovations in Facilities Management Within Higher Education Institutions. *Buildings* **2024**, *14*, 3759. [CrossRef]
43. Ahn, J. Thermal Control Strategy for the Sustainable Use of Large Classrooms Responding to User Demands in a School Building. *Buildings* **2024**, *14*, 3809. [CrossRef]
44. Romero, P.; Valero-Amaro, V.; Arranz, J.I.; Sepúlveda, F.J.; Miranda, M.T. Indoor Air Quality and Thermal Comfort in University Classrooms in Southwestern Spain: A Longitudinal Analysis from Pandemic to Post-Pandemic. *Buildings* **2025**, *15*, 829. [CrossRef]
45. Wu, S.; Zhou, P.; Xiong, Y.; Ma, C.; Wu, D.; Lu, W. Strategies for Driving the Future of Educational Building Design in Terms of Indoor Thermal Environments: A Comprehensive Review of Methods and Optimization. *Buildings* **2025**, *15*, 816. [CrossRef]
46. Ministry of Physical Planning Construction and State Assets. Long-Term Strategy for National Building Stock Renovation by 2050. Available online: https://energy.ec.europa.eu/document/download/b87dca97-b7c1-452e-ae65-fad83be5f80b_en?filename=hr_2020_ltrs_en_version.pdf (accessed on 15 October 2024).
47. Ministarstvo graditeljstva i prostornoga uređenja. Pravilnik o sustavnom gospodarenju energijom u javnom sektoru (NN 18/2015). Available online: https://narodne-novine.nn.hr/clanci/sluzbeni/2015_02_18_389.html (accessed on 15 October 2024).
48. Agencija za pravni promet i posredovanje nekretninama. Informacijski sustav za gospodarenje energijom—ISGE. Available online: https://apn.hr/gospodarenje-energijom-isge/informacijski-sustav-za-gospodarenje-energijom (accessed on 16 October 2024).
49. Ministry of Physical Planning, Construction and State Assets. *Metodologija Provođenja Energetskog Pregleda Zgrada*; Ministry of Physical Planning, Construction and State Assets: Zagreb, Croatia, 2021.
50. Ministry of Physical Planning, Construction and State Assets. *Ordinance on Property Valuation Methods (Official Gazette 105/15)*; Ministry of Physical Planning, Construction and State Assets: Zagreb, Croatia, 2015.
51. Cloud Software Group Inc. TIBCO Statistica® 14.1.0. Available online: https://docs.tibco.com/products/tibco-statistica-14-1-0 (accessed on 15 October 2024).
52. Schober, P.; Boer, C.; Schwarte, L.A. Correlation coefficients: Appropriate use and interpretation. *Anesth. Analg.* **2018**, *126*, 1763–1768.
53. Jain, R.K.; Smith, K.M.; Culligan, P.J.; Taylor, J.E. Forecasting energy consumption of multi-family residential buildings using support vector regression: Investigating the impact of temporal and spatial monitoring granularity on performance accuracy. *Appl. Energy* **2014**, *123*, 168–178.
54. Kontokosta, C.E.; Tull, C. A data-driven predictive model of city-scale energy use in buildings. *Appl. Energy* **2017**, *197*, 303–317.
55. Rana, A.; Rawat, A.S.; Bijalwan, A.; Bahuguna, H. Application of multi layer (perceptron) artificial neural network in the diagnosis system: A systematic review. In Proceedings of the 2018 International Conference on Research in Intelligent and Computing in Engineering (RICE), San Salvador, El Salvador, 22–24 August 2018; IEEE: San Salvador, El Salvador, 2018; pp. 1–6.
56. Breiman, L. Random forests. *Mach. Learn.* **2001**, *45*, 5–32.
57. Lewinson, E. *Python for Finance Cookbook: Over 50 Recipes for Applying Modern Python Libraries to Financial Data Analysis*; Packt Publishing Ltd.: Birmingham, UK, 2020.
58. Jokić, J.; Martinčić-Ipšić, S. Primjena šume slučajnih stabala za predviđanje ishoda šahovske partije reprezentirane kompleksnom mrežom. *Zbornik Veleučilišta u Rijeci* **2019**, *7*, 31–52.
59. Caruana, R.; Niculescu-Mizil, A. In An empirical comparison of supervised learning algorithms. In Proceedings of the 23rd International Conference on Machine Learning, Pittsburgh, PA, USA, 25–29 June 2006; Cohen, W., Moore, A., Eds.; Association for Computing Machinery: New York, NY, USA; Pittsburgh, PA, USA, 2006; pp. 161–168

60. Youssef, A.M.; Pourghasemi, H.R.; Pourtaghi, Z.S.; Al-Katheeri, M.M. Landslide susceptibility mapping using random forest, boosted regression tree, classification and regression tree, and general linear models and comparison of their performance at Wadi Tayyah Basin, Asir Region, Saudi Arabia. *Landslides* **2016**, *13*, 839–856.
61. Probst, P.; Boulesteix, A.-L. To tune or not to tune the number of trees in random forest. *J. Mach. Learn. Res.* **2018**, *18*, 1–18.
62. Saigal, S.; Mehrotra, D. Performance comparison of time series data using predictive data mining techniques. *Adv. Inf. Min.* **2012**, *4*, 57–66.
63. Schneider, A.; Hommel, G.; Blettner, M. Linear regression analysis: Part 14 of a series on evaluation of scientific publications. *Dtsch. Ärzteblatt Int.* **2010**, *107*, 776.
64. Wang, Z.; Bovik, A.C. Mean squared error: Love it or leave it? A new look at signal fidelity measures. *IEEE Signal Process. Mag.* **2009**, *26*, 98–117.
65. Lulić, I. The Use of Regression Analysis Method in Solving Problems from Engineering Practice. Bachelor's Thesis, University of Zagreb, Zagreb, Hrvatska, 2014.
66. Papić, M. *Primijenjena Statistika u MS Excelu za Ekonomiste, Znanstvenike i Neznalice*; Zoro d.o.o: Zagreb, Croatia, 2005.
67. Sobol, M.G. Validation Strategies for Multiple Regression Analysis: Using the Coefficient of Determination. *Interfaces* **1991**, *21*, 106–120.
68. Sailee, R. *Exploration of Variable Importance and Variable Selection Techniques in Presence of Correlated Variables*; Rochester Institute of Technology, College of Science: Rochester, NY, USA, 2019.
69. Cacabelos, A.; Eguía, P.; Febrero, L.; Granada, E. Development of a new multi-stage building energy model calibration methodology and validation in a public library. *Energy Build.* **2017**, *146*, 182–199.
70. Small, G.R.; Wong, R. The Validity of Forecasting. In Proceedings of the 8th Annual PRRES Conference in Christchurch, Christchurch, New Zealand, 21–23 January 2002; Parker, D., Ed.; PRRES Inc.: Christchurch, New Zealand, 2002; pp. 1–14.

buildings

MDPI

Article

An Investigation of Globe Temperature in Street Canyons: A Scaled Model Study Implementing Cool Materials

E. V. S. Kiran Kumar Donthu [1], Shashwat Shashwat [2], Kishor T. Zingre [2,*], Seshadhri Srinivasan [3] and Man Pun Wan [1]

1 School of Mechanical and Aerospace Engineering, Nanyang Technological University Singapore 639798, Singapore
2 Department of Architecture and Built Environment, Northumbria University, Tyne NE1 8ST, UK; s.shashwat@northumbria.ac.uk
3 Kalasalingam Academy of Research and Education, Srivilluputtur 626126, India
* Correspondence: kishor.zingre@northumbria.ac.uk

Abstract: The measurement of globe temperature (GT) is essential for investigating pedestrian thermal comfort in street canyons. The globe thermometer is the most common instrument used to measure GT; however, its application in scale models has not been thoroughly investigated to date. Therefore, this study explicitly investigates globe thermometer measurements in scale models and analyzes the need for customization of the globe thermometer for more reliable measurements. Scaling down with respect to the size of the globe thermometer and the effect of solar orientation/envelope materials are investigated in this study. The initial experiments were carried out in an outdoor setting using a typical street canyon model (scale 1:100) with an east-west street orientation. The results of the experiment are presented to compare a low solar reflectance street canyon (albedo of 0.4) and a high solar reflectance canyon (albedo of 0.6) in terms of surface temperatures, heat flux, and globe temperature. It is observed that although the wall and road surface temperatures are lower for the high solar reflectance canyon compared to those for the low solar reflectance canyon, the GT (measured at pedestrian height) is higher in a high reflectance canyon during the daytime, which could be due to the combined effect of direct radiation and short-wave reflection. However, for the hours after sunset, a reverse effect is observed, i.e., the GT becomes lower (up to 0.8 °C) in the case of a high reflectance canyon compared to that for the low reflectance canyon. Furthermore, the study emphasizes the impact of solar reflectance of canyon surfaces on GT values, due to the view factors that the globe thermometer on those surfaces.

Keywords: street canyon; globe temperature; heat capacitance; cool materials

1. Introduction

Scale model investigations can be adopted as a flexible and economically viable experimental alternative, especially for urban-scale experimental studies. The constraints of numerical modeling and the need for high performance computing also make scale models a viable substitute for software simulations [1,2]. While scale model experiments are often carried out in wind tunnels for airflow [3] and dispersion studies, the test setups are unable to replicate the interactions between solar radiation and the heat storage of building materials, which is a crucial aspect in urban microclimate studies. Performing a scale model experiment involves scaling down physical variables to mimic the actual scenario. For instance, the measurement height for human comfort is scaled down to correspond to actual human height, which in the case of a 1:5 scaled model experiment was considered to be the height of 30 cm above the ground surface [4].

Historically, urban-level scale models can be found to be used as early as the 1960s, when a 3D roughness model was used to investigate the effect of the urban climate [5].

Scale models have evolved in regard to complexity, sophistication, and realism due to advancements in methodology and technology. For instance, the modern scale model can incorporate built environment materials and other features, such as vegetation and even water features, with the possibility of studying the dynamic variation of the scale model [6]. Scale models have been used to investigate various aspects of urban microclimate and urban heat island (UHI) mitigation strategies, such as using wind tunnel experiments for European cities [7] and applying remote sensing data and geographically weighted regression models to study the relationships between different factors of UHI [8]. However, during scale model investigations, one of the primary challenges is to ensure the reliability and accuracy of the measurement equipment [9]. Additionally, the scale of the equipment is also important to ensure that it does not interfere with the airflow and measurement of other thermal properties during the investigations [10].

In thermal comfort studies, globe temperature (GT) is a key parameter, and it covers the exchange of both short- (solar and visible) and long-wave (infrared and terrestrial) radiation fluxes between the human body and the environment [11]. The globe thermometer is the most used and reliable instrument for measuring thermal comfort [12,13]. In the context of scale model studies evaluating outdoor thermal comfort, adapting the scale of the measurement equipment, such as the globe thermometer, during measurement is rarely conducted.

The primary approach for GT determination is based on measurements performed simultaneously with a globe thermometer, a dry-bulb thermometer, and an anemometer [14]. The study of Kuehn et al. reported on the theory of the globe thermometer: The temperature of the globe is governed by the combined effect of heat exchange processes through convection and radiation [15]. A temperature sensor is placed in the center of a hollow sphere to form a globe thermometer, invented by Vernon in 1932, and designed for indoor use [12]. Additionally, globe thermometers have recently been used in outdoor settings, as they can be a straightforward and less expensive alternative to net radiometers [16]. GT is therefore influenced by the values of air temperature, wind speed, and short- and long-wave radiant fluxes. Guo et al. found that air speed has a substantial influence on the mean radiant temperatures (MRTs) derived from globe thermometers [17]. Convection adjustments are essential for digital sensor-based globes because the temporal fluctuations in these data must be considered, as well as the actual influence on the globe throughout its settling period. As the differential between the air and surface temperatures in a space rises, convection correction becomes more significant [18].

In scale model studies, while it is not difficult to satisfy the geometrical similarity requirement, it is challenging to meet all the dynamic requirements related to radiation and airflow, which are critical parameters in microclimate studies [19]. In such cases, the experimental results can be complemented with numerical models or evaluated using stochastic methods [2,5]. Thus far, there have been a limited number of scale model studies related to microclimate test setups. Past studies have focused on the heat island effect and the thermal comfort aspects using scale models in outdoor conditions [19–22]. Additionally, studies have also covered the effect of cool materials and green roofs on the immediate microclimate using indoor scale models [23]. Therefore, this study aims to develop and discuss a test setup for studying the contribution of cool materials (i.e., those with high solar reflectance) on the thermal comfort parameter, such as the globe temperature in the street canyon setup, using an experimentally validated scale model.

2. Methodology

This study primarily involves the collection of experimental measurements (as shown in Figure 1) to investigate globe temperature regulation using cool materials in an urban canyon scale model setup. A series of scale model measurements were carried out, and their results were compared against real scale measurements. Specifically, the scale model measurements aimed to evaluate the effect of different globe sizes (10 mm and 50 mm) on GT, which were then compared to those obtained from the standard 150 mm globe used in the real scale setup. The real scale measurements were carried out in an industrial site in Singapore (a region characterized by a tropical climate) which represented a typical street canyon. Singapore (latitude: 1°17′ N, longitude: 103°50′ E) experiences high yearly solar insolation (1680 kWh/m^2) and moderate diurnal temperature variance (26–33 °C) throughout the year due to its proximity to the equator [24]. The site consisted of two parallel street canyons and four geometrically identical rectangular three-story buildings with matching geometrical characteristics, such as height-to-width ratio and orientation. The building walls were made of 200 mm thick bricks covered in cement plaster on both sides. The road between the two buildings was constructed of 150 mm thick asphalt concrete with a 550 mm gravel base. Further details regarding the real scale experiments can be found in the study conducted by the authors [25].

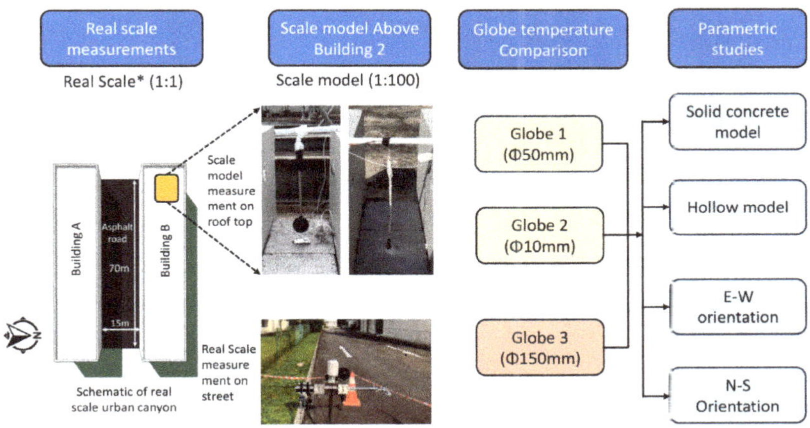

Figure 1. Schematic diagram showing workflow adopted in this study. Real Scale building [25] resembles the scale model in terms of geometry, orientation, surface reflectance and the climate of the study location.

Furthermore, the scale model measurements were performed parallelly (on one of the buildings on the same industrial site) to investigate the impact of cool materials and canyon orientation on temporal differences in GT between the cool and conventional canyons.

Experimental Design of Scale Model

The experiments were conducted outdoors using 1:100 scale models made of cement concrete (for building blocks) and asphalt concrete (for roads in between building blocks, representing four parallel residential apartment blocks of 90 m (L) × 15 m (B) × 40 m (H) in size, forming two side-by-side street canyons (as shown in Figure 2). The ration of the street width to the building height (aspect ratio) of the street canyon models is 1.3. The orientation of the street canyon is east-west, with the longer sides of the buildings facing the north and south. The materials used for building these scale models are selected carefully to replicate the processes of heat absorption, storage, and release from built-up surfaces, which is the key phenomenon in urban energy balance. Some of the previous studies involving scale model experiments [21,26] were referenced while designing the scale model for the experiments. In order to minimize errors due to scaling [27], the scaled model has been developed considering geometric similarities, i.e., the characteristic length of buildings, and kinematic similarities, i.e., the position and orientation of the buildings.

Cement concrete blocks, 900 mm × (L) 150 mm (B) × 400 mm (H) in size, were cast in the laboratory using concrete mix proportions similar to those of pre-cast concrete, with a cement, sand, and coarse aggregate weight ratio of 1:1.5:2 (SS 289, 2000), as shown in Figure 2. The density of the concrete block is assumed to be 2240 kg/m^3, with a thermal conductivity of 1.44 W/mK (U-value 3.7 W/m^2K) and a specific heat capacity of 0.75 kJ/kgK [28]. The road material between the blocks was comprised of small cubes of asphalt concrete, measuring 300 mm × 300 mm × 50 mm in size, with a density 2429 kg/m^3 and a specific heat capacity of 0.93 kJ/kgK [25]. Figure 3 shows the real images of the cement concrete and asphalt slabs prepared for the scale model setup.

Figure 2. Details of the side-by-side street canyon scale model test setup.

(a)

(b)

Figure 3. Preparation of scale models using: (**a**) cement concrete slab and (**b**) asphalt concrete slab.

In this study, a number of sensors were installed to measure surface temperature, heat flux, air velocity, and globe temperature. Two-wire PT100 sensors (Class A), 10 mm in size, and thermoelectric type heat flux sensors (gSKIN XO), 30 mm in size, were used for the surface temperature and heat flux measurements, respectively, as shown in Figure 2. Air velocity was measured using an air velocity transducer (TSI 8455). The air velocity measurements from a typical daily cycle show mostly still air conditions during the experiments (over 60% of the time, ≤0.1 m/s; average velocity, 0.2 m/s; maximum velocity, 1.6 m/s) due to parapet wall and other obstructions surrounding the model. Experimental uncertainties have been quantified in terms of sensor error for the temperature sensors as part of the calibration process, and it was determined that the systematic error was within the accepted sensor accuracy.

Hypothesizing that the size of the globe thermometer would have an effect on the GT, the first parameter investigated was the diameter of the globe. The globe sensor is customized using a 10 mm diameter ball (made of thin glass and coated with black matte paint) to suit the model scale; it is hung inside the canyon 20 mm away from the road surface to represent pedestrian height (Figure 3). A non-linear scaling was followed in this case to account for the practical constraints regarding the diameter of the globe with a temperature sensor inserted. Table 1 summarizes the instruments used for the measurements, along with the model and their accuracies.

Table 1. Details of instrumentation and sensors used in the experimental monitoring.

S. No.	Instrument	Model	Parameter	Accuracy
1	PT100 sensor (Singapore)	Class A	Temperature	±0.2–1.2 °C
2	12-bit smart sensor (Boston, MA, USA)	HOBO S-TMB-M006	Surface temperature	±0.2 °C
3	Hotwire anemometer (Singapore)	TSI 8455	Wind speed	±0.3 m/s
4	Heat flux plate (Zurich, Switzerland)	gSKIN XO	Heat flux	±3%
5	Infrared camera (Boston, MA, USA)	FLIR	Thermal image	±2 °C

In past studies using scaled models, the surface temperature was studied as a major contributing factor [2,5,28–32] when investigating the influence of canyon characteristics on the local climate. GT is chosen as a primary indicator for studying thermal interactions, as it could represent the combined effect of both radiation and convective heat transfer phenomena within the scaled street canyon. The GT has been investigated by changing the surface reflectance in various test cases. One canyon is coated with low albedo (wall 0.4, road 0.2) and the other with high albedo (wall 0.6, road 0.5) materials. Hereafter, the low albedo canyon is referred to as the low reflectance canyon and the high albedo canyon as the high reflectance canyon. The solar spectral (300–2500 nm) reflectance properties of these materials were measured in the lab in accordance with ASTM E903-12 [30] using a spectrophotometer (Perkin Elmer L950, Willingboro, New Jersey, US) [29]. Multiple rounds of experiments were conducted, each for a weeklong period during April 2019; the parameters mentioned above were logged at 30 s intervals. The test results for a selected typical non-rainy day were analyzed in terms of the difference in the GT between the low and high reflectance surfaces.

3. Results

3.1. 1 Comparison of Scale Model's Experimental Data with Real Scale Data

The results of scaling down with respect to the size of the globe and the effect of the solar orientation/envelope material are presented in this section. Figure 4 shows the difference in GT monitored between the low reflectance and high reflectance canyon models. A positive difference indicates that the low reflectance canyon resulted in a higher GT compared to that of the high reflectance canyon due to the higher absorption of solar radiation, which led to greater heat storage during the night. A negative difference during the day indicates that the high reflectance surface results in a higher GT due to exposure to solar radiation and reflection from surrounding surfaces (walls and roads). The orange and red colored lines in Figure 4 indicate a variation in the GT difference, with one measurement obtained using a standard 50 mm diameter globe and the other captured using a 10 mm globe, customized for the experiment (Figure 5). This data is compared with the real scale measurements, shown in grey color, carried out by the authors [28] in 5 min intervals during March 2019 using a 150 mm globe thermometer.

In comparing the temporal differences between the GT measurements observed for the 10 mm and the 50 mm globe thermometers, there is a negligible difference (close to zero) during the night in case of the 50mm globe. However, this difference was over 0.6–0.8 °C in the case of the 10 mm globe and up to 1 in case of the 150 mm globe used in the real scale experiment when compared with the results for the standard 50 mm globe. It is, however, important to note that the results for the 10 mm globe are more comparable with the difference in GT observed in the real scale (1:1) experiment. This was due to the obvious difference in the surface area of the globes. It is apparent that although the view factors/angle factors of the temperature sensor within the globe are same, the surface area of the 50 mm globe exposed to surrounding canyon surfaces is increased by five times in comparison with that of the 10 mm globe, which leads to higher GT values during the daytime and quicker heat dissipation (loss) during the nighttime. In addition to the dominant effect of direct solar radiation causing larger difference during the daytime, higher convective heat transfer in the case of the 50 mm globe seems to decrease the difference in GT during the nighttime. Thus, when studying the temperature of the canyon (enclosure) in scale models, the scaling of the globe is a concern.

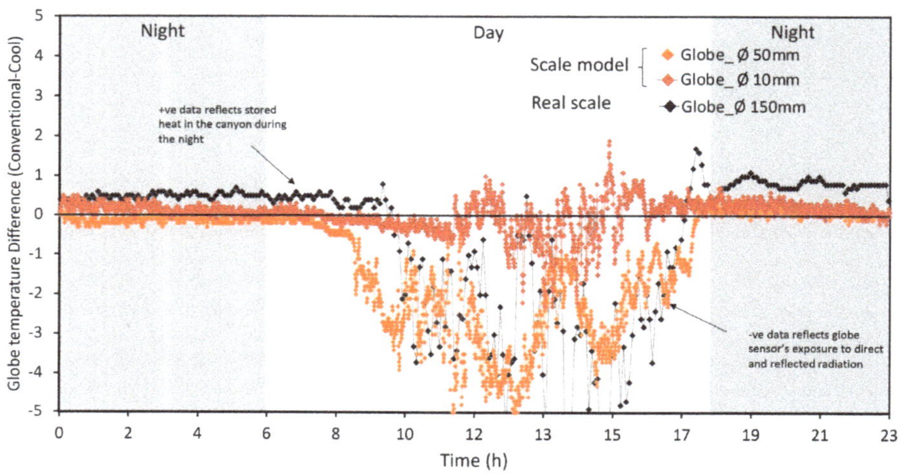

Figure 4. Comparison of globe temperature profiles of 50 mm, 10 mm, and 150 mm Ø black globes.

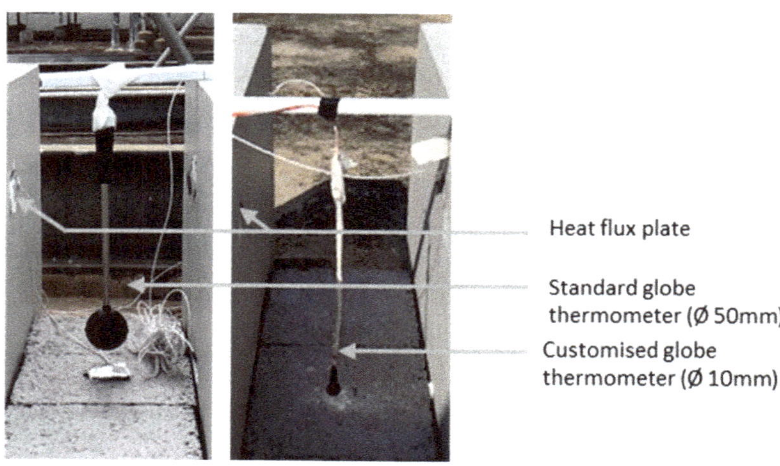

Figure 5. Scale model test setup showing globe thermometer sizes in 50 mm and 10 mm.

3.2. Relationship between Canyon Surface Temeprature and GT

Figure 6 illustrates the results collected after sunset; there is a correlation between the globe and surface temperatures in regards to the trend in radiant temperature with respect to changes in the surface temperature (long wave emission) between the wall and road surfaces. In general, the temperatures of both the wall and road surfaces of the conventional canyon are higher than those of the cool canyon. The slope of change in GT with respect to the road surface temperature is higher than that of the wall temperature. As the globe sensor is positioned closer to the road than to the wall, a change in its surface temperature seems to be more sensitive to a change in globe temperature. It is also possible that there is a higher radiant heat loss from the pavement surface compared to the wall surface due to a higher sky view factor. The change in wall temperature, however, seems to have a greater impact in terms of the magnitude of the GT due to the higher wall surface area.

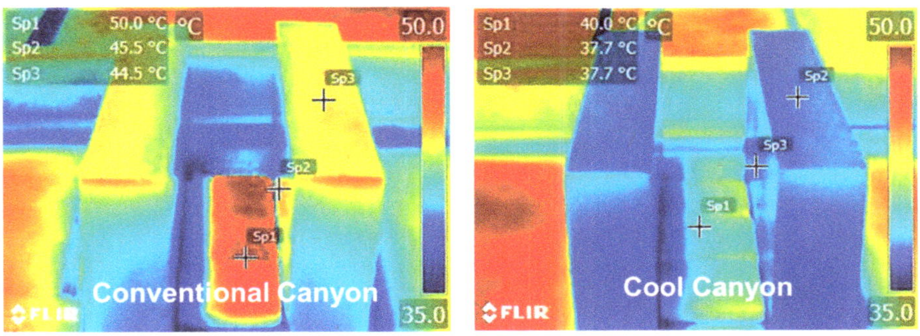

Figure 6. Variation in surface temperature of a conventional and cool canyon during a sunny afternoon.

Figure 7 shows thermal images captured for the scale model at 14:00 h, highlighting the temperature differences for the roof, wall, and road across low reflectance and high reflectance canyons. The thermography data represent qualitative results of the scale model experiment, with a lower surface temperature in the case of reflectance surfaces. The observed road surface temperature difference is up to 10 °C, whereas the wall temperature difference is up to 6.8 °C. It is evident that there is a greater difference in terms of surface reflectance for the asphalt surface (0.3) and the horizontal surface orientations compared to that of the wall surface. It can be observed that although the surface temperatures are lower in the high solar reflectance canyon compared to those in the low solar reflectance canyon, the GT is higher in the high reflectance canyon during the daytime, which could be due to the combined effect of direct radiation and short-wave reflection. However, for the hours after sunset, the GT decreases (up to 0.8 °C) in the case of the high reflectance canyon compared to the low reflectance canyon.

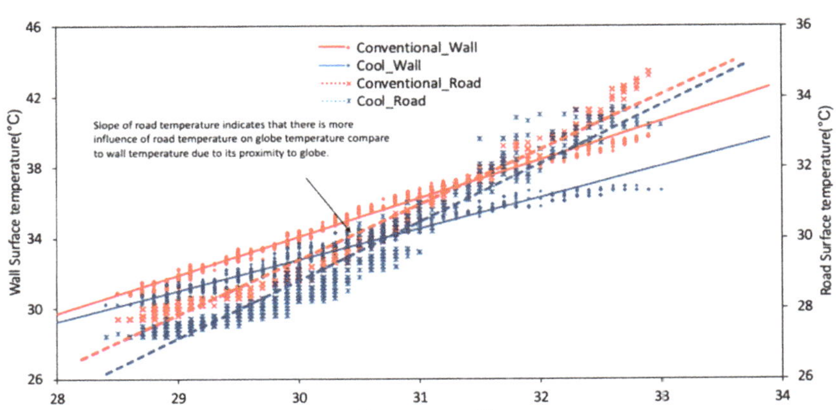

Figure 7. Correlation between nighttime globe temperature and wall/road surface temperature of conventional and cool canyons.

3.3. Effect of Wall Materials on the Globe Temperature

Considering the scaling of the wall material, a second model was created with a heat storage capacity that was scaled by $1/100^3$ times (scale factor 1/100 in all three directions of the wall, i.e., length, width, and thickness). Choosing the right material that matches the reduced scale heat storage capacity would be a challenge in this situation. In this case, 12 mm thick fiber cement boards (JamesHardie Villaboard) were used to construct the scaled geometric model, which is hollow inside. The heat capacity of the sheet (0.003135 MJ/K) matched closely with the reduced scale property of a precast concrete wall in a real scale (3276 MJ/K). The experimental results, however, show an insignificant difference between the globe temperatures measured in the conventional and the cool canyon, despite a considerable surface temperature difference during the day (Figure 8). This is possibly due to a negligible heat storage capacity and a higher heat release rate in the case of cement board compared to any real scale wall material (the thermal transmittance is over two times higher in the model than in the real scale), as the experiments were conducted in an outdoor setting. This effect could potentially be eliminated in a more controlled experimental setup by scaling the solar irradiation and adjusting the duration of the measurement.

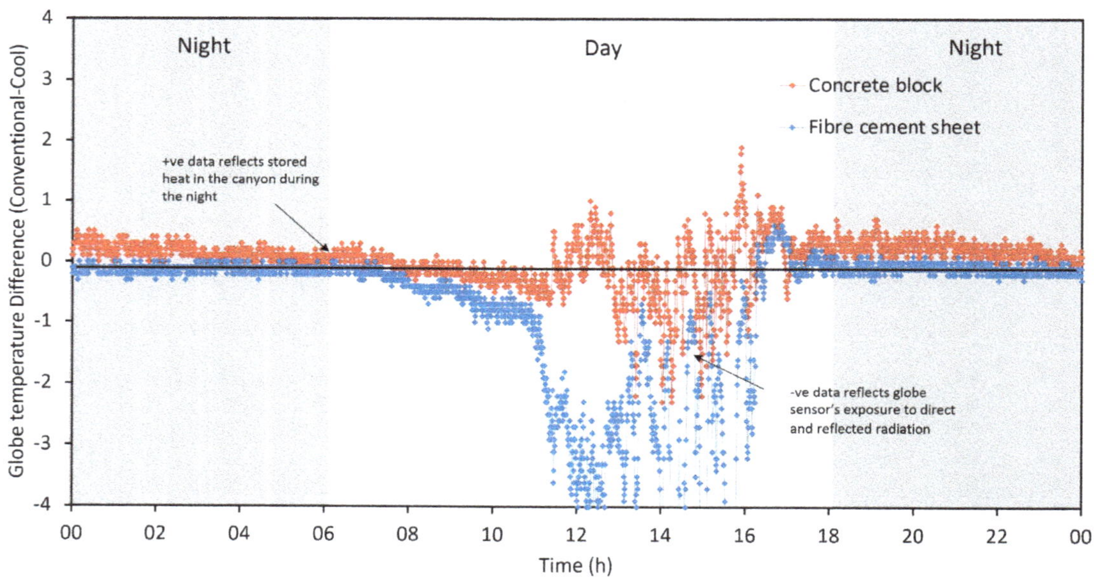

Figure 8. Temporal profile of globe temperature difference in case of concrete block and cement sheet models.

3.4. Effect of Solar Orientation on the Globe Temperature

The orientation of the canyon impacts the heat retained within the canyon. In other terms, the temperature in the canyon changes with the orientation, which has been established previously by scale model experimental investigations conducted in Italy [30]. This set of experiments shows a comparison between GT differences for two different street orientations (Figure 9). A higher positive difference can be noticed in the case of the E-W street orientation compared to the N-S street orientation due to higher solar irradiation received on E-W oriented streets.

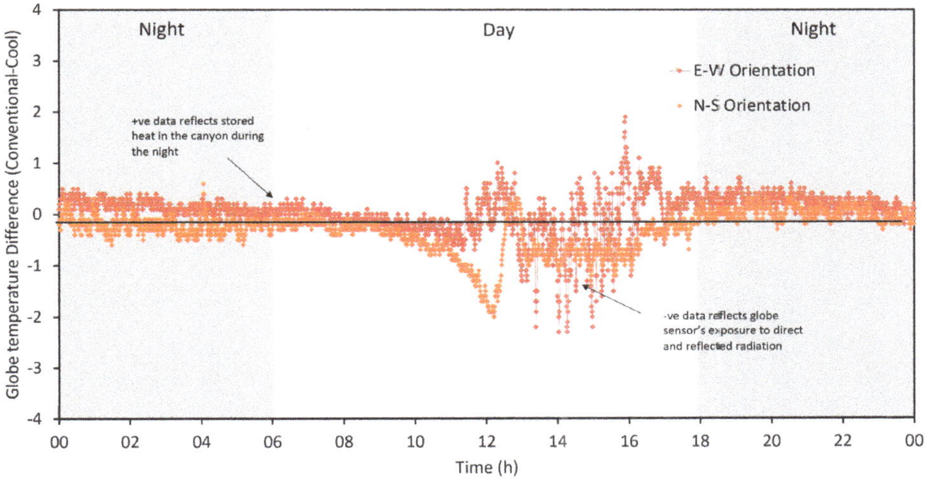

Figure 9. Temporal profile of GT difference in case of E-W and N-S canyon orientations.

3.5. Future Directions

The advantages of the scale model test setup developed in this study are not limited to the analysis of the impact of cool materials on street canyon temperatures. In practice, although this setup offers the advantages and flexibility of reduced scale models for conducting various parametric experiments (i.e., changing the surface reflectance and canyon orientations), this study could not investigate any possible air temperature differences between the two canyons. The focus of the set of experiments presented in this study has been on the GT, as well as on the surface temperature comparison between the low and high reflectance surfaces. However, to address the limitations theoretically, it is also hypothesized that the variation in canyon air temperature in the scaled model would be affected, mainly due to the change in the thermal flow field and the rate of surface heat transfer from the canyon facing surfaces. It is therefore proposed that in our future studies, mathematical thermal models should be combined with these scaled model results to further investigate the convective heat transfer and the consequent changes in canyon air and surface temperatures, based on the scale similarity theory.

4. Conclusions

This study discusses the test setups for studying a microclimate using scale models. Outdoor experiments using 1:100 reduced physical scale models were conducted to investigate the effect of building surface reflectance properties on microclimate temperatures. The temporal difference in globe temperature (GT) monitored inside the low solar reflectance and high solar reflectance canyon of the scale model was presented for three parameters, namely globe diameter, surface reflectance of the wall and street surfaces, and solar orientation. It is observed that although the surface temperatures are lower with high albedo compared to those with low albedo surfaces, the GT in the canyon is higher in the presence of high albedo materials during the daytime, which could be due to the dominance of short-wave reflection. However, for the hours after sunset, the GT becomes higher (up to 0.8 °C) in the case of a low solar reflectance albedo compared to that in a high solar reflectance canyon. It is also noted from this study that the area of the surrounding surfaces and the view factor formed by the globe thermometer are significant in regards to the GT values observed. This research might serve as a reference for policymakers as they design policies to encourage the use of cool materials to reduce urban heat by providing an evaluation of the thermal impact of cool materials prior to their large-scale adoption. The globe temperature and the surface temperature difference between the low and high reflecting surfaces were the main topics of interest for the set of tests reported in this study. However, in order to conceptually address the limitations of this study, future work will incorporate theoretical models to assess surface heat transfer rates from the canyon-facing surfaces and the fluctuation in the thermal flow field to understand the variation in canyon air temperature in the scaled model.

Author Contributions: Conceptualization, E.V.S.K.K.D.; methodology, E.V.S.K.K.D., S.S. (Shashwat Shashwat), K.T.Z., S.S. (Seshadhri Srinivasan) and M.P.W.; validation, E.V.S.K.K.D., S.S. (Shashwat Shashwat), K.T.Z., S.S. (Seshadhri Srinivasan) and M.P.W.; formal analysis, E.V.S.K.K.D., K.T.Z. and S.S. (Seshadhri Srinivasan); investigation, K.T.Z. and S.S. (Seshadhri Srinivasan); resources, E.V.S.K.K.D., S.S. (Shashwat Shashwat) and K.T.Z.; data curation, E.V.S.K.K.D. and S.S. (Shashwat Shashwat); writing—original draft, E.V.S.K.K.D. and S.S. (Shashwat Shashwat); writing—review and editing, K.T.Z.; supervision, K.T.Z. and M.P.W.; project administration, K.T.Z. and M.P.W.; funding acquisition, M.P.W. All authors have read and agreed to the published version of the manuscript.

Funding: This research was funded by the Ministry of National Development and the National Research Foundation, Prime Minister's Office, Singapore, under the Land and Liveability National Innovation Challenge (L2 NIC) Research Program (L2 NIC Award No. L2NICCFP2-2015-4). Any opinions, findings, and conclusions or recommendations expressed in this material are those of the author(s) and do not reflect the views of the Singapore Ministry of National Development and National Research Foundation, Prime Minister's Office, Singapore. Additionally, this research is also financially supported by the Faculty Fund and Research Development Fund from Northumbria University, Newcastle upon Tyne, UK.

Data Availability Statement: The raw data supporting the conclusions of this article will be made available by the authors on request.

Conflicts of Interest: The authors declare no conflicts of interest.

Abbreviations

GT	globe temperature
MRT	mean radiant temperature
UHI	urban heat island

References

1. Imbabi, M.S.-E. Scale modelling of the built environment. *Energy Build.* **1991**, *17*, 7–12. [CrossRef]
2. Lirola, J.M.; Castañeda, E.; Lauret, B.; Khayet, M. A review on experimental research using scale models for buildings: Application and methodologies. *Energy Build.* **2017**, *142*, 72–110. [CrossRef]
3. Lam, K.M.; Tam, P.W. State-of-the-art Wind Tunnel Modelling for Building Aerodynamics. *HKIE Trans.* **1996**, *3*, 17–26. [CrossRef]
4. Syafii, N.I.; Ichinose, M.; Kumakura, E.; Jusuf, S.K.; Chigusa, K.; Wong, N.H. Thermal environment assessment around bodies of water in urban canyons: A scale model study. *Sustain. Cities Soc.* **2017**, *34*, 79–89. [CrossRef]
5. Kanda, M. Progress in the scale modeling of urban climate. *Theor. Appl. Climatol.* **2006**, *84*, 23–33. [CrossRef]
6. Sharma, A.; Wuebbles, D.J.; Kotamarthi, R. The need for urban-resolving climate modeling across scales. *AGU Adv.* **2021**, *2*, e2020AV000271. [CrossRef]
7. Karlický, J.; Huszár, P.; Halenka, T.; Belda, M.; Žák, M.; Pišoft, P.; Mikšovský, J. Multi-model comparison of urban heat island modelling approaches. *Atmos. Chem. Phys.* **2018**, *18*, 10655–10674. [CrossRef]
8. Luo, X.; Peng, Y. Scale effects of the relationships between urban heat islands and impact factors based on a geographically-weighted regression model. *Remote Sens.* **2016**, *8*, 760. [CrossRef]
9. Trane, M.; Giovanardi, M.; Pejovic, A.; Pollo, R. Overview on urban climate and microclimate modeling tools and their role to achieve the Sustainable Development Goals. In *Architecture and Design for Industry 4.0: Theory and Practice*; Springer: Berlin/Heidelberg, Germany, 2023; pp. 247–267.
10. Tan, C.L.; Wong, N.H.; Jusuf, S.K. Outdoor mean radiant temperature estimation in the tropical urban environment. *Build. Environ.* **2013**, *64*, 118–129. [CrossRef]
11. Guo, H.; Aviv, D.; Loyola, M.; Teitelbaum, E.; Houchois, N.; Meggers, F. On the understanding of the mean radiant temperature within both the indoor and outdoor environment, a critical review. *Renew. Sustain. Energy Rev.* **2020**, *117*, 109207. [CrossRef]
12. Bedford, T.; Warner, C. The globe thermometer in studies of heating and ventilation. *Epidemiol. Infect.* **1934**, *34*, 458–473. [CrossRef] [PubMed]

13. Aparicio, P.; Salmerón, J.M.; Ruiz, Á.; Sánchez, F.J.; Brotas, L. The globe thermometer in comfort and environmental studies in buildings. *Rev. Construcción. J. Constr.* **2016**, *15*, 57–66. [CrossRef]
14. Oliveira, A.V.M.; Raimundo, A.M.; Gaspar, A.R.; Quintela, D.A. Globe temperature and its measurement: Requirements and limitations. *Ann. Work Expo. Health* **2019**, *63*, 743–758. [CrossRef]
15. Kuehn, L.; Stubbs, R.; Weaver, R. Theory of the globe thermometer. *J. Appl. Physiol.* **1970**, *29*, 750–757. [CrossRef]
16. Banfi, A.; Tatti, A.; Ferrando, M.; Fustinoni, D.; Zanghirella, F.; Causone, F. An experimental technique based on globe thermometers for the measurement of mean radiant temperature in urban settings. *Build. Environ.* **2022**, *222*, 109373. [CrossRef]
17. Guo, H.; Teitelbaum, E.; Houchois, N.; Bozlar, M.; Meggers, F. Revisiting the use of globe thermometers to estimate radiant temperature in studies of heating and ventilation. *Energy Build.* **2018**, *180*, 83–94. [CrossRef]
18. Teitelbaum, E.; Chen, K.W.; Meggers, F.; Guo, H.; Houchois, N.; Pantelic, J.; Rysanek, A. Globe thermometer free convection error potentials. *Sci. Rep.* **2020**, *10*, 2652. [CrossRef] [PubMed]
19. Pearlmutter, D.; Berliner, P.; Shaviv, E. Integrated modeling of pedestrian energy exchange and thermal comfort in urban street canyons. *Build. Environ.* **2007**, *42*, 2396–2409. [CrossRef]
20. Kawai, T.; Kanda, M. Urban energy balance obtained from the comprehensive outdoor scale model experiment. Part I: Basic features of the surface energy balance. *J. Appl. Meteorol. Climatol.* **2010**, *49*, 1341–1359. [CrossRef]
21. Bonamente, E.; Rossi, F.; Coccia, V.; Pisello, A.L.; Nicolini, A.; Castellani, B.; Cotana, F.; Filipponi, M.; Morini, E.; Santamouris, M. An energy-balanced analytic model for urban heat canyons: Comparison with experimental data. *Adv. Build. Energy Res.* **2013**, *7*, 222–234. [CrossRef]
22. Shashwat, S.; Zingre, K.T.; Thurairajah, N.; Kumar, D.K.; Panicker, K.; Anand, P.; Wan, M.P. A review on bioinspired strategies for an energy-efficient built environment. *Energy Build.* **2023**, *296*, 113382. [CrossRef]
23. Elnabawi, M.H.; Hamza, N.; Raveendran, R. 'Super cool roofs': Mitigating the UHI effect and enhancing urban thermal comfort with high albedo-coated roofs. *Results Eng.* **2023**, *19*, 101269. [CrossRef]
24. Devs, K.K.; Wan, M.P.; Zhou, M.; Long, Y.; Ng, B.F. Impact of solar reflectance of wall and road on outdoor thermal comfort-experimental study in a street canyon setup. In *Proceedings of the MATEC Web of Conferences*; EDP Sciences: Ulis, France, 2019; p. 02010.
25. Donthu, E.K.K.; Long, Y.P.; Wan, M.P.; Zhou, M.; Ng, B.F. Dynamics of cool surface performance on urban microclimate: A full-scale experimental study in Singapore. *Sustain. Cities Soc.* **2024**, *102*, 105218. [CrossRef]
26. Pearlmutter, D.; Berliner, P.; Shaviv, E. Physical modeling of pedestrian energy exchange within the urban canopy. *Build. Environ.* **2006**, *41*, 783–795. [CrossRef]
27. Kalloe, S.A.; Hofland, B.; Van Wesenbeeck, B.K. Scaled versus real-scale tests: Identifying scale and model errors in wave damping through woody vegetation. *Ecol. Eng.* **2024**, *202*, 107241. [CrossRef]
28. Donthu, E.K.K.; Shashwat, S.; Zingre, K.T.; Wan, M.P. Development of a simplified cool coating thermal model for predicting street canyon air temperature. *Build. Environ.* **2024**, *251*, 111207. [CrossRef]
29. Morini, E.; Castellani, B.; Presciutti, A.; Anderini, E.; Filipponi, M.; Nicolini, A.; Rossi, F. Experimental analysis of the effect of geometry and façade materials on urban district's equivalent albedo. *Sustainability* **2017**, *9*, 1245. [CrossRef]
30. *ASTM E903-2012*; Standard Test Method for Solar Absorptance, Reflectance, and Transmittance of Materials Using Integrating Spheres, American Society for Testing and Materials. ASTM International: West Conshohocken, PA, USA, 2012.
31. Shashwat, S.; Zingre, K.; Thurairajah, N. Enhancing building energy efficiency estimation by integrating microclimate conditions. In Proceedings of the 38th Annual ARCOM Conference, Glasgow, UK, 5–7 September 2022.
32. Pisello, A.L.; Piselli, C.; Cotana, F. Thermal-physics and energy performance of an innovative green roof system: The Cool-Green Roof. *Sol. Energy* **2015**, *116*, 337–356.

buildings

MDPI

Article

Research on the Influence of Open Underground Space Entrance Forms on the Microclimate: A Case Study in Xuzhou, China

Ping Chen [1], Lufeng Nie [2], Jinrun Kang [1] and Heng Liu [3,*]

[1] School of Architecture and Engineering, Huaiyin Institute of Technology, Huaian 223001, China; 11050031@hyit.edu.cn (P.C.); kangjr@hyit.edu.cn (J.K.)
[2] School of Mechanics and Civil Engineering, China University of Mining and Technology, Xuzhou 221116, China; nielufeng@cumt.edu.cn
[3] School of Architecture and Design, China University of Mining and Technology, Xuzhou 221116, China
* Correspondence: liuheng@cumt.edu.cn

Abstract: With urban development and renewal, underground space is becoming more utilized. The design and use of open underground public space entrances and exits have become more and more frequent. As a pedestrian passage connecting indoors and outdoors, the wind and thermal environment of open entrances have a great impact on human comfort. This paper investigates the open underground space entrances and exits in Xuzhou. Physical environments such as temperature and wind speed were measured. Through numerical simulation, the influence relationships between the spatial form elements of open entrances and exits and microclimate and thermal comfort were investigated. This study showed that there are four common spatial morphological elements of open entrances and exits. The physiologicafl equivalent temperature (PET) of the outdoor part of the entrance is the highest in summer, and the lowest in winter, and the PET is most affected by the shape of the opening plane and the aspect ratio, which are linearly related. The trends of the spatial morphology elements were not consistent when seeking the optimal situation of PET in summer and winter, respectively. The relationship between the spatial form elements of entrances and PET established in this study provides technical guidance for the design of open entrances, which can help improve environmental quality and enhance human comfort.

Keywords: underground space; entrance forms; microclimate; thermal comfort; PET; numerical simulation

1. Introduction

With the development of cities and the increase in population density, urban land is gradually strained. Underground space has a huge development prospect as urban stock land [1,2]. According to statistics, the cumulative floor area of urban underground space in China is 2.962 billion square meters, and the average per capita size of underground space is 3.8 square meters. Japan has built at least 146 underground streets in 26 cities, and the number of people entering and leaving the underground streets has reached 12 million. The New York City subway in the United States has the longest operating line in the world (443 km), the largest number of stations (504), and receives 5.1 million daily trips, or nearly 2 billion trips per year. As urban underground spaces are being further utilized, it is becoming increasingly important to study the thermal environment of underground spaces, considering people's health and comfort [3].

As a pedestrian passage connecting indoors and outdoors, the entrance of the underground space can interact with the outdoor environment and, thus, change the physical environment of the indoor. Their morphological design and physical environment are essential. Some scholars have studied the relationship between PM10 concentration in parking lots, underground hotels' spatial depth, and the surrounding thermal and humid environments [4,5]. This kind of underground space is deeper and has a weaker correlation with the outdoor environment. Other scholars have investigated the relationship between atrium morphology, spatial form proportions, materials and vents, and underground spaces' thermal and humid environments [6–8]. Their studies take advantage of the characteristics of underground buildings interacting with the ground and the outdoors and naturally regulate them by changing the spatial forms, materials, systems, or other passive facilities at the entrances and exits. In addition, many scholars have studied the relationship between the morphology of entrances and exits of subway spaces such as subways and commercial exhibition buildings and human comforts, such as ventilation, temperature, air quality, and visual preference, with regard to the national conditions and climatic conditions of each country [9–13]. With underground space development, the form of entrances also tends to diversify gradually. The spatial form is divided into three forms: open entrance and exit, independent foyer-type entrance and exit, and sunken plaza-type entrance and exit. Since underground space construction is later than the general urban development, the newly developed underground space is subject to the limitations of norms, land use rights, and existing building protection, and only open entrances and exits can be used in some areas. In addition, the short construction period and lower impact on the surrounding environment are also advantages of open entrances and exits.

There is no transition space as open entrances differ from separate foyer and sunken plaza entrances. This situation leads to a direct connection between indoor and outdoor environments, which has a greater impact on human thermal comfort. Many scholars have extensively researched underground space environments for different space types. Wen Yueming et al. proposed design strategies to enhance comfort and health based on solar radiation and air pollution to retrofit sunken plaza-type entrances and exits in underground spaces [14]. Guo Haoxu et al. found that the choice of the orientation of a sunken plaza has a significant influence on the thermal environment and human comfort of underground space [15]. The above studies mainly focused on sunken plaza-type entrances and exits. Liu Yanan et al. investigated the effects of outdoor wind speed, flow resistance of the subway station, height and length of the shelter, and other parameters on the airflow at the entrance of the subway station through numerical simulation of the horizontal entrance of the subway station [16]. Guan Bowen et al. found that the larger the entrance opening, the better the indoor air quality [17]. These studies explored the effects of wind and thermal environments on comfort. It provides a theoretical basis for the design of entrances and exits. Therefore, studying the microclimate and human comfort of open entrances and exits is important. However, these studies have certain limitations. On the one hand, few studies have focused on field measurements; on the other hand, few studies have determined the spatial morphology elements of underground space entrances and exits through a large amount of research, as well as few studies have quantified the effects of multiple morphology features of entrances and exits on the thermal environment and human comfort in underground spaces.

Therefore, this study aims to explore the effects of spatial morphological changes of open entrances and exits on microclimate and human comfort. The spatial morphology characteristics of general open underground space entrances and exits were collected and organized through field research. A representative open underground space entrance in Xuzhou, China, was selected. The microclimate physical environment and wind speed of the entrance and exit were measured. The relationship between spatial morphology parameters and PET was mathematically modeled by modeling various spatial patterns of entrances and exits and software simulation, and the general pattern of the influence of spatial patterns of open entrances and exits on microclimate was finally derived. Finally, the general rule of the influence of open entrance space pattern on microclimate was found, and the optimization design strategy of entrance space pattern for human comfort was proposed. The results of this study provide data support and technical guidance for the design of open entrances and exits in underground spaces, which can help improve the environmental quality of underground entrances and exits and enhance human comfort.

2. Materials and Methods

2.1. Overview

In order to explore the influence of underground public entrances and exits on the surrounding wind and thermal environment, this study selected the open underground public space entrances and exits and applied the response surface method (RSM). This statistical method uses reasonable experimental design methods and obtains certain data through experiments, adopts multiple quadratic regression equations to fit the functional relationship between factors and response values, and seeks the optimal parameters through the analysis of regression equations to solve the multivariable problem [18]. The response surface method finds the relationship between the output response value and the input design parameter by conducting a limited number of experiments or simulations [19,20]. This type of relational model is called the response surface model. It reveals the effect of design parameters on the response value. In this way, according to the influence relationship shown by the model, the design parameters can be adjusted to optimize and even achieve the best response effect, significantly reducing the experiment and repetitive work [21]. In addition, the response surface method has been applied to the design of indoor air quality and thermal comfort in the built environment. In 2010, Tomas Norton conducted a quantitative study on the spatial form and eave opening conditions of livestock farms and improved the uniformity of the indoor wind environment of livestock farms by changing these design variables [22]. Xiong Shen used different DOE experimental design methods in 2012 to explore the relationship between wind speed and wind direction changes in buildings and ventilation rates [23]. All of them have explored ways to improve the natural ventilation performance of buildings. Zhou Liang studied the distribution of underfloor air distribution (UFAD) in 2008 and used response surface design to establish appropriate experimental simulation, aiming to improve thermal comfort and indoor air quality without increasing the energy consumption of mechanical ventilation [24]. In 2013, Chen Huijuan used a response surface method and CFD to study the influence of different configurations and flow parameters of impact jet ventilation on local thermal discomfort [25]. Both of them have explored ways to improve the mechanical properties of buildings. There are other studies on thermal comfort [26]. In order to analyze and explore the relationship between the underground public space entrance and exit forms that interact with the ground and outdoor climate and the surrounding microclimate environment, this paper generates a polynomial relationship model through RSM to describe the relationship between multiple parameterized architectural features and climate environment.

The research process is shown in Figure 1. Firstly, the paper conducts a field investigation on the open entrance and exit, summarizes the spatial characteristics of the open underground public space entrance and exit, and selects the corresponding quantitative and categorical form elements. Secondly, the current wind and thermal environment of typical entrances and exits is measured to clarify the characteristics of urban wind and thermal environment of the entrances and exits of open underground public Spaces. The typical inlet and outlet air thermal environment is simulated and verified with the actual situation. Thirdly, the response surface method was used to set the simulation experimental group and simulate the experimental group with the obtained quantitative spatial morphological elements as the design parameters. Fourthly, the regression analysis and data verification of the simulation results are carried out to obtain the prediction model. Finally, the corresponding design optimization method of the entrance and exit space form is proposed.

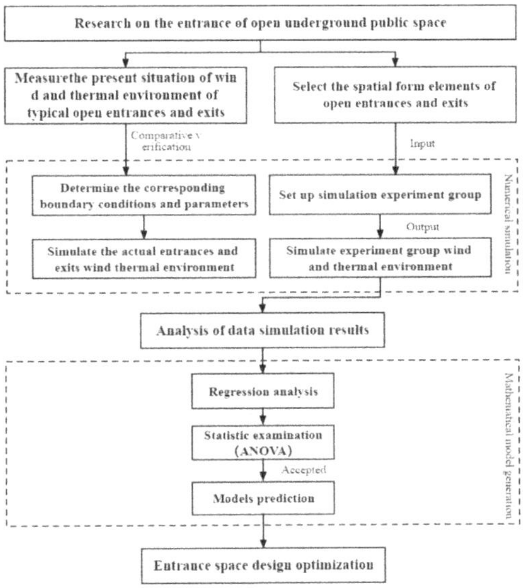

Figure 1. Research framework of influence of entrance and exit shape on wind and thermal environment.

2.2. Measurement Method

2.2.1. Measured Time

Xuzhou is located in the cold area of China's building thermal engineering zone, but its climate is both hot in summer and cold in winter [27]. In the process of outdoor climate design, it is necessary not only to consider the contradiction of winter cold protection but also to pay attention to summer heat protection and ventilation [28]. Therefore, this study fully considered the above conditions in the field measurement process, selected the more typical natural days in summer and winter, and observed the seasonal characteristics of the thermal environment at the entrances and exits of urban underground public Spaces. In addition, considering the influence of rainfall and direct sunlight on the measured results, a cloudy natural day with no rainfall was selected [29]. The specific time of summer measurement is from 20 August to 26 August 2021; The actual time of winter measurement was from 20 December to 26 December.

For the selection of actual time, it is necessary to consider the opening time of public places and the high frequency of people's use time. The venue is open from 6 a.m. to 11 p.m., which overlaps with the residents' main travel and outdoor activities [30,31]; 7:00–10:00 and 16:00–19:00 are the rush hours, and the main dining, shopping, and leisure hours, which can better represent the main usage feelings of the population. Therefore, it is of great significance to pay attention to the environmental comfort of this period. Therefore, the specific recording time is set to 7:00–10:00 and 16:00–19:00.

2.2.2. Tool Selection

The actual measuring instrument is JT2020(Beijing Shijijiantong Technology, Beijing, China) multifunctional environmental tester. The environmental tester can simultaneously collect air temperature, black globe temperature, relative humidity, wind speed, and other climatic environment data. The physical parameters to be collected in this paper are air temperature, black globe temperature, relative humidity, and breeze wind speed, and the measuring instruments all meet the measurement accuracy and error requirements specified in the literature.

2.2.3. Measured Site

The test site is located in Pengcheng Square, Xuzhou City, Jiangsu Province. Firstly, Xuzhou is located in the "cold region" in China's building climate zoning and building thermal design zoning, which has certain special climate characteristics. Secondly, Xuzhou has abundant underground space. Moreover, this area is located in the center of Xuzhou city, which is the largest city center square in the province, with huge underground space. Covering a variety of functions such as shopping malls, pedestrian streets, museums, subways, and underground passages, the flow of people is large. Finally, the case has a rich type of underground space entrances and exits. There is a certain amount of urban construction, and there is a rich urban neighborhood form [32]. The surrounding climate and environmental conditions are more consistent, which is conducive to actual measurement and analysis and comparison. Therefore, the two open underground public space entrances and exits with the largest flow of people among them are selected [31]. They are Exit 5 of Pengcheng Square Station of Metro Line 1 (Case 1) and the West Entrance of Golden Eagle Underground Supermarket (Case 2). In Case 1, the material is ordinary concrete; in Case 2, the structural part of the opening baffle is ordinary concrete with a layer of perforated aluminum plate, and the rest is the same as in Case 1. The specific actual measurement points are shown in Figure 2. The measurement points are located at the edges of the entrance and the walkway, avoiding the middle part where people move around as much as possible. Measurement points are located at the edges of entrances and walkways, avoiding the middle part where people move frequently. The effect of crowd flow on the actual results will be taken into account, and the measurement error will be corrected in the following statistics.

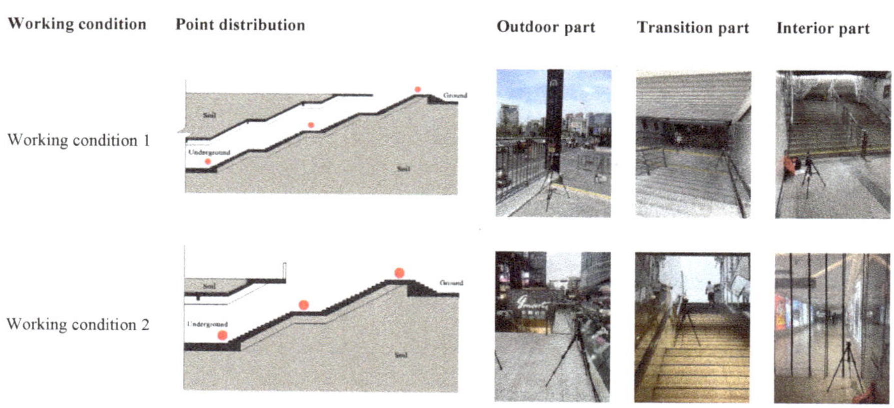

Figure 2. Open entrance and exit measured point arrangement.

2.2.4. Measured Process

The measurement follows three parts: outdoor slow walking, transition space slow walking, and indoor slow walking. The outdoor preparation stage is mainly for the installation preparation of the instrument, and the instrument is used to collect data every 15 s in the three parts of the outdoor, transition, and indoor space.

2.3. Design Experiment Method

In order to judge the influence of the spatial form index of the entrance and exit of underground public space on the wind and thermal environment around the entrance and exit, effectively control the shape parameters with greater influence and maximize the comfort level of the wind and thermal environment of the entrance and exit. Therefore, the response surface method (RSM) was used to analyze the correlation between the quantitative indicators of the spatial morphology of the entrance and exit and the existing wind thermal physical environment data, and the main influencing indicators were found. The response surface method is a statistical method that uses a reasonable experimental design method and obtains certain data through experiments, adopts multiple quadratic regression equations to fit the functional relationship between factors and response values, and seeks the optimal parameters through the analysis of regression equation to solve the multivariable problem [33]. The experimental design of the response surface method includes central composite design CCD, Box-Behnken Design BBD, Doptimality, orthogonal design OA, uniform design UN, etc. [34]. This study will use Design Expert as the main test design and result analysis output software to complete the main body of test design, sample extraction, data result fitting, equation establishment, trend analysis, target optimization, and other work. Select the Box-Behnken Design BBD method in the response surface method module to operate [35]. Through the correlation and regression analysis of different elements, the influence of the spatial form elements of the entrance and exit on the air and thermal environment of the underground public space is determined.

2.4. CFD Software Simulation Process

The CFD simulation software used in this article is FLUENT 2020 R2. The main steps of CFD software simulation are divided into three parts. The first part is to import the completed block model into the preprocessor, set the border inspection conditions and calculation conditions, and divide the calculation grid. The second part is to simulate the existing model, load the model and data of the previous part, and calculate and output the calculation result. The third part is to generate the corresponding simulation analysis results and output them in the form of visualization. The specific basic flow of numerical simulation of wind and thermal environment is shown in Figure 3.

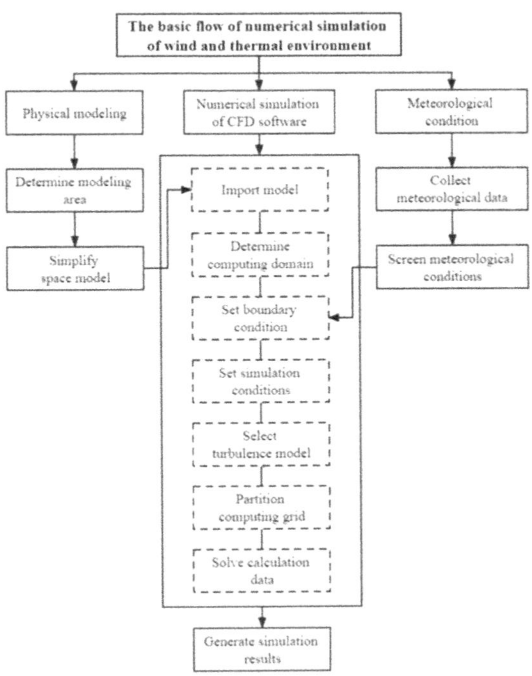

Figure 3. The basic process of numerical simulation of the urban wind-heat environment.

Two of the most important aspects are the setting of external conditions and the physical modeling of the simulated area. The setting of boundary conditions is mainly to determine the simulated wind and thermal environment [36]. Meteorological data, including temperature, humidity, wind speed, etc., are determined by actual measurement, and the corresponding data parameters are set as boundary conditions according to software requirements. For simulation area modeling, the key morphological features of the model should be retained first, and the hardware requirements of the computer should be taken into account to reduce irregular model details, reduce the modeling workload, and the complexity of calculation and solution so as to reduce the problem of increasing calculation time or calculation errors due to too small grid-scale [37,38]. Therefore, targeted deletion should be carried out, integrated, and simplified into regular spatial models.

3. Results

3.1. Analysis of Measured Results

3.1.1. Field Research

The types of entrances and exits of underground public Spaces are rich, and the design of space forms are also different. Therefore, the author conducted an in-depth investigation of the existing open underground public space entrances and exits, totaling 10. It covers the entrances and exits of underground public Spaces around Xuzhou Suning Square, Sanpower Square, Yunlong Campus of Normal University, Nanhu Campus of China University of Mining and Technology, and City Wall Museum.

Research site 1: The entrance and exit of the underground public space is located at the north side of Huaihai East Road and the west side of Pengcheng Road, Gulou District, Xuzhou City, Jiangsu Province, at the intersection of the two roads, which is the exit 5 of Pengcheng Square Station of Xuzhou Metro Line 1. From the ground, directly through the steps into the underground space, there is no shelter roof above.

Research site 2: The entrance and exit of the underground public space are located on the north side of Heqing Road and the west side of Pengcheng Road, Gulou District, Xuzhou City, Jiangsu Province. It is the entrance of the underground supermarket of Golden Eagle International Shopping Center and also exit 19 of Pengcheng Square Station of Metro Line 1. From the ground, directly through the steps into the underground space, there is no shelter roof above.

Research site 3: The entrance and exit of the underground public space is located at the north side of Heqing Road and the west side of Pengcheng Road, Gulou District, Xuzhou City, Jiangsu Province. It is the entrance and exit of the central Fashion Avenue and the underground pedestrian corridor at the intersection of the two roads. From the ground directly through the steps into the underground space, there is a semishielded roof above.

Research site 4: The entrance of the underground public space is located north of Huaihai East Road, Gulou District, Xuzhou City, Jiangsu Province, and south of Suning Square B1 south entrance. There is no shelter roof above from the ground directly through the steps into the underground space.

Research site 5: The entrance and exit of the underground public space are located near Sanpower Square, Tongshan District, Xuzhou City, Jiangsu Province, and the southwest entrance of the underground Food Street of Hi Street. From the ground through the steps, experience two turns into the underground food city, with no shelter above.

Research site 6: The entrance and exit of the underground public space are located near Sanpower Square, Tongshan District, Xuzhou City, Jiangsu Province, and the north entrance of the underground Food Street, Hi Street. There is an escalator above the shelter from the ground directly through the steps into the underground food city.

Research site 7: The entrance and exit of the underground public space is located at the north side of Heqing Road and the west side of Pengcheng Road, Gulou District, Xuzhou City, Jiangsu Province, at the intersection of the two roads, and is exit No. 18 in Pengcheng Square Station, Metro Line 1. There is no shelter roof above the ground directly through the steps into the underground space.

Research site 8: The entrance and exit of the underground public space is located in the underground city site Museum, Gulou District, Xuzhou City, Jiangsu Province. It is the north exit of the downtown Museum. There is no shelter roof above the ground directly through the steps into the underground space.

Research site 9: The entrance and exit of the underground public space is located in the City Wall Museum, Yunlong District, Xuzhou City, Jiangsu Province. It is the east exit of the City Wall Museum. There is no shelter roof above from the ground directly through the steps into the underground space.

Research site 10: The entrance and exit of the underground public space is located opposite the west gate of Wenchang Campus, Tongshan District, Xuzhou City, Jiangsu Province. Mine West Food City south entrance. There is no shelter roof above from the ground directly through the steps into the underground space.

According to the basic spatial characteristics of the research objects above, the entrances and exits of open underground public spaces are classified as shown in Table 1. The following table lists the profile types of each open entrance and exit investigated and summarizes its basic spatial prototype.

Table 1. Open underground public space entrances and exits.

Research Site	Location	Site Photo	Profile Prototype
Research site 1	Xuzhou Metro Line 1 Pengcheng Square Station, Exit 5		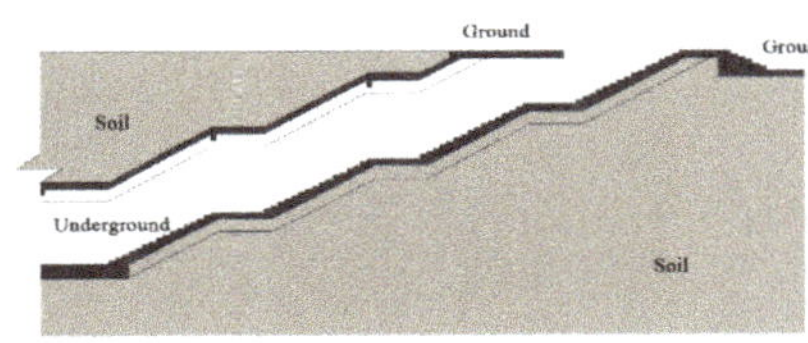
Research site 2	Xuzhou Metro Line 1 Pengcheng Square Station, Exit 19		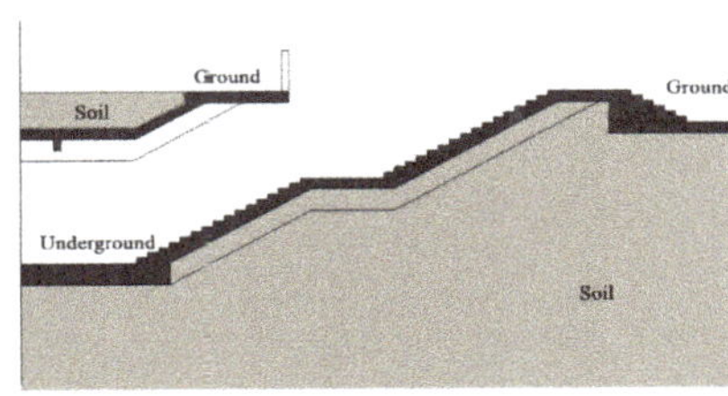
Research site 3	Xuzhou city center fashion Avenue and underground walkway north entrance		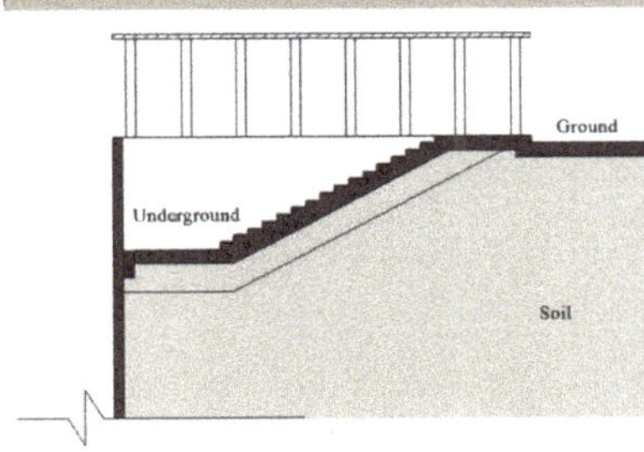
Research site 4	Xuzhou Suning Plaza B1 south entrance		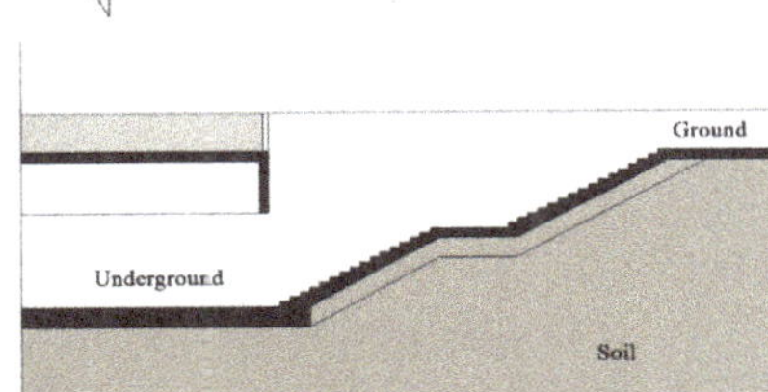

Table 1. *Cont.*

Research Site	Location	Site Photo	Profile Prototype
Research site 5	Sanpower Square hi Street underground food street southwest entrance		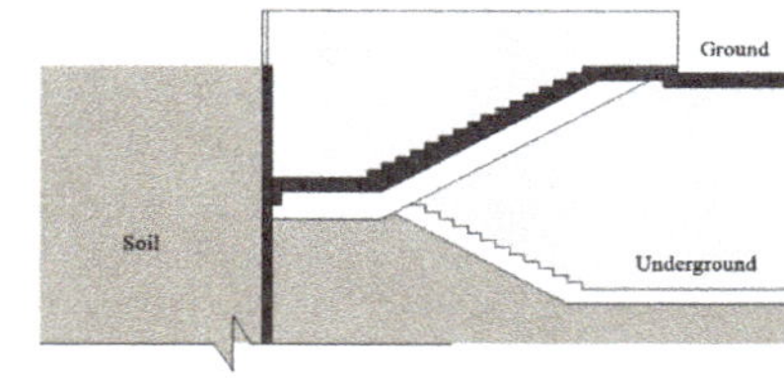
Research site 6	Sanpower Square hi Street underground food street north entrance		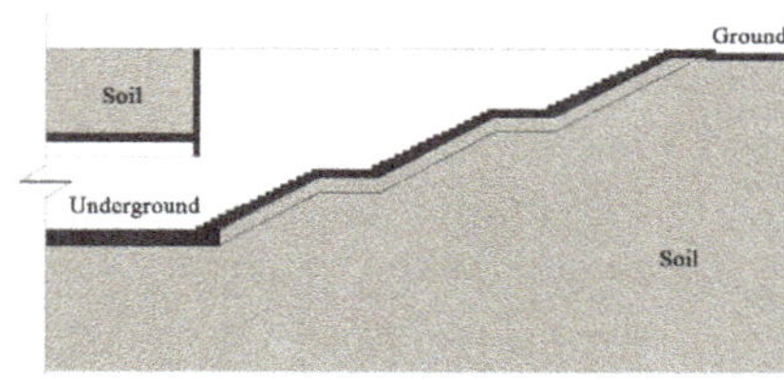
Research site 7	Xuzhou Metro Line 1 Pengcheng Square Station, Exit 18		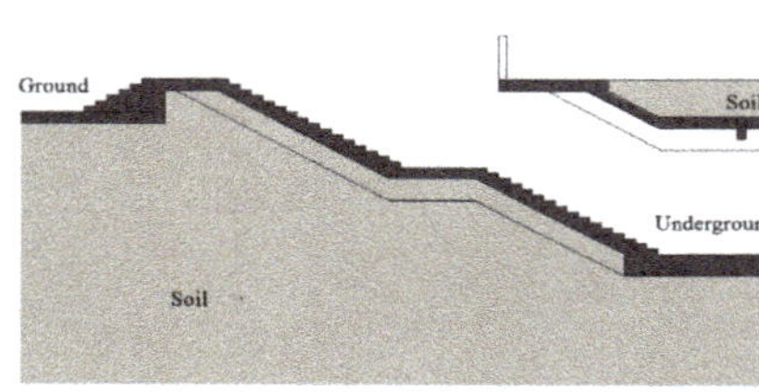
Research site 8	Xuzhou underground city site Museum north exit		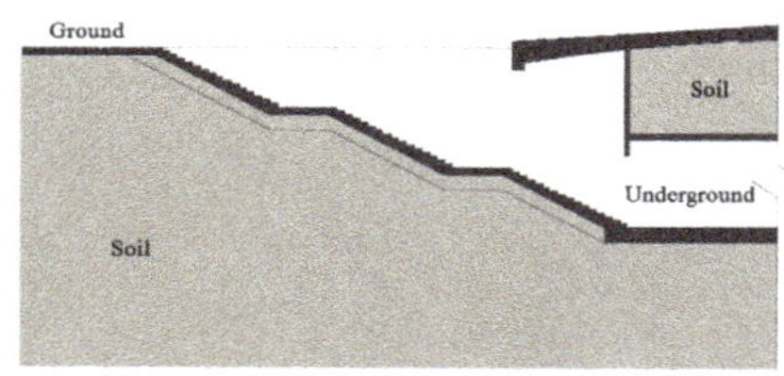
Research site 9	Xuzhou City wall Museum east entrance		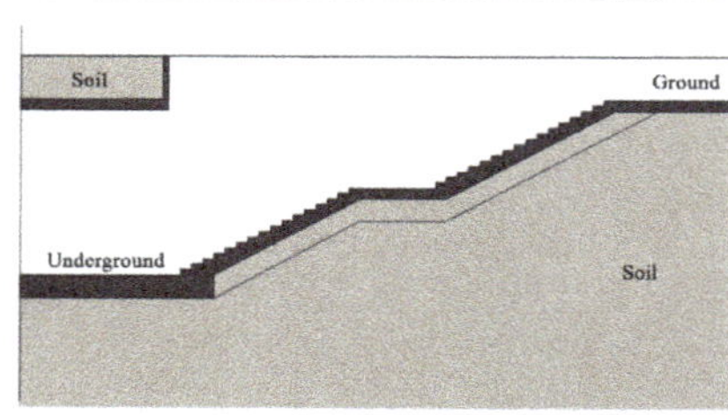
Research site 10	Xuzhou mineral west food city		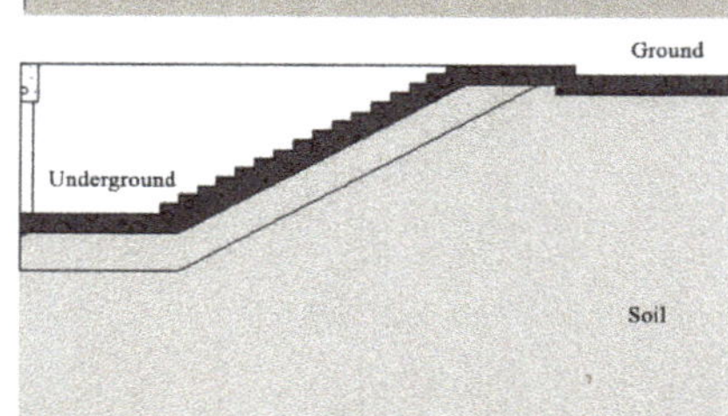

3.1.2. Typical Inlet and Outlet Air Thermal Environment Measurement

(1) Summer

The environmental and physical parameters of the outdoor part, the transition part, and the indoor part in the three stages of the first group of open inlet and outlet conditions 1 and 2 in summer are shown in Table 2. You can see from the table.

Table 2. Environmental physical parameters in each stage of summer conditions 1 and 2.

	Location	Index	Mean Value	Standard Deviation	Maximum Value	Minimum Value
Working condition 1	Outdoor part	Air temperature (°C)	35.49	0.07	35.59	35.31
		Black globe temperature (°C)	35.95	0.07	36.06	35.81
		Relative humidity (%)	59.48	1.12	61.33	58.06
		Breeze speed (m/s)	1.11	0.87	4.03	0.22
	Transition part	Air temperature (°C)	33.27	0.35	33.72	32.49
		Black globe temperature (°C)	33.68	0.28	34.13	33.25
		Relative humidity (%)	65.78	1.06	67.73	63.63
		Breeze speed (m/s)	0.18	0.12	0.53	0.00
	Interior part	Air temperature (°C)	29.05	0.28	29.80	28.60
		Black globe temperature (°C)	29.47	0.52	30.90	29.00
		Relative humidity (%)	75.29	1.56	78.10	71.60
		Breeze speed (m/s)	0.91	0.29	1.65	0.46
	Location	**Index**	**Mean Value**	**Standard Deviation**	**Maximum Value**	**Minimum Value**
Working condition 2	Outdoor part	Air temperature (°C)	34.44	0.05	34.54	34.36
		Black globe temperature (°C)	34.53	0.04	34.63	34.44
		Relative humidity (%)	63.01	1.11	64.36	58.63
		Breeze speed (m/s)	0.24	0.04	0.53	0.10
	Transition part	Air temperature (°C)	32.64	0.21	33.06	32.40
		Black globe temperature (°C)	32.94	0.26	33.44	32.63
		Relative humidity (%)	51.61	0.34	52.63	51.05
		Breeze speed (m/s)	0.18	0.57	3.11	0.01
	Interior part	Air temperature (°C)	30.92	0.25	31.40	30.60
		Black globe temperature (°C)	31.18	0.33	31.90	30.80
		Relative humidity (%)	52.47	0.31	53.00	51.90
		Breeze speed (m/s)	0.24	0.10	0.63	0.12

In condition 1, compared with the air temperature and the black globe temperature, the black globe temperature is generally higher than the air temperature, and the outdoor to indoor air temperature drops from 35.49 °C to 29.05 °C, with a more obvious drop.

The standard deviation of relative humidity in outdoor, indoor, and transitional parts is 1.12, 1.06, and 1.56, respectively, which are all higher than 1 and significantly higher than the standard deviation of other groups of data. In addition, the maximum and minimum relative humidity values of the three parts from outdoor to transition and then to indoor are 61.33 and 58.06%, 78.1 and 71.6%, 67.73 and 63.63%, respectively, and the fluctuation is also relatively obvious, indicating that the overall relative humidity of the underground public space entrance part of the working condition 1 has a large change range.

In terms of breeze wind speed, the outdoor wind speed fluctuates between 0.22 and 4.03 m/s, with an average of about 1.11 m/s; the indoor wind speed fluctuates between 0.46 and 1.65 m/s, with an average of about 0.91 m/s; the maximum wind speed in the transition part is 0.53 m/s, the minimum is 0 m/s, and the standard deviation is 0.12. It is possible that due to the relatively stable space of the transition part and the human environment, the wind speed is too small to be measured, and the overall wind speed change is not large.

Condition 2 is slightly different from condition 1. The overall black globe temperature is higher than the air temperature, and the outdoor to indoor air temperature drops from 34.44 °C to 30.92 °C. Compared with the same type of working condition 1, the temperature drop is small, and the temperature difference is more moderate. The standard deviation of the overall physical environment of the outdoor, transition, and indoor parts is 85% below 0.5, and only the relative humidity of the outdoor part is, with a maximum value of 64.36%, a minimum value of 58.63% and the standard deviation of 1.11, indicating a large overall fluctuation. Considering that there are some factors in the outdoor part of the working condition 2 that will change with changes in the outside world, it has an impact on the overall humidity.

In terms of breeze wind speed, the outdoor wind speed fluctuates between 0.1 and 0.53m/s, with an average of about 0.24m/s, and the indoor wind speed fluctuates between 0.12 and 0.63 m/s, with an average of about 0.24 m/s. The maximum wind speed in the transition part is 3.11 m/s, the minimum wind speed is 0.1 m/s, and the standard deviation is 0.57. The difference between the maximum wind speed and the minimum wind speed is large, but the standard deviation is not too high compared with the overall data, so it can be judged that the measured maximum wind speed should be an individual case.

Figure 4 shows the whole process change trend of working conditions 1 and 2 in summer. The data in the figure is the data change record of the whole process on a certain day under different working conditions in the same season. In the figure, 1–5 min represents the outdoor part, 6–10 min represents the transition part, and 11–15 min represents the indoor part. Combined with the above chart comprehensive analysis.

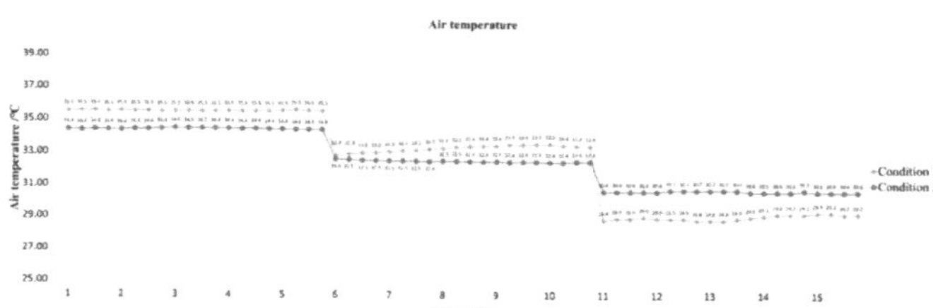

(**a**) Air temperature

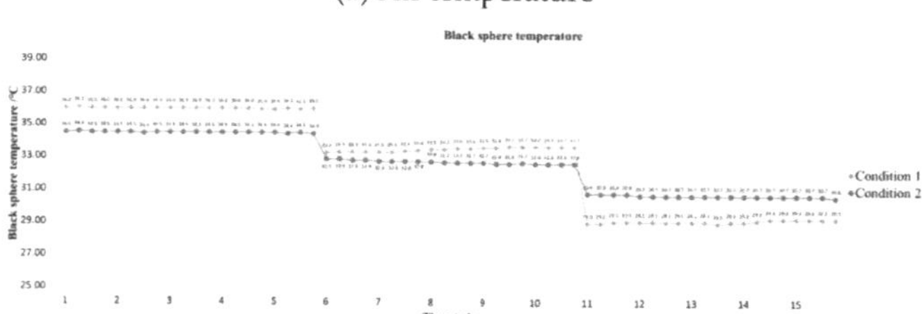

(**b**) Black globe temperature

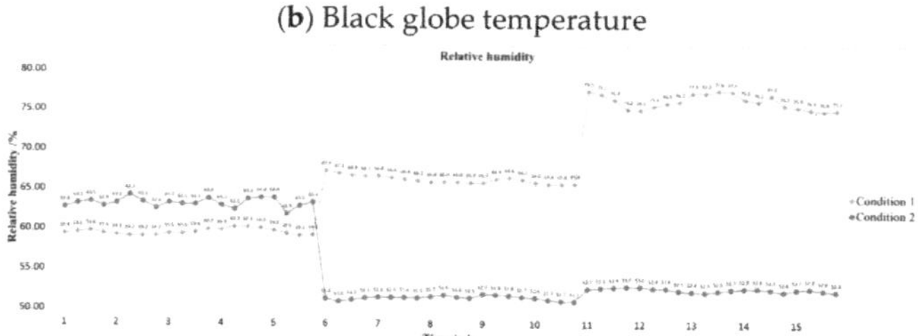

(**c**) Relative humidity

Figure 4. *Cont.*

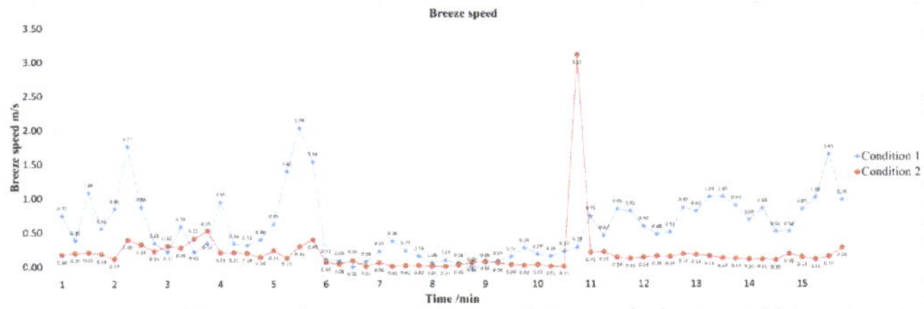

(**d**) Breeze speed

Figure 4. Analysis of air temperature, black ball temperature, relative humidity and breeze velocity in the whole process of summer condition 1 and condition 2.

From the point of view of air temperature, the overall air temperature of working condition 1 and working condition 2 is relatively stable in each part, especially in working condition 2, there is almost no change. Compared with working condition 2, working condition 1 has a gradual upward trend in the transition part, and there are small fluctuations in the room within a certain range.

From the perspective of relative humidity, in the process of turning from the outdoor part to the transition part and then to the indoor part, the humidity of working conditions 1 and 2 showed a completely opposite trend; the humidity of working condition 1 showed an upward trend, while the humidity of working condition 2 showed a significant downward trend from the outdoor part to the transition part, and the overall average relative humidity from the transition part to the indoor part began to rise.

From the point of view of air velocity, the air velocity of working conditions 1 and 2 has the largest fluctuation in the outdoor part, and the overall wind environment is unstable, especially in working condition 1, which has great fluctuation. Not only that, the air velocity in the indoor part of working condition 1 also has a large fluctuation. Working condition 2 is generally smooth, and the wind environment is relatively stable, while working condition 1 only shows little fluctuation of the measured data in the transition part, and the wind environment is relatively stable.

Through the calculation of software, the PET values of the environment around the underground public space entrance and exit conditions 1 and 2 are obtained. It can be seen from the change of PET value with the slow walking process (Figure 5) that the PET value in the outdoor part of working condition 1 is about 37.14 °C on average; when entering the transition part, the PET value drops to 34.92 °C; when entering the indoor part, the PET value drops to 29.88 °C and tends to be stable. The PET value of working condition 2 is about 30.58 °C on average; when entering the transition part, the PET value drops to 27.87 °C; when entering the indoor part, the PET value drops to 26.39 °C and tends to be stable. In the space from outdoor to indoor, the outdoor part has the highest PET.

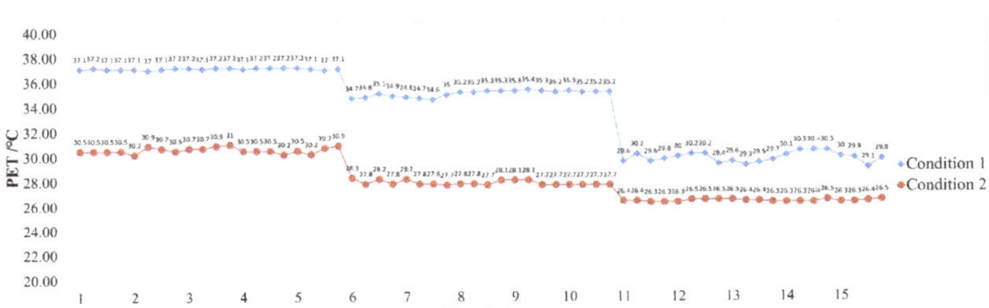

Figure 5. Analysis of the change of physiological equivalent temperature PET value with time in summer conditions 1 and 2.

(2) Winter

The environmental physical parameters of the outdoor part, the transition part, and the indoor part of the working conditions 1 and 2 in winter are shown in Table 3. You can see from the table.

Table 3. Environmental, physical parameters in each stage of winter conditions 1 and 2.

	Location	Index	Mean Value	Standard Deviation	Maximum Value	Minimum Value
Working condition 1	Outdoor part	Air temperature (°C)	5.18	0.08	5.30	5.10
		Black globe temperature (°C)	5.97	0.07	6.10	5.90
		Relative humidity (%)	30.14	0.26	30.70	29.70
		Breeze speed (m/s)	0.79	0.41	1.81	0.34
	Transition part	Air temperature (°C)	9.46	0.26	9.70	9.00
		Black globe temperature (°C)	9.43	0.41	9.90	8.60
		Relative humidity (%)	33.12	1.47	34.80	30.60
		Breeze speed (m/s)	0.19	0.06	0.35	0.12
	Interior part	Air temperature (°C)	10.39	0.19	10.70	10.00
		Black globe temperature (°C)	10.46	0.27	10.80	10.00
		Relative humidity (%)	33.28	0.62	34.20	32.30
		Breeze speed (m/s)	0.36	0.05	0.48	0.28
	Location	**Index**	**Mean Value**	**Standard Deviation**	**Maximum Value**	**Minimum Value**
Working condition 2	Outdoor part	Air temperature (°C)	6.30	0.09	6.40	6.10
		Black globe temperature (°C)	7.20	0.09	7.30	7.10
		Relative humidity (%)	28.07	0.11	28.30	27.90
		Breeze speed (m/s)	1.45	1.01	3.66	0.30
	Transition part	Air temperature (°C)	7.63	0.13	7.80	7.40
		Black globe temperature (°C)	8.28	0.18	8.60	8.00
		Relative humidity (%)	24.89	0.19	25.20	24.70
		Breeze speed (m/s)	0.53	0.25	1.26	0.21
	Interior part	Air temperature (°C)	8.61	0.04	8.70	8.50
		Black globe temperature (°C)	9.04	0.05	9.10	9.00
		Relative humidity (%)	23.36	0.08	23.50	23.20
		Breeze speed (m/s)	0.66	0.18	1.18	0.39

In condition 1, compared with the air temperature and the black globe temperature, the black globe temperature is generally higher than the air temperature, and the outdoor to indoor air temperature rises from 5.18 to 8.61 °C, with a significant increase.

The standard deviations of outdoor and indoor relative humidity are 0.26 and 0.62, respectively, while the standard deviation of the transition part is 1.47, which is much higher than the other two sets of data. In addition, the maximum and minimum values of relative humidity from the transition part are 34.8 and 30.6%, respectively, and the fluctuation is also relatively obvious, indicating that the relative humidity of the underground public space entrance and exit of the working condition 1 changes greatly in the transition part, and the environmental performance of the other two parts is relatively stable.

In terms of breeze wind speed, the outdoor wind speed fluctuates between 0.34 and 1.81 m/s, with an average of about 0.79 m/s; the indoor wind speed fluctuates between 0.28 and 0.48 m/s, with an average of about 0.36 m/s; the maximum wind speed in the transition part is 0.35 m/s, the minimum wind speed is 0.12 m/s, and the standard deviation is 0.06. Considering that the space of the transition part and the man-made environment are relatively stable, the overall wind speed changes little.

Condition 2 is slightly different from condition 1. Except that the overall black globe temperature is higher than the air temperature, the outdoor to indoor air temperature rises from 6.3 to 9.04 °C, which is smaller than the same type of working condition 1, and the temperature change is more moderate. The standard deviation of the physical environment of the overall relative humidity of outdoor, transition, and indoor parts is almost below 0.2, and the highest is only 0.19. Considering that the space and external environment of working condition 2 are relatively stable, the overall impact on relative humidity is small.

In terms of wind speed, the outdoor wind speed fluctuates between 0.3 and 3.66 m/s, with an average of about 1.45 m/s, and the indoor wind speed fluctuates between 0.39 and 1.18 m/s, with an average of about 0.66 m/s. The maximum wind speed in the transition part is 1.26 m/s, the minimum wind speed is 0.21 m/s, and the standard deviation is 0.25. The maximum wind speed differs greatly from the minimum wind speed, but the standard deviation is not too high compared to the overall data.

Figure 6 shows the whole process change trend of operating conditions 1 and 2 in winter. The data in the figure is the change record of the whole process on a certain day under different working conditions in the same season. In the figure, 1–5 min represent the outdoor part, 6–10 min represent the transition part, and 11–15 min represent the indoor part. Combined with the above chart comprehensive analysis.

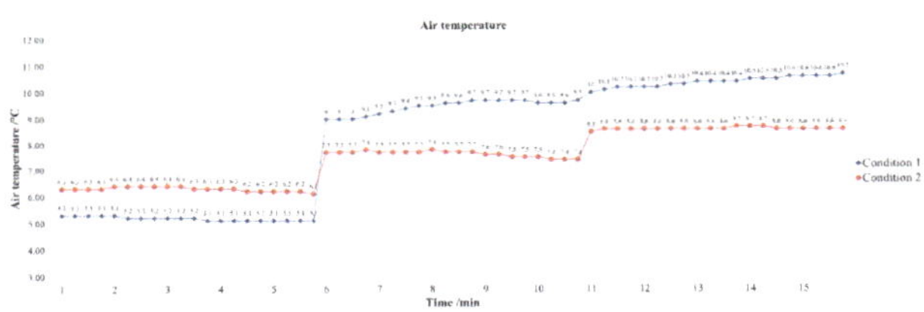

(**a**) Air temperature

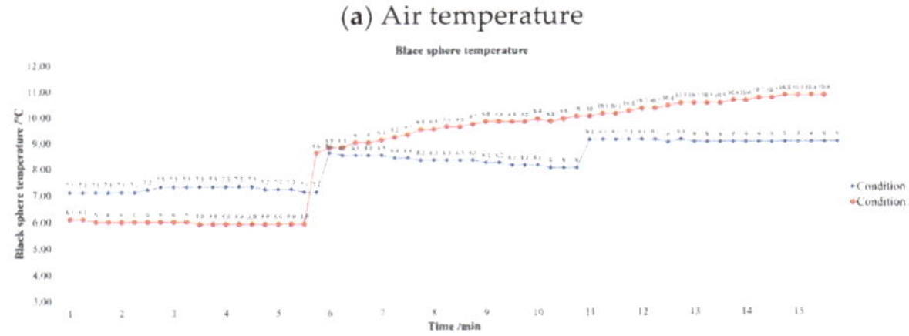

(**b**) Black globe temperature

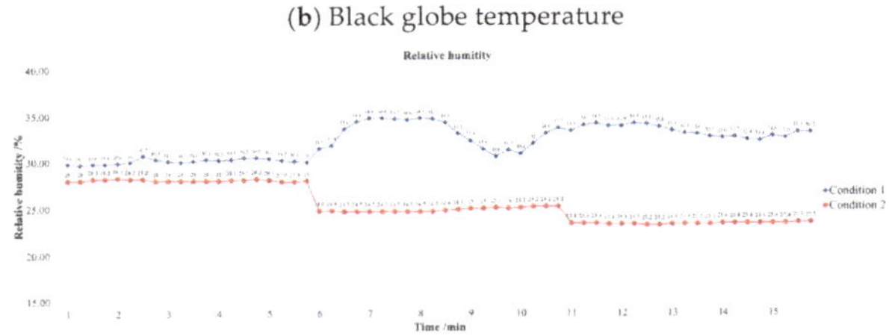

(**c**) Relative humidity

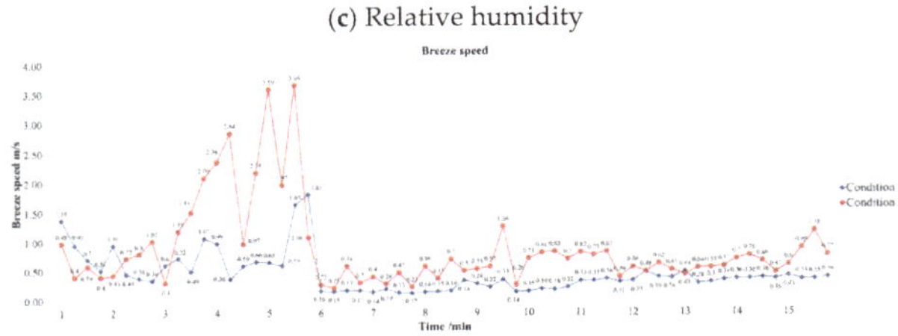

(**d**) Breeze speed

Figure 6. Analysis of air temperature, black ball temperature, relative humidity and breeze velocity in the whole process of winter condition 1 and condition 2.

From the point of view of air temperature, the overall air temperature of working conditions 1 and 2 in winter from indoors to outdoors is contrary to that in summer, showing a gradual upward trend. The temperature is relatively stable in each part, especially in working condition 2, where there is almost no big change. Compared with condition 2, condition 1 has a slight gradual upward trend in the transition part and the indoor part, and there are slight fluctuations within a certain range. In addition, compared with the transition part to the indoor part, the air temperature drop from the outdoor part to the transition part in condition 1 is larger and more sudden.

The blacksphere temperature is generally consistent with the air temperature, but there are differences. In condition 1, the outdoor black globe temperature is significantly higher than the air temperature, and the black globe temperature is almost the same as the air temperature after entering the transition part and the indoor part, with a difference of about 0.2 °C.

From the perspective of relative humidity, in the process of turning from the outdoor part to the transition part and then to the indoor part, the humidity of working conditions 1 and 2 is still the same as in summer, showing a completely opposite trend. The humidity of working condition 1 shows an overall upward trend, while the humidity of working condition 2 shows a significant downward trend from outdoor to indoor part.

From the point of view of air velocity, the air velocity of working conditions 1 and 2 has the largest fluctuation in the outdoor part, and the overall wind environment is unstable, especially in working condition 1, which has great fluctuation. Compared with the outdoor part, the other two parts change less. Working condition 2 is generally smooth, and the wind environment is stable.

Through the calculation of software, the PET values of the environment around the underground public space entrance and exit conditions 1 and 2 are obtained. It can be seen from the changes in PET value with the slow walking process (Figure 7) that the physiological equivalent temperature PET of the outdoor part of working conditions 1 and 2 is not particularly stable, and the change is large. After entering the transition and indoor part, the physiologically equivalent temperature PET of both conditions has a great increase, with an increase of about 5–7 °C. In addition, the PET value of the outdoor part of working condition 1 is about 2 °C on average. When entering the transition part, the PET value rises to 8.61 °C, and when entering the indoor part, the PET value drops to 8.55 °C and tends to be stable. The PET value of working condition 2 is about 2 °C on average; when entering the transition part, the PET value rises to 4.73 °C; when entering the indoor part, the PET rises to 5.24 °C and tends to be stable. In the space from outdoor to indoor, the outdoor part has the lowest PET.

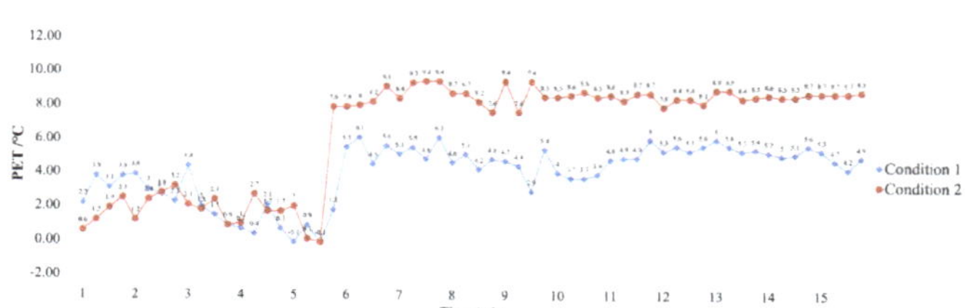

Figure 7. Analysis of the change of physiological equivalent temperature PET value with time in winter conditions 1 and 2.

3.2. *Test and Simulation Results Verify*

In practical applications, the accuracy of simulation software parameter Settings is very important because the different values of these parameters will have a huge impact on the final simulation results. Therefore, verifying the reliability of simulation software parameter Settings is an essential step before the wind and thermal environment simulation. Two groups of adjacent open underground public space entrances are measured, analyzed, and compared. The built environment model of 1 km around Pengcheng Square was established.

3.2.1. Construction of Measurement Model

The simulation object of this part is Exit 5 of Pengcheng Square Station of Xuzhou Metro Line 1. By surveying and mapping with the laser rangefinder, the author established the overall model of the entrance and exit space of No. 5. The red part in the figure is the entire entrance and exit space. In addition, due to the complexity of the entrance and exit, the modeling will be difficult, and the Mesh mesh division in Fluent will deteriorate the value of the minimum orthogonal quantity in the later stage. Too fine mesh division of the entire model will also lead to poor computer operation, and the interior of the entrance and exit will be controlled by the underground ventilation system. Therefore, the model of the entrance and exit will be simplified [34]; the underground part also only involves an extension of 2 m behind the groundwater level. The simplified model retains the main influencing factors, such as the depth of the entrance and exit, the area of the opening, and the size of the section of the channel, as shown in Figure 8. The actual location distribution of measuring points is consistent with the measured position.

(**a**) Model axis side diagram (**b**) Model profile

Figure 8. Simplified model of Exit 5 of Pengcheng Square Station on Xuzhou Metro Line 1.

3.2.2. Air Temperature, Breeze Wind Speed Simulation and PET Calculation

Meteorological data of temperature and wind speed in Gulou District, Xuzhou City, from 20 August to 26 August, were collected and input into Fluent for solving. Fluent software was used to simulate the physical environment at a height of 1.5 m in the outdoor part, the transition part, and the indoor part of Gate No.5 [39], The temperature and wind speed of the corresponding seven days are obtained, as shown in the following table: Table 4 shows the cloud image of computer-simulated air temperature and breeze wind speed.

Table 4. Simulate air temperature cloud chart.

Date	Air Temperature Cloud Map	Wind Speed Cloud Picture
8.20	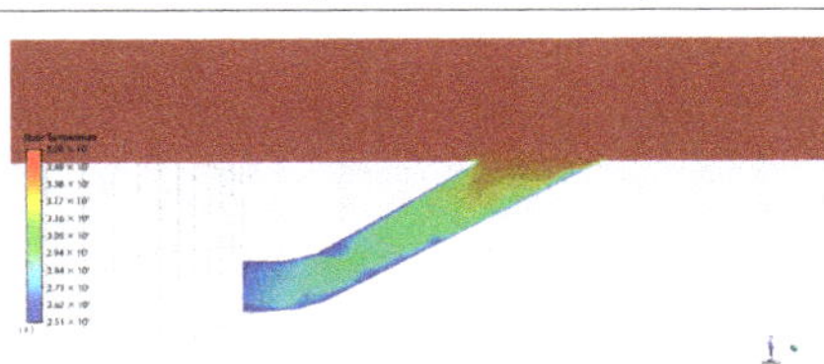	
8.21	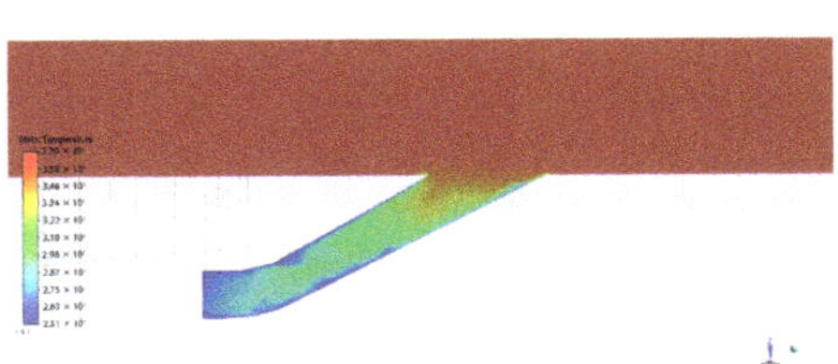	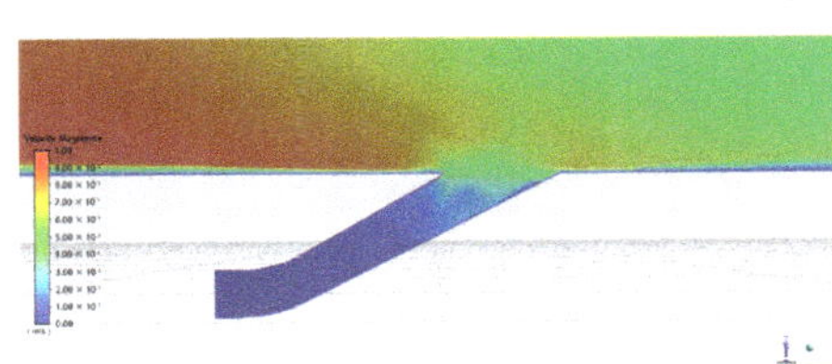

Table 4. *Cont.*

Date	Air Temperature Cloud Map	Wind Speed Cloud Picture
8.22	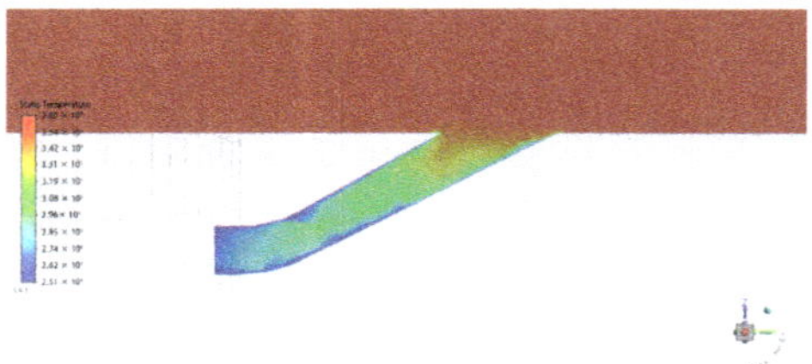	
8.23	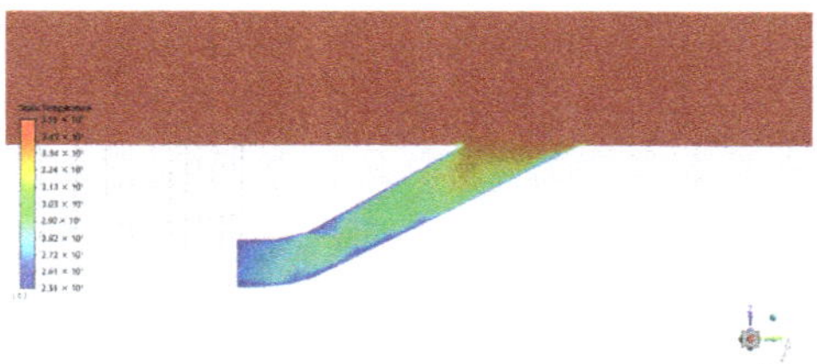	
8.24	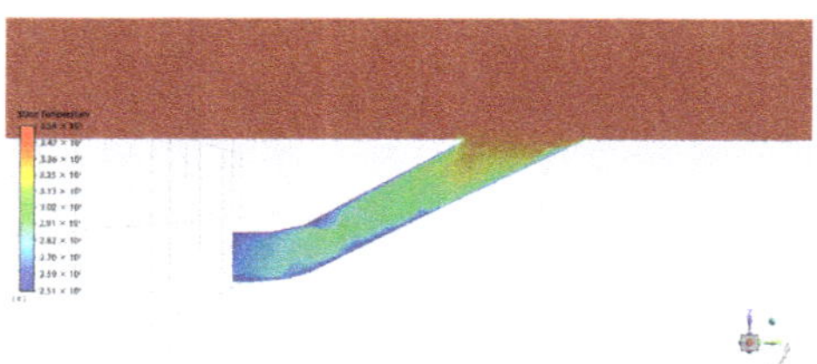	
8.25	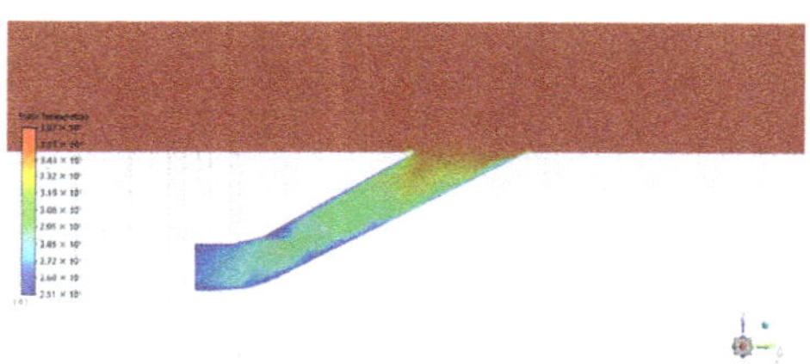	
8.26	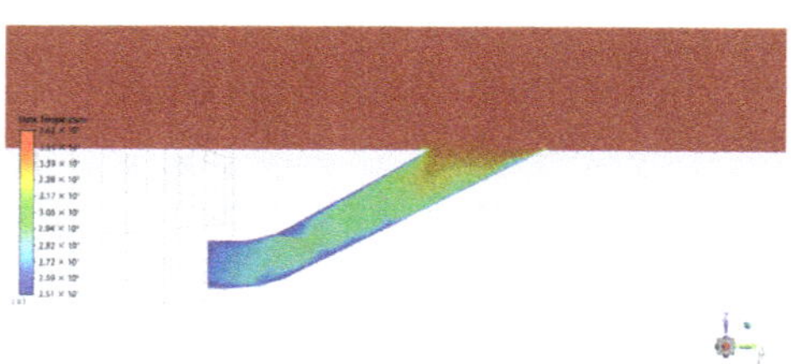	

Through the above-simulated air temperature and breeze wind speed, the software Rayman 1.2 can directly calculate the corresponding measurement point Tmrt and PET values [40]. Thus, the inlet and outlet air thermal environment can be compared and analyzed.

3.2.3. Simulation Verification Analysis

As shown in Figures 9–11, PET measured and simulated data analysis of air temperature, breeze wind speed, and physiological equivalent temperature at a height of 1.5 m from outdoors to indoors. After comparing the measured daily average values with simulated values, it is found that the temperature changes from the 20th to the 26th are not large, indicating that the weather conditions are stable. The difference between the measured temperature and the simulated temperature is basically controlled around 0.1–0.2 °C, and the general trend is that the measured temperature is higher than the simulated temperature. Considering that other surrounding green plants and ground texture materials are not added, certain errors will be caused. The variation of breeze wind speed fluctuates obviously, but the difference between the actual and simulated wind speed is also controlled at about 0.2 m/s on average. Physiological equivalent temperature PET is the same as air temperature.

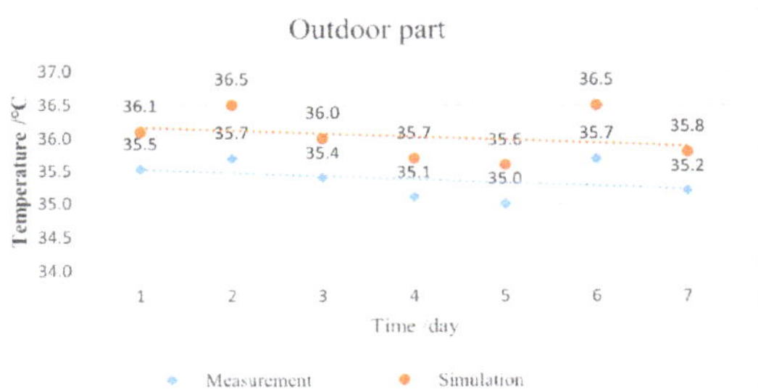

(**a**) Outdoor part

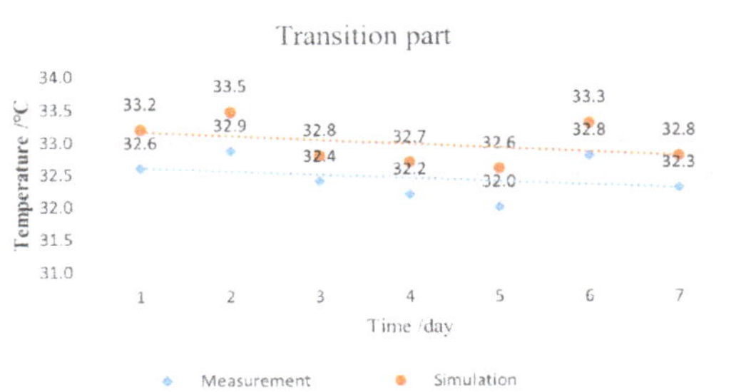

(**b**) Transition part

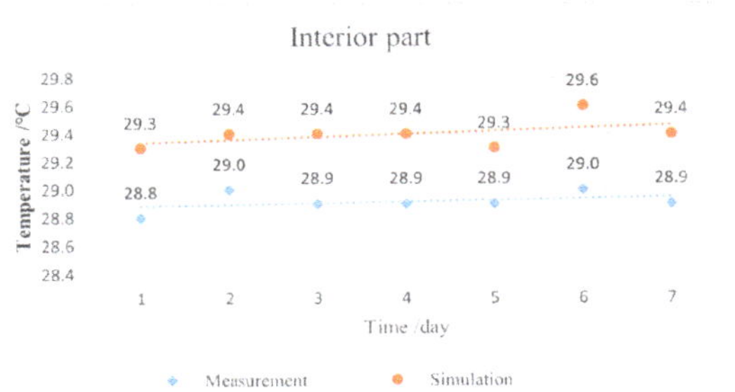

(**c**) Interior part

Figure 9. Correlation analysis of measured and simulated air temperature.

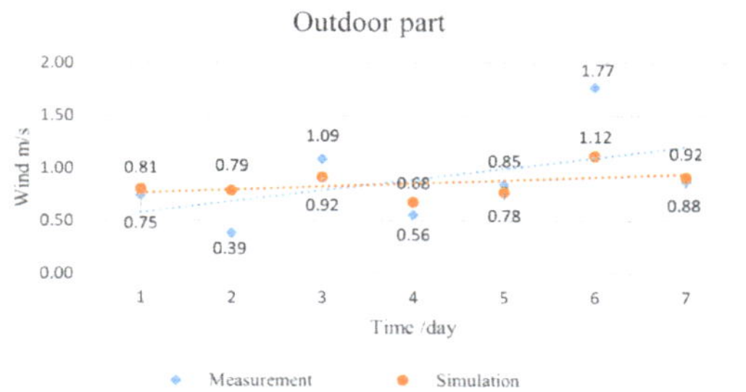

(**a**) Outdoor part

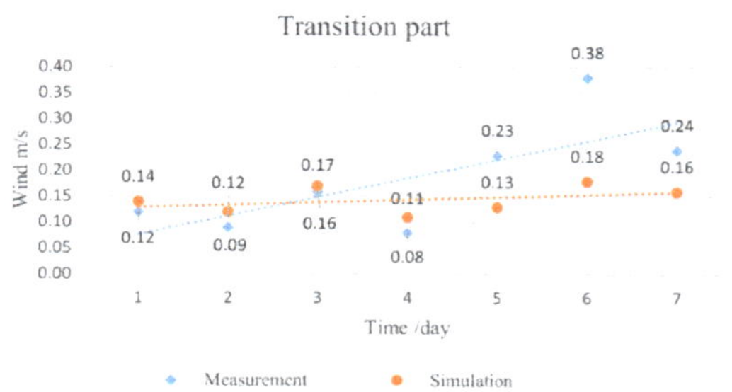

(**b**) Transition part

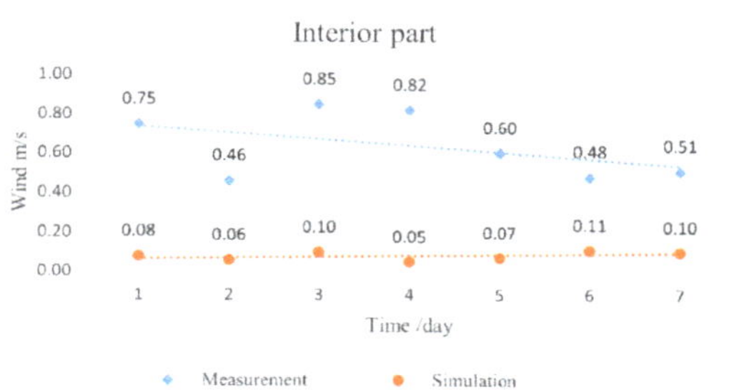

(**c**) Interior part

Figure 10. Correlation analysis of measured and simulated breeze speed.

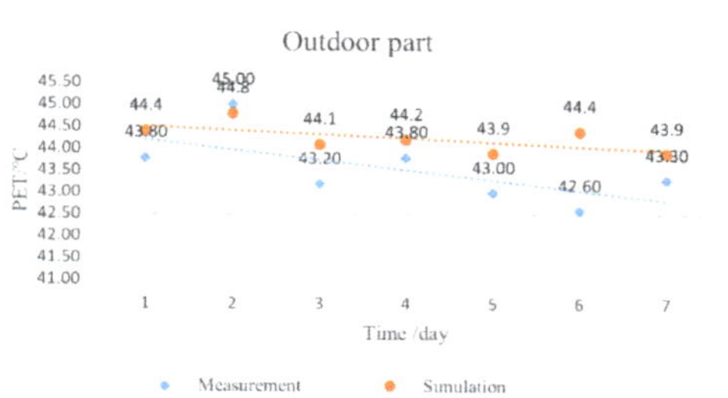

(**a**) Outdoor part

Figure 11. *Cont.*

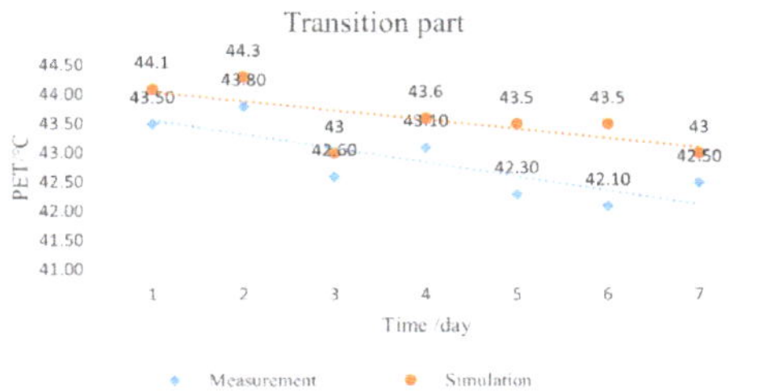

(**b**) Transition part

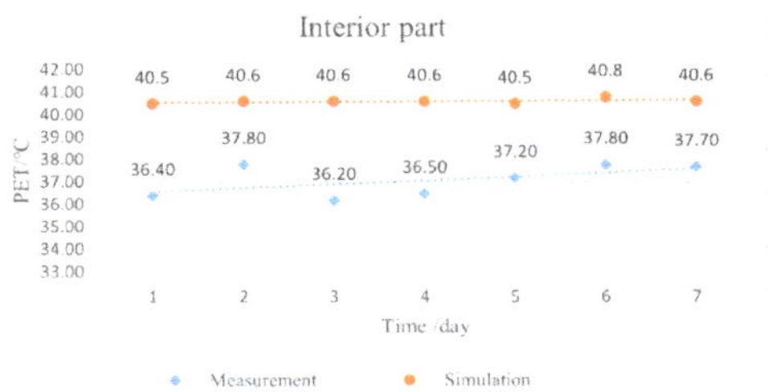

(**c**) Interior part

Figure 11. Correlation analysis between measured and simulated PET.

In the analysis process, it was found that there are some deviations between the simulation results and the measured data of the underground public space entrance of Pengcheng Square, and the overall thermal environment has a high degree of fitting, while the wind environment is low.

In the actual situation, part of the wind and thermal environment in the entrance and exit room of underground public space is affected by many factors, such as pedestrians and air conditioning equipment, and the above situation is not truthfully considered in the simulation. However, after comparing the measured data with the simulation results, it can be found that there is still a high consistency between the two. Therefore, it can be considered that using Fluent software to simulate the wind and thermal environment of the underground public space entrance and exit is reliable.

3.3. Design and Determination of Simulated Experimental Group

In order to quantitatively analyze the influence of the spatial shape characteristics of the entrance and exit on the wind and thermal environment of the entrance and exit space, the shape characteristics of the entrance and exit are described from three aspects as follows: plane form, section form, and interface form, and various index elements are sorted out and selected. The spatial form index elements of the entrance and exit of the open underground public space are shown in Table 5. See Appendix A for the nomenclature and explanation of the indicators.

Table 5. Open underground public space entrance space form factor index.

Sort		Spatial Form Index Name	Units	Data Nature	Data Source and Method of Determination
Import and export index	Planar form	Opening aspect ratio		Quantification	Field investigation
		Open plane shape		Quantification	Field investigation
		Opening orientation		Categorization	Field investigation
	Profile form	Opening depth	M	Quantification	Field investigation
		Open baffle height	M	Quantification	Field investigation

In order to judge the influence of the spatial form index of the entrance and exit of underground public space on the wind and thermal environment around the entrance and exit, one must effectively control the shape parameters with greater influence, and maximize the comfort level of the wind and thermal environment of the entrance and exit. Therefore, the correlation analysis between the quantitative indicators of the spatial form of the entrance and exit and the existing wind thermal physical environment data is carried out to find the main influencing indicators. Through the correlation and regression analysis of different elements, the influence of the spatial form elements of the entrance and exit on the air and thermal environment of the underground public space is determined. It mainly uses software Design-Expert [41].

After determining the value range of single-factor variables, with wind speed, temperature, and physiological equivalent temperature PET as response values, the Box-Behnken four-factor and four-level response surface Design was carried out by Design-Expert10 software with plane form, section form, and interface form as influencing factors for the open entrance and exit. The actual variable model is selected by response surface analysis. Based on the above research and analysis, the single-factor value range of each influencing factor of the spatial form of the entrance and exit is selected, as shown in Table 6 below.

Table 6. Selection of boundary range of each element of open underground public space entrance and exit space form.

Name	Units	Minimum Value	Maximum Value
Open plane shape	edge	4	8
Opening aspect ratio		1	3
Opening depth	m	2.2	5.1
Open baffle height	m	0	3

A total of 29 groups of open entrance and exit experiments were designed, of which 5 groups had the same factors and conditions, and repeated group experiments were needed to reduce the error of experimental results. Since the same boundary conditions were used in the simulation experiments, the final results of the same experiments were consistent, so the number of simulations could be reduced. According to the response surface analysis requirements of the experimental group, 25 groups of experimental models under different working conditions were established (Table 7). When determining the parameter variables, the width of each outlet is controlled to be 6m unchanged, and the length of the outlet varies according to the proportion of the parameters. In addition to the differences in the shape of the opening plane, the ratio of length to width of the opening, the depth of the opening and the height of the baffle, the length of the ground part around the underground opening, and the length of the extension part into the underground plane remain the same when the model is built.

The comparison and verification of the measured and simulated data of Fluent software are carried out above, and it is believed that the software simulation has a certain degree of reliability. Therefore, the experimental method of software simulation was adopted for 25 groups of experiments.

The core of numerical simulation is to set up accurate boundary conditions. ANSYS FLUENT 2020 R2 provides several types of boundary conditions. This study conducted detailed field tests on the physical environment around Pengcheng Square and combined them with the obtained network data, including air temperature, black ball temperature, site breeze wind speed, and relative humidity. Considering the difference between numerical simulation and actual measurement, only part of the measured data was used as numerical calculation parameters.

Table 7. 25 groups of experimental models of open underground public space entrance and exit conditions.

Serial Number	Open Plane Shape (Edge)	Opening Aspect Ratio	Opening Depth (m)	Open Baffle Height (m)
1	4	1	3.65	1.5
2	4	2	2.2	1.5
3	4	2	3.65	3
4	4	2	5.1	1.5
5	4	2	3.65	0
6	4	3	3.65	1.5
7	6	1	5.1	1.5
8	6	1	2.2	1.5
9	6	1	3.65	3
10	6	1	3.65	0
11	6	2	5.1	3
12	6	2	2.2	0
13	6	2	5.1	0
14	6	2	2.2	3
15	6	2	3.65	1.5
16	6	3	3.65	3
17	6	3	5.1	1.5
18	6	3	2.2	1.5
19	6	3	3.65	0
20	8	1	3.65	1.5
21	8	2	5.1	1.5
22	8	2	3.65	3
23	8	2	3.65	0
24	8	2	2.2	1.5
25	8	3	3.65	1.5

3.4. Analysis of Experimental Group Simulation Results

3.4.1. Summer Design Experiment Simulation

Under summer conditions, the main research is to study the changes in wind speed, temperature, and physiological equivalent temperature PET value at the same position under different spatial shape factors of open inlet and outlet. According to the above-measured data, the size of the outflow field is set to 200 m × 100 m × 50 m, the velocity is set to 1.5 m/s, and the wind field temperature is set to 36 °C. The entrance and exit of the open underground public space are set as free-flow boundary conditions.

According to the above analysis of changes in physiological equivalent temperature PET data, for the entrance and exit of open underground public space, the value of PET shows a downward trend with the change of outdoor to indoor space. Therefore, it is necessary to control the physiologically equivalent temperature of PET outdoors within a reasonable range. Therefore, in this paper, only the distribution of wind speed and temperature in the outdoor part at a height of 1.5 m is captured.

Model building and significance analysis: PET value was calculated by Rayman. In the experimental design, the plane shape of the opening, the aspect ratio of the opening, the depth, and the opening baffle's height are set as the horizontal factors, and the horizontal factors are independent of each other. The response values are set to the physiologically equivalent temperature PET, temperature, and wind speed. Combined with the temperature and wind speed of the measuring point, the response surface experiment results were obtained, and the response surface design was carried out according to them, as shown in Table 8 below.

Table 8. Correlation analysis table of quantitative spatial morphology index with wind speed, temperature, and PET value in summer.

Number	Horizontal Factor				Response Value		
	Open Plane Shape (Number of Plane Edges) A	Opening Aspect Ratio B	Opening Depth (m) C	Open Baffle Height (m) D	PET (°C)	Temperature (°C)	Wind Speed (m/s)
1	4	1	3.65	1.5	41.7	35.11	0.12
2	4	2	2.2	1.5	41.4	35.11	0.18
3	4	2	3.65	3	41.4	35.13	0.20
4	4	2	5.1	1.5	41.1	35.03	0.28
5	4	2	3.65	0	41.2	35.13	0.26
6	4	3	3.65	1.5	40.9	34.98	0.44
7	6	1	5.1	1.5	41.3	34.66	0.10
8	6	1	2.2	1.5	41.4	34.07	0.03
9	6	1	3.65	3	40.8	35.28	0.61
10	6	1	3.65	0	40.7	34.95	0.52
11	6	2	3.65	1.5	40.9	34.82	0.31
12	6	2	5.1	3	40.9	34.80	0.31
13	6	2	2.2	0	41.1	34.97	0.27
14	6	2	5.1	0	41.3	34.95	0.25
15	6	2	3.65	1.5	40.7	35.00	0.32
16	6	2	3.65	1.5	41.2	34.47	0.29
17	6	2	3.65	1.5	41	34.24	0.28
18	6	2	2.2	3	41	34.89	0.30
19	6	2	3.65	1.5	40.5	35.25	0.33
20	6	3	3.65	3	40.8	34.74	0.30
21	6	3	5.1	1.5	40.3	33.64	0.40
22	6	3	2.2	1.5	40.8	34.52	0.16
23	6	3	3.65	0	40.8	34.73	0.28
24	8	1	3.65	1.5	40.7	33.90	0.22
25	8	2	5.1	1.5	40.6	35.92	1.51
26	8	2	3.65	3	40.9	35.07	0.30
27	8	2	3.65	0	40.8	34.54	0.25
28	8	2	2.2	1.5	40.5	35.33	0.87
29	8	3	3.65	1.5	40.8	34.71	0.27

Using the Design-Expert 10 software, variance analysis was performed on the PET simulation experiment results of wind speed, temperature, and physiological equivalent temperature at the entrance and exit of open underground public space in summer. The PET fitting function of physiological equivalent temperature at the entrance and exit was obtained as a first-order function, and the first-order polynomial regression equations were, respectively, as follows:

$$Y = 40.95 - 0.28 * A - 0.18 * B - 0.058 * C - 8.33 * 10^{-3} * D \quad (1)$$

In the formula:

Y—PET
A—Open plane shape
B—Opening aspect ratio
C—Opening depth
D—Open baffle height

The results of variance analysis of regression model are shown in Table 9.

Table 9. Regression model analysis of variance in summer.

Source	Sum of Squares	df	Mean Square	F Value	*p* Value	Significance
Model	1.41	4	0.35	5.77	0.0021	**
A	0.96	1	0.96	15.79	0.0006	**
B	0.40	1	0.40	6.61	0.0168	*
C	0.041	1	0.041	0.67	0.4213	
D	8.333×10^{-4}	1	8.333×10^{-4}	0.014	0.9079	
Residual	1.46	24	0.061			
Lack of fit	1.17	20	0.059	0.80	0.6762	
Pure error	0.29	4	0.073			
Cor Total	2.87	28				
R^2	0.4903					
R^2_{Adj}	0.4053					
R^2_{Pred}	0.2582					

Note: * was significant difference ($p < 0.05$); ** was very significant difference ($p < 0.01$).

As can be seen from Table 9, $p < 0.01$ indicates that the model is extremely significant. The missing item was not significant ($p = 0.6762 > 0.05$), indicating that the repeated test results were reasonable. $R^2 = 0.4903$, indicating the degree of equation fitting is not very good [42]. However, after comparing Linear, 2FI, Quadratic and Cubic, it is found that Linear's R^2_{Adj} is positive and has higher predictability than the other three fitting methods, so Linear is selected for fitting. In addition, the difference between R^2_{Pred} and R^2_{Adj} is less than 0.2, indicating a high correlation between the actual and predicted values. Therefore, the model can properly reflect the relationship between various form factors and response value PET during the use of open underground public space entrances and exits in summer, and predict the best design form factors.

According to the *p* value, the PET value of the entrance and exit of the open underground public space in summer was significantly affected by the plane shape of the opening ($p < 0.01$), and significantly affected by the aspect ratio of the opening ($p > 0.05$). According to the F value, the order of influence of four factors on the PET value of the entrance and exit of the open underground public space in summer is as follows: A > B > C > D, that is, the plane shape of the opening > the aspect ratio of the opening > the depth of the opening > the height of the opening baffle (Figure 12). The single factor opening plane shape has the most obvious influence on the outdoor PET value, followed by the opening length-width ratio, and the other two factors have little influence. In addition, the interaction relationship among the four factors A, B, C and D had little effect on PET outside the entrance and exit.

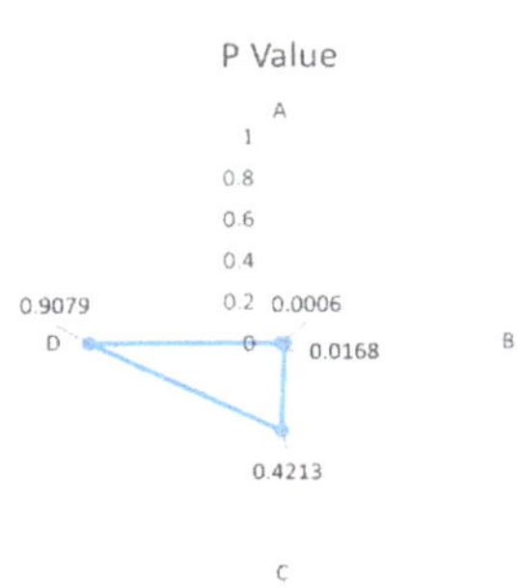

Figure 12. Importance of factors for open entrances and exits in summer.

In addition, according to Figure 13, the trend relationship of the influence of single factors on PET values can be found. The physiological equivalent temperature PET has a linear negative correlation with the opening plane shape. The more the opening plane shape edges, the smaller the PET value. There is also a linear negative correlation with the opening length-width ratio. The larger the opening length-width ratio, the smaller the PET value. With the change of the depth of the opening and the height of the opening baffle, the difference between the maximum and the minimum value of the physiological equivalent temperature PET is small, indicating that these two factors have little influence on the PET value.

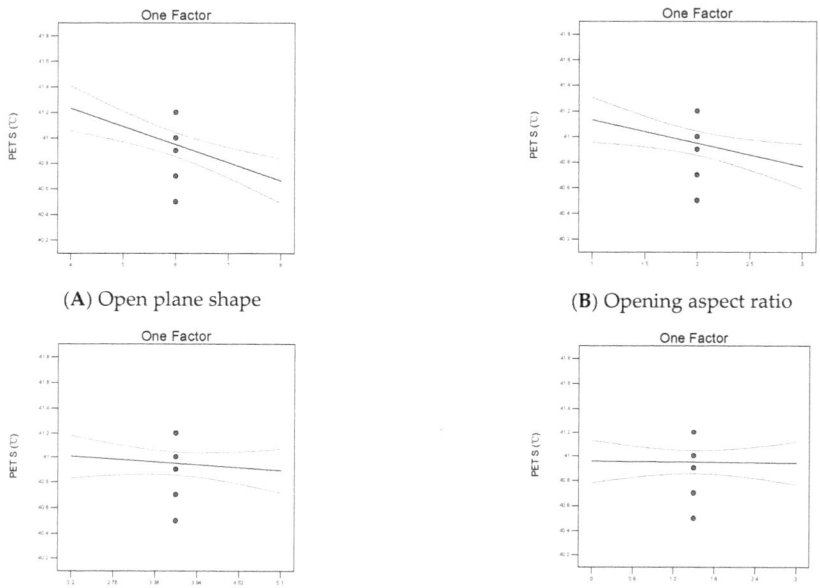

(**A**) Open plane shape (**B**) Opening aspect ratio

(**C**) Opening depth (**D**) Open baffle height

Figure 13. The influence of single factor on the open entrance and exit relationship in summer.

3.4.2. Winter Design Experiment Simulation

Under winter conditions, the main research is also the change law of wind speed, temperature, and physiological equivalent temperature PET value at the same position under different spatial shapes influencing factors of open inlet and outlet. The size of the outflow field is set to 200 m × 100 m × 50 m, the velocity is set to 1.5 m/s, and the wind field temperature is set to 6.3 °C. The entrance and exit of the open underground public space are set as free-flow boundary conditions.

According to the above analysis of changes in physiological equivalent temperature PET data, for the entrance and exit of open underground public space, the value of PET shows an upward trend with the change of outdoor to indoor space. Therefore, it is necessary to control the physiologically equivalent temperature of PET outdoors within a reasonable range. Therefore, in this paper, only the distribution of wind speed and temperature in the outdoor part at a height of 1.5 m is captured.

Model establishment and significance analysis: The value of PET was calculated by Rayman, and the response surface experimental results were obtained by combining the temperature and wind speed of the measuring point, and the response surface design was carried out according to it, as shown in Table 10 below.

Table 10. Correlation analysis table of quantitative spatial morphology index with wind speed, temperature, and PET value in winter.

Number	Horizontal Factor				Response Value		
	Open Plane Shape (Number of Plane Edges) A	Opening Aspect Ratio B	Opening Depth (m) C	Open Baffle Height (m) D	PET (°C)	Temperature (°C)	Wind Speed (m/s)
1	4	1	3.65	1.5	4.8	6.75	0.15
2	4	2	2.2	1.5	5.4	6.76	0.23
3	4	2	3.65	3	4.8	6.77	0.26
4	4	2	5.1	1.5	5.4	6.81	0.24
5	4	2	3.65	0	4.8	6.76	0.26
6	4	3	3.65	1.5	7.8	6.60	0.04
7	6	1	5.1	1.5	3.1	6.42	0.80
8	6	1	2.2	1.5	3.2	6.44	1.38
9	6	1	3.65	3	3	6.32	1.31
10	6	1	3.65	0	3.2	6.32	0.63
11	6	2	3.65	1.5	4.4	6.91	0.31
12	6	2	5.1	3	4.9	6.91	0.33
13	6	2	2.2	0	4.8	6.84	0.26
14	6	2	5.1	0	4.8	6.84	0.25
15	6	2	3.65	1.5	4.9	6.91	0.33
16	6	2	3.65	1.5	5.1	6.91	0.33
17	6	2	3.65	1.5	5.2	6.91	0.33
18	6	2	2.2	3	4.9	6.88	0.30
19	6	2	3.65	1.5	4.6	6.91	0.33
20	6	3	3.65	3	4.9	6.95	0.32
21	6	3	5.1	1.5	5	6.96	0.32
22	6	3	2.2	1.5	5.5	7.01	0.20
23	6	3	3.65	0	5	6.95	0.30
24	8	1	3.65	1.5	3.9	6.80	0.18
25	8	2	5.1	1.5	3.4	6.30	0.74
26	8	2	3.65	3	4.3	7.13	0.11
27	8	2	3.65	0	5.8	7.31	0.15
28	8	2	2.2	1.5	3.2	6.39	1.30
29	8	3	3.65	1.5	5	6.97	0.27

Using Design-Expert 10 software, variance analysis was performed on the PET simulation experiment results of wind speed, temperature, and physiological equivalent temperature at the entrance and exit of open underground public space in summer. The PET fitting function of physiological equivalent temperature at the entrance and exit was obtained as a first-order function, and the first-order polynomial regression equations were, respectively, as follows:

$$Y = 4.66 - 0.62 * A + B - 0.033 * C - 0.13 * D \quad (2)$$

In the formula:

Y—PET
A—Open plane shape
B—Opening aspect ratio
C—Opening depth
D—Open baffle height

The results of variance analysis of regression model are shown in Table 11.

As can be seen from Table 11, $p < 0.01$ indicates that the model is extremely significant. The missing item was not significant ($p = 0.0694 > 0.05$), indicating that the repeated test results were reasonable. $R^2 = 0.5981$, although the fit is not excellent, after comparing Linear, 2FI, Quadratic, and Cubic, it is found that Linear's R^2_{Adj} is positive, and its value is higher than the other three, and the fitting degree is better. Therefore, Linear is selected for fitting. In addition, the difference between R^2_{Pred} and R^2_{Adj} is less than 0.2, indicating a high correlation between the actual and predicted values. Therefore, the model can properly reflect the relationship between various form factors and response value PET during the use of open underground public space entrances and exits in winter and predict the best design form factors.

Table 11. Regression model analysis of variance in winter.

Source	Sum of Squares	df	Mean Square	F Value	*p* Value	Significance
Model	16.79	4	4.20	8.93	0.0001	**
A	4.56	1	4.56	9.71	0.0047	**
B	12.00	1	12.00	25.53	<0.0001	*
C	0.013	1	0.013	0.028	0.8677	
D	0.21	1	0.21	0.45	0.5069	
Residual	11.28	24	0.47			
Lack of fit	10.83	20	0.54	4.79	0.0694	
Pure error	0.45	4	0.11			
Cor Total	28.07	28				
R^2	0.5981					
R^2_{Adj}	0.5312					
R^2_{Pred}	0.3811					

Note: * was significant difference ($p < 0.05$); ** was very significant difference ($p < 0.01$).

According to the P value, the PET value of the open underground public space in winter was significantly affected by the opening plane shape A and the opening aspect ratio ($p < 0.01$). According to the F value, the influencing sequence of four factors on the PET value of the entrance and exit of the open underground public space in winter is as follows: B > A > D > C, that is, the aspect ratio of the opening > the plane shape of the opening > the height of the opening baffle > the depth of the opening (Figure 14). Among the single factors, the ratio of opening length to width has the most obvious effect on PET value, followed by the opening plane shape, and the other two factors have little effect. In addition, the interaction relationship among the four factors A, B, C, and D had little effect on PET outside the entrance and exit.

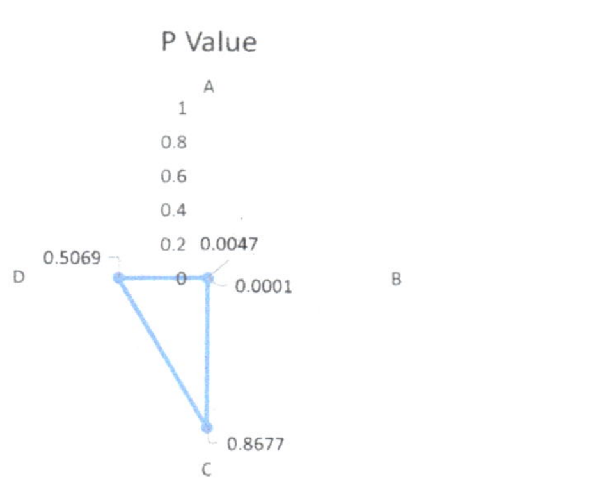

Figure 14. Importance of factors for open entrances and exits in winter.

In addition, according to Figure 15, the trend relationship of the influence of single factors on PET values can be found. The physiological equivalent temperature PET has a linear negative correlation with the opening plane shape. The more the opening plane shape edges, the smaller the PET value. The linear relationship with the opening aspect ratio is positive. The larger the opening aspect ratio is, the larger the PET value is. With the change of the opening depth and the height of the opening baffle, the physiolcgically equivalent temperature PET in its value range changes little, indicating that these two factors have little influence on the PET value.

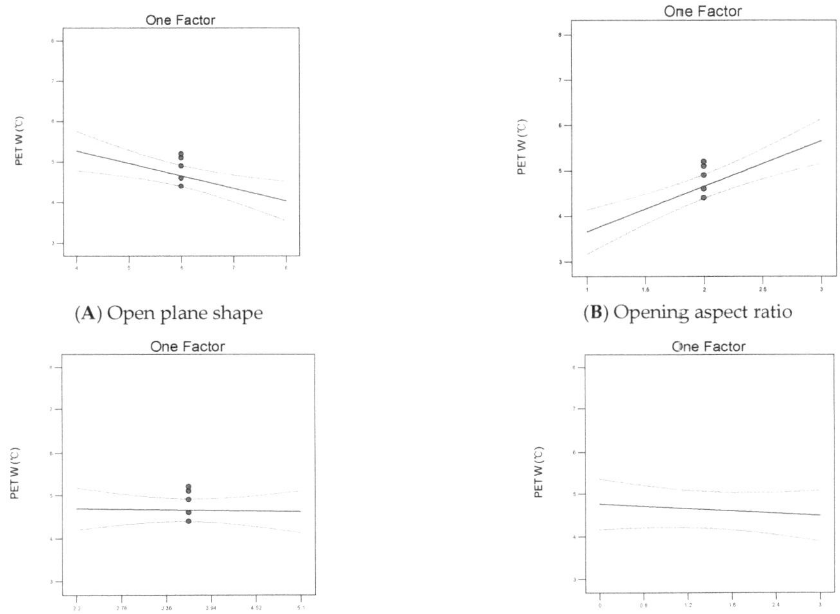

(**A**) Open plane shape (**B**) Opening aspect ratio

(**C**) Opening depth (**D**) Open baffle height

Figure 15. The influence of single factor on the open entrance and exit relationship.

4. Discussion

4.1. Air Thermal Environment Characteristics of Inlet and Outlet of Xuzhou Underground Public Space

In the summer, the entrance and exit of the open underground public space, from the outdoor part through the transition part to the indoor of this complete stage, the relative temperature, black globe temperature, and physiological equivalent temperature PET values are constantly decreasing; the relative temperature, black globe temperature, and PET values of the outdoor part are the highest. From outdoor to indoor, the relative humidity of working condition 1 shows an increasing trend, the humidity of the outdoor part is the highest, the relative humidity of working condition 2 shows a decreasing trend, and the humidity of the outdoor part is the lowest. The data variations of relative temperature, black sphere temperature, and humidity for Case 1 agree with Liu Xiaobin et al. [31]. Compared with condition 2, the air temperature in condition 1 drops more sharply and suddenly from the transition part to the indoor part. The wind speed of the breeze is more unstable in the outdoor and indoor parts, and the transition part is more stable.

In winter, from outdoor to indoor, the relative temperature, black globe temperature, and physiologically equivalent temperature PET values continue to rise, and the outdoor relative temperature, black globe temperature, and PET values are the lowest. The trend of relative humidity change in condition 1 and condition 2 is opposite to that in summer. The outdoor part of the breeze wind speed is the most unstable. In contrast, Liu Yanan [16] et al. found that the airflow in the outdoor portion of an open entrance was more stable. Their findings are not consistent with the results obtained in this study. This may be due to the fact that the wind tunnel tests did not consider the effects of buildings, vegetation, air-conditioning equipment, and pedestrians at the site.

Consider the situation of summer and winter. The trend of relative humidity change in condition 1 and condition 2 is not consistent, which is related to the adjustment function of the air conditioning system [43,44]. Working condition 1 is the subway entrance with an air conditioning system in summer, while working condition 2 is the Golden Eagle underground supermarket entrance without an air conditioning system to adjust the entrance. In addition, the indoor part and outdoor parts of working condition 1 are in a state of no door, no curtain, and no shelter indoor and outdoor connectivity, which is different from working condition 2. In addition, the number of users of working condition 1 during the test is larger, which may have a certain impact on the results of the measured data [12]. The variation of outdoor breeze wind speed in winter is larger than that in summer, the overall wind environment is unstable, and the overall wind speed is higher than that in summer. The preliminary speculation may be due to the withered plants around the entrance and exit, which do not have the functions of blocking the urban wind, shading and cooling [45].

4.2. Optimal Design of Human Thermal Comfort at the Entrance and Exit of Open Underground Public Space

4.2.1. Summer

In order to meet the thermal comfort requirements of the underground public space entrance and exit in summer, the physiologically equivalent temperature PET value of the outdoor part needs to reach a reasonable range, and the minimum value of physiological equivalent temperature PET should be selected. Through the multi-factor coupling model established above, the simulation data is optimized and predicted. When all factors are within the range and PET is the lowest, the plane shape of the opening is 8 sides, the aspect ratio of the opening is 3, the depth of the opening is 5.1 m, and the height of the opening baffle is 2.937 m. Under this condition, the physiologically equivalent temperature PET of the outdoor part of the entrance and exit of the open underground public space in summer is 40.415 °C. According to the actual achievable situation, the plane shape of the opening is 8 sides, the aspect ratio of the opening is 3, the depth of the opening is 5.1 m, and the height of the opening baffle is 2.95 m. Finally, the PET of the outdoor part of the open entrance and exit in summer is 40.4 ± 0.2 °C, which is close to the predicted value of the model, indicating that the model can be used to optimize the shape design of the open underground public space entrance and exit in summer.

4.2.2. Winter

In order to meet the human thermal comfort requirements of the underground public space entrance and exit in winter, the physiologically equivalent temperature PET value of the outdoor part needs to reach a reasonable range, and the maximum physiological equivalent temperature PET should be selected. Through the multi-factor coupling model established above, the simulation data is optimized and predicted. When all factors are within the range, and PET is the largest, the plane shape of the opening is 4 sides, the aspect ratio of the opening is 2.999, the depth of the opening is 2.2 m, and the height of the opening baffle is 0.006 m. In this case, the physiologically equivalent temperature PET of the outdoor part of the entrance and exit of the open underground public space in winter is 6.441 °C. According to the practical situation, the plane shape of the opening is 4 sides, the aspect ratio of the opening is 3, the depth of the opening is 2.2, the height of the opening baffle is 0m, and the model is simulated again. Finally, the PET of the outdoor part of the open entrance and exit in winter is 6.4 ± 0.2 °C, which is close to the predicted value of the model, indicating that the model can be used to optimize the shape design of the open underground public space entrance and exit in winter.

4.2.3. Combine Summer and Winter

In order to meet the most suitable physiological equivalent temperature PET in both winter and summer, the four single-factor variables of the opening plane shape, opening aspect ratio, opening depth, and opening baffle height cannot be agreed upon. Combined with the relationship model established in summer and winter, when the physiologically equivalent temperature PET in the two seasons is in the most suitable condition, the plane shape of the opening is 5.330 sides, the opening aspect ratio is 2.959, the opening depth is 5.1 m, and the opening baffle height is 0m. In this case, the physiologically equivalent temperature PET of the outdoor part of the entrance and exit in summer is 40.817 °C, and the physiologically equivalent temperature PET of the outdoor part of the entrance and exit in winter is 5.924 °C. According to the actual situation, the plane shape of the opening is 5 sides, the aspect ratio of the opening is 3, the depth of the opening is 5.1 m, the height of the opening baffle is 0m, and the model is simulated again. Finally, the PET of the outdoor part of the entrance and exit in summer is 40.8 ± 0.2 °C, and the breeze wind speed is 0.354 ± 0.02 m/s. The PET of the outdoor part of the entrance and exit in winter is 5.9 ± 0.2 °C, and the breeze wind speed is 0.152 ± 0.02 m/s, which is close to the predicted value of the model. The average wind speed ratio in summer is 0.4, and the average wind speed ratio in winter is 0.1, which is within the range of 0.1–1.2, meeting the specification "Standard for Wind Tunnel Experiment Method of Construction Engineering", and can achieve a better wind environment comfort. At this time, it is the optimal design target for the open underground public space entrance and exit form.

The four factors influencing the PET value in summer are the plane shape of the opening, the ratio of length to width of the opening, the depth of the opening, and the height of the baffle. The order of influence in winter is opening length-width ratio > opening plane shape > opening baffle height > opening depth. In summer and winter, single-factor opening plane shape and opening aspect ratio have the most obvious influence on the outdoor PET value of the entrance and exit, while the other two factors have little influence. The interaction relationship between A, B, C, and D has almost no influence on the outdoor PET value of the entrance and exit.

The regression model of the spatial morphological characteristics of the open inlet and outlet in summer and winter and the physiologically equivalent temperature PET is established. The specific relationship equation is as follows.

In the formula:

$$Y = 40.95 - 0.28 * A - 0.18 * B - 0.058 * C - 8.33 * 10^{-3} * D \tag{3}$$

$$Y = 4.66 - 0.62 * A + B - 0.033 * C - 0.13 * D \tag{4}$$

Y—PET
A—Open plane shape
B—Opening aspect ratio
C—Opening depth
D—Open baffle height

5. Conclusions

This study first measured the wind and thermal environment characteristics of typical open underground public space entrances and exits in Xuzhou city, then sorted out the basic spatial form characteristics of open underground public space entrances and exits through investigation, and finally explored the relationship between the spatial form of underground public space entrances and exits, wind and thermal environment and human comfort. The research shows the following:

1. The air temperature, black globe temperature, and physiological equivalent temperature PET values of the outdoor part of the entrance and exit of the underground public space in summer are higher than those of the transition part and indoor part of

the entrance and exit space of the entire underground public space, and the relative humidity is lower than the other two parts; In winter, the air temperature, black globe temperature and physiologically equivalent temperature PET values of the outdoor part are lower than the transition part, and the indoor part of the whole underground public space entrance and exit space, and the relative humidity is higher than the other two parts.

2. There are four main quantitative spatial form elements of the entrance and exit of Xuzhou open underground public space: the plane shape of the opening, the aspect ratio of the opening, the depth of the opening, and the height of the baffle at the opening.
3. In Xuzhou and other cold areas of Jiangsu Province, the human thermal comfort of the open entrance is most affected by the plane shape of the opening, followed by the aspect ratio of the opening, while the influence of the depth of the opening and the height of the baffle at the opening can be ignored.
4. The relationship model between the spatial form of open underground public space and physiological equivalent temperature PET in cold regions of Jiangsu, such as Xuzhou, was established, respectively, in summer and winter. The planar shape of the opening and the length-width ratio of the opening is linear with the physiologically equivalent temperature.

Generally speaking, in cold areas, the shape design of underground public space entrances and exits greatly affects the thermal comfort of the human body. The relationship between each factor and thermal comfort is not simply a single variable relationship. On the contrary, human thermal comfort and multiple spatial morphologic variables interact. In addition, when seeking the optimal physiological equivalent temperature in summer and winter, the changing trend of some spatial form elements with great influence is not consistent. Therefore, when designing for the human comfort of the entrance and exit, the situation of summer and winter should be considered comprehensively. Architects should mainly focus on the physical environment data and human comfort of the outdoor part when designing. In the design of spatial form, more attention should be paid to the opening plane shape and the opening length-width ratio.

This study still has some limitations. On the one hand, we cannot guarantee that all parameters of the simulation are exactly the same as reality. The effects of equipment and pedestrians were not considered. On the other hand, we only built a limited number of 25 sets of experimental models for simulation. There is a certain bias in modeling the mathematical relationship with the results of this simulation. Therefore, in our future work we will consider more factors comprehensively and use more data results for analysis and generalization. Currently, underground space is developing rapidly. In addition to open entrances and exits, there are separate foyers and sunken plaza entrances and exits. Their physical environment and human comfort are also future research directions.

Author Contributions: Conceptualization, H.L. and L.N.; methodology, H.L.; software, H.L.; validation, H.L., J.K., P.C. and L.N.; formal analysis, P.C. and H.L.; investigation, H.L., P.C. and L.N.; resources, P.C.; data curation, J.K.; writing—original draft preparation, H.L.; writing—review and editing, P.C. and J.K.; visualization, H.L., P.C. and J.K.; supervision, P.C.; project administration, J.K.; funding acquisition, J.K. All authors have read and agreed to the published version of the manuscript.

Funding: Multi-factor coupling mechanism and regulation method of heat and humid environment in underground atrium space of the National Natural Science Foundation of China, grant number 51778611, Jiangsu Social Science Foundation Provincial City Collaboration Project: Research on Huai'an Village Residential Model and Renewal Strategies Based on Rural Revitalization, grant number 22XZB020 and the Assistance Program for Future Outstanding Talents of China University of Mining and Technology Research on the Optimal Design of Wind Environment at the Entrance and Exit of Underground Public Space—Taking Xuzhou City as an Example, grant number 2022WLJCRCZL314.

Data Availability Statement: The data presented in this study are available on request from the corresponding author. The data are not publicly available due to the large amount of data, it is inconvenient to upload.

Acknowledgments: We would like to express our gratitude to those who helped us while writing this article. Our most profound appreciation goes to Ji Xiang, from the China University of Mining and Technology in China, for his encouragement and inspiration.

Conflicts of Interest: The authors declare no conflicts of interest.

Appendix A

PET: Physiological Equivalent Temperature. It is an indicator of thermal comfort. It is defined as the temperature at which the human skin and body temperatures in an indoor or outdoor environment reach a thermal state equivalent to that of a typical indoor environment.

Air temperature (°C): The physical quantity of how hot or cold the air is called the air temperature.

Black globe temperature (°C): Indicates the actual sensory temperature expressed in terms of temperature when a person or object in a radiant heat environment is subjected to the combined effects of radiant heat and convective heat.

Relative humidity (%): Refers to the water vapor pressure in air as a percentage of the saturated water vapor pressure at the same temperature.

Breeze speed (m/s): Rate of airflow.

RSM: Response Surface Method. This method is a statistical method that uses reasonable experimental design methods and obtains certain data through experiments, adopts multiple quadratic regression equations to fit the functional relationship between factors and response values, and seeks the optimal parameters through the analysis of regression equations to solve the multivariable problem.

Opening aspect ratio: It means the ratio of the lengths of the long and short axes of an open plane. Opening aspect ratio = b/a (Figure A1).

Open plane shape: It means the number of sides that is the shape of the open plane (Figure A2).

Opening orientation: It means the angle between the normal to the long axis of the open plane and the direction due to the south.

Opening depth: It means the vertical distance of the opening from the level at which it is located, below that elevation, to the next stabilizing level (Figure A3).

Open baffle height: It means the height of the baffle at the opening (Figure A4).

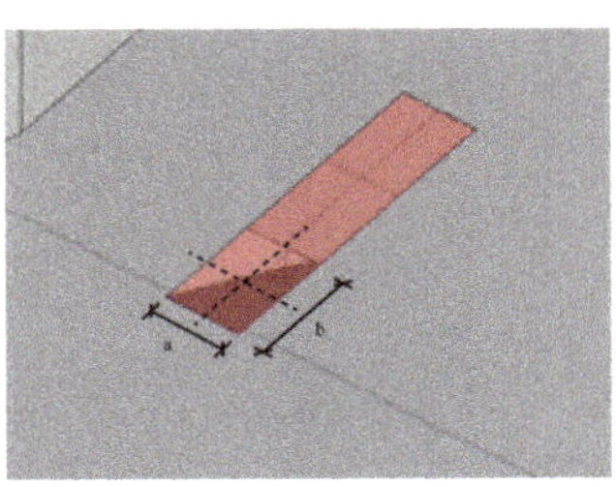

Figure A1. Opening aspect ratio.

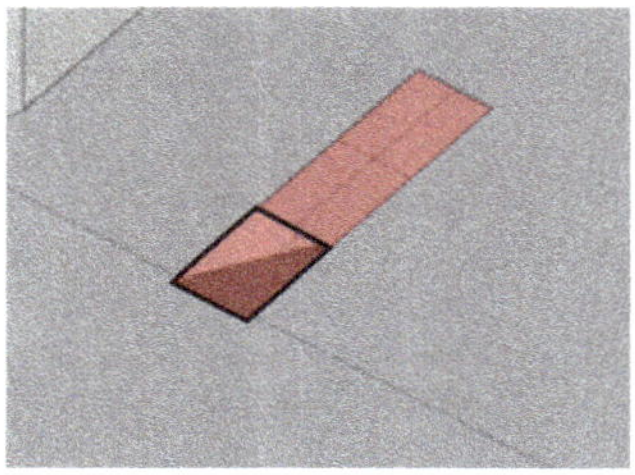

Figure A2. Open plane shape.

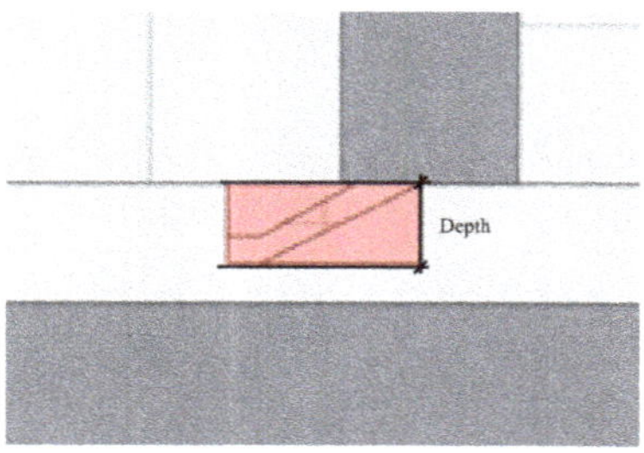

Figure A3. Opening depth.

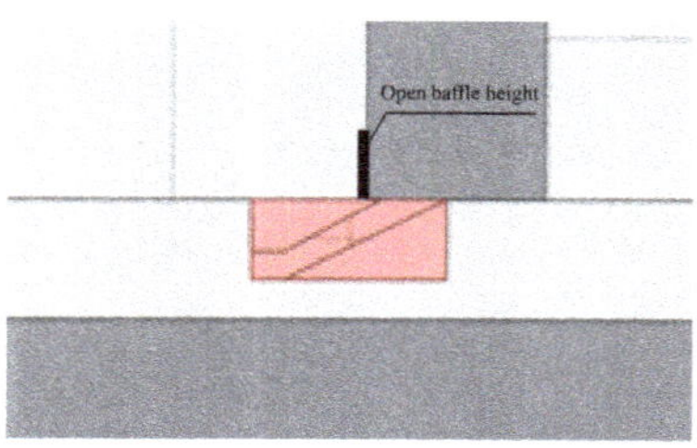

Figure A4. Open baffle height.

References

1. Cao, S.; Leng, J.; Qi, D.; Kumar, P.; Chen, T. Sustainable underground spaces: Design, environmental control and energy conservation. *Energy Build.* **2022**, *257*, 111779. [CrossRef]
2. Peng, Y.; Liu, H.; Wang, Z.; Fu, J.; Zhang, H.; Wang, J.; Yang, Q. Development of urban underground space in coastal cities in China: A review. *Deep. Undergr. Sci. Eng.* **2023**, *2*, 148–172. [CrossRef]
3. Besner, J. Cities Think Underground—Underground Space (also) for People. *Procedia Eng.* **2017**, *209*, 49–55. [CrossRef]
4. Hwang, S.; Park, W.M. Radon and PM10 concentrations in underground parking lots and subway stations with health risks in South Korea. *Environ. Sci. Pollut. Res.* **2018**, *25*, 35242–35248. [CrossRef]
5. Xueyuan, H.; Yu, S. The urban underground space environment and human performance. *Tunn. Under-Ground Space Technol.* **1988**, *3*, 193–200. [CrossRef]
6. Shafiei Fini, A.; Moosavi, A. Effects of "wall angularity of atrium" on "buildings natural ventilation and thermal performance" and CFD model. *Energy Build.* **2016**, *121*, 265–283. [CrossRef]
7. Vaccarini, M.; Giretti, A.; Tolve, L.C.; Casals, M. Model predictive energy control of ventilation for underground stations. *Energy Build.* **2016**, *116*, 326–340. [CrossRef]
8. Samant, S. Atrium and its adjoining spaces: A study of the influence of atrium façade design. *Archit. Sci. Rev.* **2011**, *54*, 316–328. [CrossRef]
9. Liang, X.; Lu, Z.; Ye, F.; Zhang, W. Investigation of design of independent metro station entrances in China. *Proc. Inst. Civ. Eng.—Munic. Eng.* **2023**, *176*, 10–31. [CrossRef]
10. Afshin, M.; Sohankar, A.; Manshadi, M.D.; Esfeh, M.K. An experimental study on the evaluation of natural ventilation performance of a two-sided wind-catcher for various wind angles. *Renew. Energy* **2016**, *85*, 1068–1078. [CrossRef]

11. Choi, C.; Morita, M. Characteristics of Universal Design Environmental Installation in Subways' Exits and Entrances in Osaka City (1). *Des. Stud.* **2002**, *49*, 19–28. [CrossRef]
12. Chen, T.; Cao, S.; Wang, J.; Nizamani, A.G.; Feng, Z.; Kumar, P. Influences of the optimized air curtain at subway entrance to reduce the ingress of outdoor airborne particles. *Energy Build.* **2021**, *244*, 111028. [CrossRef]
13. Li, G.; Han, C.; Shen, J.; Wang, X.; Gu, T.; Yang, Z.; Zhang, L. A Demographic Characteristics-Based Study on the Visual Impact Assessment of the External Form of Entrance Pavilions to the Underground Stations of China's Subway. *Appl. Sci.* **2023**, *13*, 4030. [CrossRef]
14. Han, G.; Wen, Y.; Leng, J.; Sun, L. Improving Comfort and Health: Green Retrofit Designs for Sunken Courtyards during the Summer Period in a Subtropical Climate. *Buildings* **2021**, *11*, 413. [CrossRef]
15. Guo, H.; Deng, M.; Li, Y. Influence of sunken plaza on ventilation performance of underground commercial space. *J. South China Univ. Technol. (Nat. Sci. Ed.)* **2014**, *42*, 114–120.
16. Liu, Y.; Hu, Y.; Xiao, Y.; Chen, J.; Huang, H. Effects of different types of entrances on natural ventilation in a subway station. *Tunn. Undergr. Space Technol.* **2020**, *105*, 103578. [CrossRef]
17. Guan, B.; Zhang, T.; Liu, X. Performance investigation of outdoor air supply and indoor environment related to energy consumption in two subway stations. *Sustain. Cities Soc.* **2018**, *41*, 513–524. [CrossRef]
18. Simpson, T.W.; Peplinski, J.D.; Koch, P.N.; Allen, J.K. Metamodels for Computer-based Engineering Design: Survey and recommendations. *Eng. Comput.* **2001**, *17*, 129–150. [CrossRef]
19. Shen, X.; Zhang, G.; Bjerg, B. Assessments of experimental designs in response surface modelling process: Estimating ventilation rate in naturally ventilated livestock buildings. *Energy Build.* **2013**, *62*, 570–580. [CrossRef]
20. Geyer, P.; Schlüter, A. Automated metamodel generation for Design Space Exploration and decision-making—A novel method supporting performance-oriented building design and retrofitting. *Appl. Energy* **2014**, *119*, 537–556. [CrossRef]
21. Yu, F.; Leng, J. Multivariable interactions in simulation-based energy-saving glass roof designs. *Sol. Energy* **2020**, *201*, 760–772. [CrossRef]
22. Norton, T.; Grant, J.; Fallon, R.; Sun, D. Optimising the ventilation configuration of naturally ventilated livestock buildings for improved indoor environmental homogeneity. *Build. Environ.* **2010**, *45*, 983–995. [CrossRef]
23. Shen, X.; Zhang, G.; Bjerg, B. Investigation of response surface methodology for modelling ventilation rate of a naturally ventilated building. *Build. Environ.* **2012**, *54*, 174–185. [CrossRef]
24. Zhou, L.; Haghighat, F. Optimization of ventilation system design and operation in office environment, Part I: Methodology. *Build. Environ.* **2009**, *44*, 651–656. [CrossRef]
25. Chen, H.; Moshfegh, B.; Cehlin, M. Computational investigation on the factors influencing thermal comfort for impinging jet ventilation. *Build. Environ.* **2013**, *66*, 29–41. [CrossRef]
26. Cetin, K.S.; Manuel, L.; Novoselac, A. Thermal comfort evaluation for mechanically conditioned buildings using response surfaces in an uncertainty analysis framework. *Sci. Technol. Built Environ.* **2016**, *22*, 140–152. [CrossRef]
27. Cetin, K.S.; Manuel, L.; Novoselac, A. Effect of technology-enabled time-of-use energy pricing on thermal comfort and energy use in mechanically-conditioned residential buildings in cooling dominated climates. *Build. Environ.* **2016**, *96*, 118–130. [CrossRef]
28. Zhou, H.; Tao, G.; Nie, Y.; Yan, X.; Sun, J. Outdoor thermal environment on road and its influencing factors in hot, humid weather: A case study in Xuzhou City, China. *Build. Environ.* **2022**, *207*, 108460. [CrossRef]
29. Liu, H.; Liu, X.; Nie, L.; Hong, X.; Ji, X. Research Progress of Urban Wind and Thermal Environment Based on CiteSpace and China National Knowledge Infrastructure Database. *Sustainability* **2022**, *14*, 13108. [CrossRef]
30. Medeiros, B.; Clement, A.C.; Benedict, J.J.; Zhang, B. Investigating the impact of cloud-radiative feedbacks on tropical precipitation extremes. *npj Clim. Atmos. Sci.* **2021**, *4*, 18. [CrossRef]
31. Liu, X. *Research on Thermal Comfort Optimization of Subway Entrance Transition Space Based on Measured and Simulat-Ed.*; China University of Mining and Technology: Beijing, China, 2022. [CrossRef]
32. Liu, Q.; Liu, B.T. Sustainable Urban Revitalization in Historic Districts: A Case Study of Huilongwo Area in Xuzhou. *Adv. Mater. Res.* **2014**, *919–921*, 1567–1572. [CrossRef]
33. Cheng, J.; Cai, C.S.; Xiao, R.-C. Probabilistic shear-lag analysis of structures using Systematic RSM. *Struct. Eng. Mech.* **2005**, *21*, 507–518. [CrossRef]
34. Ge, Y. *Experimental Design Method and Design-Expert Software Application*; Harbin Institute of Technology Press: Harbin, China, 2015.
35. Kuznik, F.; Brau, J.; Rusaouen, G. A RSM Model for the Prediction of Heat and Mass Transfer in a Ventilated Room. In Proceedings of the Building Simulation 2007, Beijing, China, 3–6 September 2007.
36. Gao, X.; Wang, S.; Wang, P. The evaluation method of PID controller parameter tuning based on Fluent. In Proceedings of the 27th Chinese Control and Decision Conference (2015 CCDC), Qingdao, China, 23–25 May 2015; pp. 4850–4854.
37. Li, L.; Zhang, L.J.; Zhang, N.; Hu, F.; Jiang, Y.; Xuan, C.Y.; Jiang, W.M. Study on the micro-scale simulation of wind field over complex terrain by rams/fluent modeling system. *Wind Struct. Int. J.* **2010**, *13*, 519–528. [CrossRef]
38. Zhang, T. Study on Coupling of Wind Environment and Spatial Morphology in Urban Central Area. Master's Thesis, Southeast University, Nanjing, China, 2015.
39. Han, J.; Liao, L.; Zhang, G. Review and trend prospect of the development process of China's green building evaluation standard system. *Constr. Sci. Technol.* **2017**, 10–13. [CrossRef]

40. Lee, H.; Mayer, H. Validation of the mean radiant temperature simulated by the RayMan software in urban environments. *Int. J. Biometeorol.* **2016**, *60*, 1775–1785. [CrossRef]
41. Uddin, A.; Khan, M.K.; Campean, F. Function-Based Conceptual Design Expert (CDE) Systems: Development Trend and Gaps Identification. *Appl. Mech. Mater.* **2014**, *564*, 590–596. [CrossRef]
42. Ozili, P.K. The Acceptable R-Square in Empirical Modelling for Social Science Research. *SSRN Electron. J.* **2022**. [CrossRef]
43. Zhang, Y.; Li, X. Numerical analysis on the condenser inlet air temperature of train-mounted air conditioner when a train stops in subway station tunnel. *Sustain. Cities Soc.* **2021**, *69*, 102793. [CrossRef]
44. Wang, C.; Li, X. Measurement-based investigation of subway station tunnel thermal environment. *J. Build. Eng.* **2022**, *57*, 104757. [CrossRef]
45. Wu, L. Research on the Influence Mechanism and Evaluation of Built Environment Characteristics on the Design of Subway Station Transfer Space. Ph.D. Thesis, Beijing Jiaotong University, Beijing, China, 2021. [CrossRef]

MDPI AG
Grosspeteranlage 5
4052 Basel
Switzerland
Tel.: +41 61 683 77 34

Buildings Editorial Office
E-mail: buildings@mdpi.com
www.mdpi.com/journal/buildings

www.ingramcontent.com/pod-product-compliance
Lightning Source LLC
Chambersburg PA
CBHW040747250926
51850CB00007BB/1430
9783725874422